FRUIT CULTURE:
Its Science and Art

*An Old Italian woodcut
(artist anon.)*

FRUIT CULTURE:

Its Science and Art

KAY RYUGO
University of California
Davis, California

John Wiley & Sons
New York • Chichester • Brisbane •
Toronto • Singapore

Copyright © 1988, by John Wiley & Sons, Inc.

All rights reserved. Published simultaneously in Canada.

Reproduction or translation of any part of
this work beyond that permitted by Sections
107 and 108 of the 1976 United States Copyright
Act without the permission of the copyright
owner is unlawful. Requests for permission
or further information should be addressed to
the Permissions Department, John Wiley & Sons.

Library of Congress Cataloging-in-Publication Data:
Ryugo, Kay.
   Fruit culture.

   Includes bibliographies and indexes.
   1. Fruit-culture. 2. Fruit. I. Title.
SB355.R96 1988   634   88-5540
ISBN 0-471-89191-6

Printed in the United States of America

10 9 8 7 6 5 4 3 2 1

Preface

This textbook is intended primarily for horticulture and plant science majors who are interested in pursuing a career in some phase of pomology, the science and art of fruit culture. The reader is assumed to have taken botany and courses allied to horticulture, such as economic entomology, plant pathology, soils, and irrigation.

One of the dilemmas of teaching fruit culture is the varied backgrounds of our contemporary students. A majority of them have little or no experience in commercial agriculture. Even those who were raised on farms have not had wide exposure to the various facets of orcharding because farms are often large and specialized. Thus, students from an apple-growing district are not well versed in the cultural practices of peach and grape growers. Furthermore, the enrollment of foreign students, eager to expand the fruit-growing potential of their native lands, has added to the diversity of students.

This book consists of three parts. The first four chapters in Part I cover the morphology of vegetative and reproductive organs and the physiology of fruits, trees, and vines. Chapter 5 addresses the reproductive processes, beginning with a juvenile seedling tree and ending with flowering and seed formation. Chapter 6 continues with postbloom development of the fruit, proceeding from fruit set to physical and chemical changes, maturity, harvesting techniques, and physiological disorders.

Part II, which includes chapters on various orchard management and nursery practices, serves as an introduction to pomology. Although these subjects are covered more extensively in other horticultural textbooks, they are included here to complement or reinforce the information covered in Part I. Brief histories of some manipulations of fruit trees and vines that are virtually unchanged from ancient times are given. The ancient Assyrians and Babylonians cross-pollinated date palms; the Romans knew how to prune (see frontispiece) and graft trees and how to thin fruits, but they did not comprehend why these practices succeeded. Even today, pomologists do not have an adequate explanation for the mechanisms behind such phenomena as graft incompatibility and disease resistance.

In Part III, pomological species and cultivars are described. Old apple and pear cultivars are included because of their history and because some are still grown today. Minor crops such as the Asian pear, pomegranate, kiwifruit, and persimmon are also included, although these crops are not yet widely grown in the United States. With their increased popularity, more and more acres are being planted to them in the Pacific Coast states, Chile, New Zealand, and Europe.

By integrating the principles in Part I with the practices and special characteristics of species described in Parts II and III, students should gain a good understanding of fruit tree and vine behavior and culture. With this background, students should be better prepared to solve pomological problems, thereby lowering production costs to growers and making high-quality fruits available to more consumers.

Appendix A is a short history of growth regulators, a subject which could not be adequately covered in any one chapter. Because this textbook is expected to be used in conjuction with a laboratory course, several outdoor and indoor exercises are listed in Appendix D. These exercises, which cover field practices and reproductive and vegetative characteristics of different species, were deemed necessary in a recent survey of American instructors and students of horticulture. A glossary of botanical terms is given in Appendix E.

The lists of references cited at the ends of chapters are by no means complete. They represent only findings that offer evidence to strengthen a hypothesis or a horticultural axiom or to demolish it as a myth. Articles that are no less important than those cited were regretfully omitted to keep the book from becoming a review of current pomological progress.

I have found the profession of a pomologist not only challenging but also interesting and highly satisfying, enabling me to meet innumerable students, fruit growers, and scientists, here and abroad. This book is dedicated to the late Luther Dent Davis, Professor Emeritus of the University of California and a grandson of a humble Indiana homesteader. He guided me, a son of a Japanese immigrant, to my horticultural career. Without Professor Davis's help and the understanding of the late Warren P. Tufts, former chairman of the Pomology Department during the post-World War II period, my career in horticulture would probably not have been realized.

I am very grateful to Julian C. Crane and Silviero Sansavini who encouraged me to write this book. Dr. Crane contributed to the parts dealing with juvenility and species descriptions, and to the glossary, and he reviewed some chapters. Special thanks go to Frank G. Dennis, Jr., for his excellent editorial review of the manuscript and to my wife, Masako, for her patience in editing the manuscript and in proofreading the text. I am indebted to P. J. Breen, S. D. Seeley, W. C. Stiles, K. Uriu, and J. A. Wolpert for taking their valuable time to review parts of the manuscript before it went to press, and especially to Kiyoto Uriu and the Division of Agricultural Science, University of California, for permission to reproduce the eight color plates depicting mineral deficiency symptoms of French prunes. These plates and those of chimeras are placed in Appendix F.

My sincere gratitude is extended to the numerous individuals, publishers, and equipment manufacturers who graciously authorized the use of their data and photographs, and to my American and foreign colleagues and former graduate students who helped me gather the much needed information for this book.

KAY RYUGO

Davis, California
April 1988

Contents

PART I   1
Structures and Functions of Cultivated Fruit Plants

CHAPTER 1
Anatomy and Morphology of Fruit Plants and Their Parts   3

Stem   4

ONTOGENY 5
BUD CLASSIFICATION 5

Leaf   5

ONTOGENY AND MORPHOLOGY 5

Root 8
> MORPHOLOGY AND ORIGIN OF ROOTS 8

Flower 10
> MORPHOLOGY 10
> CLASSIFICATION 14
> CLASSIFICATION OF INFLORESCENCES 18

Fruit 19
> MORPHOLOGY 19
> FRUIT TYPES 20
> FRUIT CLASSIFICATION 20
> PLACENTATION 26

CHAPTER 2
Physiology and Functions of Buds and Stems ... 27

Control of Bud Development 27
> DORMANCY AND REST OF BUDS 27
> THE NECESSITY FOR REST AND CHILLING 28
> OTHER ENVIRONMENTAL FACTORS GOVERNING DORMANCY AND REST 30
> CORRELATIVE INHIBITION AND APICAL DOMINANCE 31

Bud, Shoot, and Spur Development 32
> DETERMINATE AND INDETERMINATE GROWTH 32
> SPUR FORMATION 32
> GROWTH IN GIRTH 33
> TENSION WOOD 34
> SAPWOOD AND HEARTWOOD FORMATION 35

Carbohydrate Storage 35

Mutations, Bud Sports, and Chimeras 37

CHAPTER 3
Physiology and Functions of Leaves 41

Photosynthesis 41
> NET PHOTOSYNTHESIS AND NET ASSIMILATION RATE 44
> INFLUENCE OF LIGHT AND TEMPERATURE ON PHOTOSYNTHESIS 44
> LIGHT SATURATION AND COMPENSATION POINTS 44

Transpiration 46
> WATER POTENTIAL 46
> RELATIONSHIP BETWEEN PN AND T 47
> DAILY TRENDS OF PN AND G 47
> ENDOGENOUS AND ENVIRONMENTAL FACTORS THAT INFLUENCE PN, G, AND T 49
> CULTURAL MEANS OF MAINTAINING OPTIMAL RATES OF PN AND T 53

Other Photochemical and Related Leaf Functions ... 56
> SOURCE OF FLOWERING STIMULUS 56
> NITRATE REDUCTION AND AMINO ACID SYNTHESIS 56
> BIOSYNTHESIS OF TERPENOIDS, INCLUDING HORMONES 56
> SYNTHESIS OF PHENOLIC SUBSTANCES AND TANNINS 56
> FOOD AND MINERAL NUTRIENT STORAGE 57

CHAPTER 4
Physiology and Functions of Roots   61

Growth Periodicity   62

Functions   62

ABSORPTION AND CONDUCTION OF WATER AND MINERAL NUTRIENTS   62
STORAGE AND METABOLISM OF RESERVE CARBOHYDRATES   63
ANCHORAGE AND SUPPORT   64
ROOT PRESSURE AND XYLEM EXUDATION   65

Root Distribution   65

SPECIES CHARACTERISTICS   65
ENVIRONMENTAL EFFECTS   67

CHAPTER 5
The Flowering Process and Seed Formation   69

Juvenility   69

CHARACTERISTICS OF JUVENILE PLANTS   69
LENGTH AND HERITABILITY OF THE JUVENILE PERIOD   70

Flower Bud Formation   71

FOUR PHYSIOLOGICAL AND MORPHOLOGICAL STAGES   71
INVESTIGATIONS ON FLOWER INITIATION   72
  History of the Nutritional Concept   72
  History of the Hormonal Concept   72
  The Nutritional and Nutrient Diversion Concept   73
  The Florigen Concept   74
TIME OF FLORAL INITIATION AND SOME PREREQUISITES   75
FLOWERING IN STRAWBERRIES   80
FLOWERING IN GRAPEVINES   81
FLOWERING IN KIWIFRUIT VINES   82
CULTURAL PRACTICES THAT INFLUENCE FLOWER BUD FORMATION   82
SUMMARY OF TREATMENTS INHIBITORY TO FLOWER INITIATION   85

Flower Differentiation   85

Abscission of Buds, Flowers, and Fruits   87

BASES FOR FLOWER BUD ABSCISSION   87

Anthesis and Zygote Formation   89

ANTHESIS   89
POLLINATION   89
GAMETOPHYTIC INCOMPATIBILITY   90
SPOROPHYTIC INCOMPATIBILITY   91
POLLEN TUBE GROWTH AND FERTILIZATION   91
THE EFFECTIVE POLLINATION PERIOD   92
FACTORS INFLUENCING POLLINATION AND FERTILIZATION   94
ENVIRONMENTAL CONDITIONS   95
COMPETITION FROM WEEDS   95
BEE POPULATION   96
CULTURAL PRACTICES THAT FAVOR FRUIT SET   96

POLLINATION-RELATED
PHENOMENA 97
   Apomixis or Parthenogenesis
   and Polyembryony 97
   Xenia and Metaxenia 97
PARTHENOCARPY 98
SEED DEVELOPMENT
AND PHYSIOLOGY 99
   Ovule and Embryo Ontogeny 99
   Developmental Physiology 100
SEED GERMINATION
AND STORAGE 101
   Factors Influencing
   Germination 101
   Storage of Seeds 103

CHAPTER 6
Fruit Growth and Development   107

Fruit Set   107

Cell Division and Tissue Differentiation   108

   GROWTH PATTERNS 109
   ENDOGENOUS FACTORS
   AFFECTING FRUIT GROWTH 113
   ENVIRONMENTAL FACTORS 116

Blossom and Fruit Thinning   120

   MEANS OF RELIEVING
   THE CROP LOAD 120
   GRAPE THINNING 124

Source–Sink Relationships   125

Morphological, Physical, and Chemical Changes   128

   DRUPES 128
   POMES 131
   BERRIES 133

   STRAWBERRY, AN
   AGGREGATE FRUIT 136
   NUTS 136

Ripening and Maturation Processes   137

   CARBOHYDRATE
   METABOLISM 138
   ORGANIC ACID
   METABOLISM 139
   PIGMENT FORMATION
   AND DEGRADATION 140
   RESPIRATION AND
   ETHYLENE EVOLUTION 143
   FLAVOR AND
   AROMATIC COMPOUNDS 146

Harvest Technology   146

   HARVESTING TECHNIQUES 146
   MATURATION AND
   HARVESTING INDICES 149

Abscission and Preharvest Drop   153

Physiological Disorders   154

   TEMPERATURE-RELATED
   DISORDERS 155
   WATER-RELATED
   DISORDERS 157
   GROWTH-RELATED
   DISORDERS 158
   DISORDERS OF IMMATURE
   AND OVERMATURE FRUITS 161
   ROOTSTOCK-RELATED
   DISORDERS 162

PART II
Orchard Management and Nursery Practices   169

CHAPTER 7
Establishing and Managing an Orchard 171

Site Selection and Preparation 171

REGIONAL AND LOCAL CONSIDERATIONS 171
PREPARATION OF THE SITE 173

Soil–Plant–Water Relationships 174

IRRIGATION PRINCIPLES 174
IRRIGATION SYSTEMS 175

Orchard Design 179

PLANTING SYSTEMS 179
TREE PLANTING 182
POLLINIZERS, THEIR SELECTION AND PLACEMENT 182

Vegetation Management 184

CLEAN CULTIVATION 184
SOD CULTURE 184

Protection Against Cold Injury 184

WINTER INJURY AND ACQUIRED COLD HARDINESS 184
FACTORS FAVORING FROST 185
MEANS OF REDUCING FROST DAMAGE 186
CARE OF FREEZE-DAMAGED TREES 189

CHAPTER 8
Soil and Soil Fertility 191

Soil Classification 191

Fertilizers 192

MANURING 192
CHEMICAL FERTILIZERS 193

Mineral Nutrition 193

BRIEF HISTORY OF ESSENTIAL ELEMENTS 193
MACRONUTRIENTS AND THEIR ROLES 194
MICRONUTRIENTS AND THEIR ROLES 196

Mineral Deficiencies and Excesses 196

SYMPTOMS AND TREATMENTS 196
MINERAL ANALYSIS 200

Soil Amendments 201

CHAPTER 9
Pruning and Training of Fruit Trees 203

Principles of Pruning 204

MECHANICS AND TYPES OF PRUNING CUTS 204
ESSENTIALS OF A STRONG TREE FRAME 205

Pruning During the Formative Years 206

TIME OF PRUNING AS AN INVIGORATING OR DEBILITATING PROCESS 206

Pruning of Mature Trees 208

DORMANT PRUNING 208
SUMMER PRUNING 209

Training Systems 210

Pruning of Bearing Vines, Brambles, and Bushes 216

GRAPEVINES 216
KIWIFRUIT VINES 218
BRAMBLES AND BUSH BERRIES 220

CHAPTER 10
Nursery Practices and Management 223

Propagation 223

SEXUAL PROPAGATION OF PLANTS 224
ASEXUAL, VEGETATIVE, OR CLONAL PROPAGATION 224

The Need for Specific Rootstocks 225

THE FUNCTION OF ROOTSTOCKS 225
ECONOMICALLY IMPORTANT TRAITS OF ROOTSTOCKS 225

Methods of Vegetative Propagation 228

BUDDING 228
GRAFTING 230
CUTTINGS 237
LAYERAGE OR LAYERING 239
PROPAGATION BY CROWN DIVISION, SUCKERS, AND STOLONS 241
IN VITRO CULTURE, OR MICROPROPAGATION 241

PART III
Species Characteristics 245

CHAPTER 11
Pomes 247

Apples, *Malus* spp. 247

Pears, *Pyrus* spp. 252

EUROPEAN OR FRENCH PEAR, *P. COMMUNIS* 252
ASIAN PEAR, *P. SEROTINA*, SYN. *PYRIFOLIA* 255
EURASIAN HYBRIDS 256

Quince, *Cydonia oblonga* 256

Loquat, *Eriobotrya japonica* 257

Pomegranate, *Punica granatum* 257

CHAPTER 12
Stone Fruits 259

Almond, *Prunus dulcis* 259

Peach and Nectarine, *Prunus persica* 261

Apricot, *Prunus armeniaca* 263

Plums 264

EUROPEAN OR FRENCH PLUM, *PRUNUS DOMESTICA* 264
JAPANESE PLUM, *PRUNUS SALICINA* 266

Cherries 267

SWEET CHERRY, *PRUNUS AVIUM* 267
SOUR CHERRY, *PRUNUS CERASUS* 270

CHAPTER 13
Nut Crops 271

Walnuts, *Juglans* spp. 271

PERSIAN WALNUT, *J. REGIA* 271

Pecan, *Carya illinoensis* 274

Chestnuts, *Castanea* spp. — 276

Filbert, *Corylus avellana* — 277

Pistachio, *Pistacia vera* — 278

CHAPTER 14
Vine Crops and Bushberries — 281

Grapes, *Vitis* spp. — 281
- OLD WORLD GRAPE, *V. VINIFERA* 281
- FOX OR EASTERN GRAPE, *V. LABRUSCA* 283

Kiwifruit or Chinese Gooseberry, *Actinidia deliciosa* — 283

Cranberry and Blueberries, *Vaccinium* spp. — 286
- CRANBERRY, *V. MACROCARPON* 286
- BLUEBERRIES 286

Currants and Gooseberries, *Ribes* spp. — 287
- CURRANTS 287
- GOOSEBERRIES 287

Raspberries, Blackberries, and Dewberries, *Rubus* spp. — 288

CHAPTER 15
Miscellaneous Fruit Species — 291

Persimmons, *Diospyros* spp. — 291
- ORIENTAL PERSIMMON, *D. KAKI* 291
- AMERICAN PERSIMMON, *D. VIRGINIANA* 293

Fig, *Ficus carica* — 293
- CAPRIFIG AND CAPRIFICATION 295
- SMYRA FIG 295
- COMMON FIG 295
- SAN PEDRO FIG 295

Mulberries, *Morus* spp. — 296

Feijoa, *Feijoa sellowiana* — 296

Zizyphus, *Zizyphus jujuba* — 297

APPENDIXES — 299

APPENDIX A
Natural Hormones and Synthetic Growth Regulators — 301

Natural Hormones — 301
- AUXINS 301
- GIBBERELLINS 302
- CYTOKININS 304
- ABSCISIC ACID 304
- ETHYLENE 305

Synthetic Growth Regulators — 306

APPENDIX B
Table of Equivalent Weights and Measures — 309

APPENDIX C
Some Common Units of Expression — 311

APPENDIX D
Laboratory Exercises
and Objectives 313

APPENDIX E
Glossary 321

APPENDIX F
Color Plates
Showing Chimeras
and Nutritional
Deficiency Symptoms 333

INDEX 335

PART I

Structures and Functions of Cultivated Fruit Plants

Students who have studied elementary botany and plant physiology will find portions of Part I to be a review. The main difference between this work and a general botany textbook is that, whereas botanists and physiologists usually select herbaceous plants as the most suitable to illustrate a specific structure, function and/or physiological function, here members of pomological species are given. Although fruit trees have been cultivated from prehistoric times, botanists rarely use them as examples because until recently little basic botanical information had been collected on fruit trees, partly on account of their large size and unwieldiness. The wide range of horticultural specimens in this text will introduce you to the intricacies and infinite beauty of fruit plants and their flowers, fruits, and seeds.

Pomology is a branch of horticulture. The name is derived from the Latin word *pomum*, for fruit, and the Greek word *logos*, for discourse. The term *horticulture* comes from the Latin words *hortus* and *cultura*, meaning, respectively, garden and cultivation. Horticulture was not given an impetus in the United States until 1862 when President Abraham Lincoln signed the Morrill Act providing for the establishment of a land-grant college within each state.

With this brief introduction, the author invites you to step into the realm of pomology.

CHAPTER 1

Anatomy and Morphology of Fruit Plants and Their Parts

A tree or a vine is made up of two parts: the aerial portion or shoot that consists of the trunk and branches and their appendages and the subterranean part or roots. Each thrives in an entirely unique environment and performs vastly different functions. Branches make up the basic scaffolding of a tree that supports the leaves, flowers, and fruit. Water, dissolved minerals, and some organic substances are translocated upward in the xylem to the upper extremities, while products of photosynthesis elaborated in leaves move downward in the phloem to the roots. Roots anchor the tree to the soil from which they absorb water and minerals. These absorbed minerals and some organic compounds such as amino acids and hormones that are synthesized in root tips are drawn into the xylem and are moved upward in the transpiration stream. Horticulturists refer to tissues external to the lateral cambium as the bark and all that is within it as wood. Xylem ray cells and phloem parenchyma cells in the wood and bark store the food, most of which is utilized for shoot growth in the spring.

A tree growing in its natural state possesses shoot and root systems that are genetically alike because the plant originated from a seed or a root sucker. Plants derived from seeds may look alike but are genetically different. Similarly, fruit plants raised from seeds collected from the same tree are likely to be quite different from each other because they, being heterozygous, do not breed true. Consequently, to establish an orchard bearing a desirable cultivar (cultural variety), its scions, short pieces of dormant shoot, are normally grafted onto rootstocks. These rootstocks may be sexually propagated from seeds, in which case they are different from one another, or they may be vegetatively propagated by cuttings or layering, in which case the rootstocks would be genetically alike. Whether the rootstock is propagated from seeds or vegetatively, the grafted scion top will be genetically dissimilar from the rootstock. Horticulturists refer to a grafted tree as a scion/rootstock combination, for example, Golden Delicious/M9, or Redhaven/Nemaguard. Vegetatively propagating a series of plants from a single mother plant, a clone, to retain its desirable traits is known as cloning. Grafting is not necessary if scion cultivars are propagated on their own roots by cuttings or layerage; in such cases the top and root are genetically alike.

Stem

ONTOGENY

The stem or shoot grows in length because of cell division in the apical meristem followed by cell enlargement in the subapical region. The protoderm, ground meristem, and procambium (Fig. 1.1) derived from the apical meristem give rise to the epidermis, cortex, and vascular cambium, respectively. The epidermis is usually a cutinized single layer of cells that protects the stem from desiccation. The cortex is beneath the epidermis and encircles an inner core of vascular column or the stele; the primary phloem is separated from the primary xylem by the cambium (Esau, 1958).

The tissues of the apical meristem of a stem, unlike those of roots, differentiate into nodes, where buds and leaves are located, and internodes. Nodes become separated by internodes of varying lengths depending on shoot vigor. The spacial arrangement of leaves about the stem axis is known as phyllotaxy. Leaves are arranged helically about the stem axis so that each succeeding leaf is offset from the previous one by a fraction of the stem circumference. This positioning of leaves allows a distance of three to eight nodes before one leaf is directly above another, thus affording better light exposures to basal leaves. The arrangement of leaves about a single node is categorized as (1) alternate, if there is one leaf per node as in most fruit tree species; (2) opposite, if two leaves are opposite each other as in the case of pomegranate; or (3) whorl, if there are three or more leaves per node as in the *Catalpa* which has three.

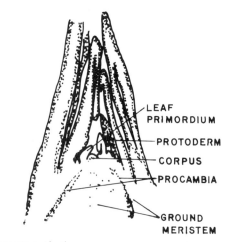

Figure 1.1
Longitudinal section of a peach shoot apex with its meristematic tissues.

BUD CLASSIFICATION

Buds are classified by their (1) activity: active or dormant (latent); (2) function: vegetative, reproductive, or both; or (3) position or origin on the stem: terminal, lateral, or adventitious. Active buds on shoots may (a) extend shoot length by terminal growth, (b) grow into short lateral branches or spurs, or (c) develop flowers and inflorescences. Dormant buds that do not grow into shoots or spurs within a few years eventually become overgrown with bark tissue and remain latent, growing just enough to remain beneath the bark surface (Fig. 1.2). These latent buds may be forced to grow by severe pruning.

A bud is further classified as (1) a leaf or vegetative bud if it gives rise to a shoot, (2) a flower bud if it produces a flower or an inflorescence, or (3) a mixed bud if it possesses both leaf and flower primordia. A flower bud is (1) simple, if it possesses only flowers, for example, cherry, peach, and plum; (2) mixed, if a bud gives rise to a shoot with floral or inflorescence primordia, for example, apple, kiwifruit, pear, persimmon, and walnut; or (3) compound, if a single bud contains two or more vegetative and/or mixed buds, as in the grape. Shoots arising from mixed buds have flowers that are borne terminally as in the apple, pecan, and pomegranate, or laterally as in the grape, kiwifruit, and persimmon.

Adventitious buds are formed at sites other than nodes, such as on roots, which give rise to suckers, and occasionally from callus, resulting in a chimera (see Mutations, Ch. 2).

Figure 1.2
Cross section of a six-year-old branch revealing a latent bud trace.

Leaf

ONTOGENY AND MORPHOLOGY

Fitzpatrick (1934) found that the terminal 1.5 millimeters of the peach leaf does not mature until the blade is about 6 centimeters in length. Veins and stomata in the terminal region then become functional while cells in the basal portion of the leaf are still dividing and enlarging (Fig. 1.3). When fully expanded and mature, a leaf may be 20 centimeters in length and 4 centimeters in width.

MacDaniels and Cowart (1944) found that a dormant, vegetative bud of apple contains an apical meristem and leaf primordia (Fig. 1.4). Thus, a preformed shoot is developed before budbreak. As the bud elongates into a shoot, the apical meristem continues to initiate additional leaf and bud primordia while the basal leaves expand. This expansion is the result of cell division and enlargement.

In apple leaves, cell division initially occurs throughout the entire leaf blade or lamina; cells near the large veins soon cease dividing, but those of the marginal meristem (Fig. 1.5) remain active, adding cells to the lamina surface area but not to its thickness. Cell division ceases first in the upper epidermis and then in the palisade layer, but cell enlargement continues until the leaf is fully expanded. The diameter of epidermal cells in the plane of the leaf surface is three to four times greater than that of palisade cells, but the vertical axis of the palisade cells is two to three times longer than the thickness of epidermal cells (Fig. 1.6). Cell division ceases in the lower spongy mesophyll layer before that in the palisade layer and lower epidermis. As cells begin to enlarge, the spongy parenchyma cells separate, leaving large intercellular air spaces and substomatal chambers. The ratio of epidermal cells to spongy parenchyma cells is nearly 2:1. The lower epidermal cells of the apple leaf, where the stomata are located, are interlocked much like pieces of a jigsaw puzzle.

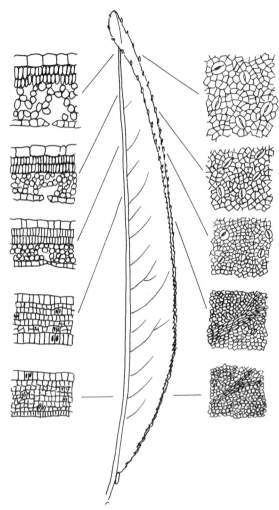

Figure 1.3
Cross sections (left) and surface views (right) of lower epidermis of a developing peach leaf about 6 centimeters in length. Lines indicate approximate positions of the leaf sections. (From Fitzpatrick, 1934)

Leaf maturation is initially basipetal but becomes acropetal as the leaf approaches full size. Hence, when young leaves are marked with a grid, the grid lines are farther apart in the midsection than at the apical or basal ends after the leaves mature (Fig. 1.7).

The first basal leaves to unfold on shoots are

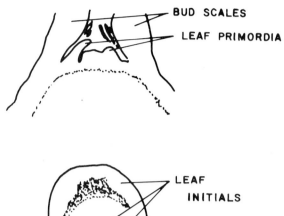

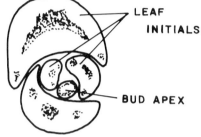

Figure 1.4
Longitudinal (upper) and transverse (lower) sections of apple leaf buds. (Adapted from MacDaniels and Cowart, 1944)

generally smaller than those that emerge later during the period of rapid shoot elongation. Terminal leaves that develop later in the season as shoot growth rate slows are smaller, usually as a result of water stress and a diminishing supply of nitrogenous compounds.

An important morphological feature of leaves is the presence of the stoma (pl. stomata or stomates). A stoma, a Greek word for mouth, is a tiny pore surrounded by a specialized pair of guard cells on the epidermis. Stomata are located predominantly on the lower epidermis of leaves, but they are also present on the epidermis of stems, flowers, and fruit. Roots lack stomata.

The primary function of stomata is to govern the intake of carbon dioxide and transpiration of water. The two guard cells and adjacent subsidiary or accessory cells make up the stomatal apparatus. In this textbook, the stomatal apparatus and aperture together will be considered the stoma.

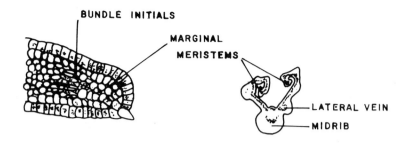

Figure 1.5
Cross section of a young apple leaf (right) and an enlarged section of a meristematic leaf margin (left). (Adapted from MacDaniels and Cowart, 1944)

Stomatal density differs greatly from species to species, ranging from 125 to over 1000 per square millimeter (Table 1). The sum of the areas of stomatal apertures on a leaf may be less than 1 percent of the total leaf area. However, stomatal positions and arrangements on the leaf surface are such that water loss through them can approach 50 percent of that lost by evaporation from a moistened filter paper having an area equivalent to the leaf.

Leaves of fruit trees found in the temperate zone are dorsiventral in that the external morphology and internal arrangement of mesophyll cells differ from the upper and lower sides of the leaf blade. The leaves form a mosaic allowing little light to penetrate through the canopy. A notable exception is the leaf of the pistachio of commerce, *Pistacia vera*. Leaves of this species are nearly isolateral in that the two sides are almost indistinguishable. Their leaflets are randomly oriented with respect to the position of the sun (Lin et al., 1984), so that the orchard floor is well lit even when the tree has a full canopy. This xeromorphic characteristic of avoiding the direct rays of the sun is not unexpected, considering that the species originated in the hot, dry climate of the Middle East.

Trichomes or epidermal hairs on the leaf surface are believed to reduce transpiration, especially if the stomata are sunken and covered with these trichomes. The hairs reduce the rate of air movement adjacent to the leaf surface so that the boundary layer of humid air is not disturbed. The lower epidermis of kiwifruit leaves is covered with stellate hairs borne on short stalks. Chestnut leaves have both stellate and glandular hairs;

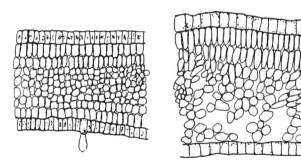 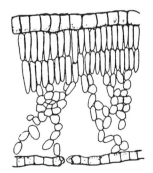

Figure 1.6
Transverse sections of a developing apple leaf in a stage of rapid cell division (left); in an early stage of intercellular space formation (center); and when fully mature (right). (Adapted from MacDaniels and Cowart, 1944)

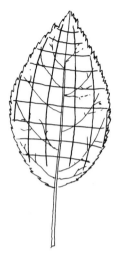

Figure 1.7
Mature apple leaf marked with grid lines while the leaf was young and expanding. The greater distances between lines at the lower midsection of the leaf blade reveal that most expansion occurred in that part. (Adapted from MacDaniels and Cowart, 1944)

the densities, kinds, and locations of these hairs are used as a taxonomic key to distinguish the four major species.

Root

MORPHOLOGY AND ORIGIN OF ROOTS

Roots are unlike shoots because they do not have leaves and nodes with buds, but they possess other morphologically distinct features (Fig. 1.8): (1) the hard root cap at the tip which is forced ahead between soil particles by expanding cells behind it; (2) the absence of the pith; (3) the presence of an endodermis, a primary tissue whose cells possess the Casparian strip (Fig. 1.9), a band of lignified or suberized substances encircling each cell tangentially; and (4) the pericycle, a layer of primary tissue immediately adjacent and interior to the endodermis. These features are confined to the root apex where there are three distinct regions: (1) cell division, (2) cell elongation, and (3) cell differentiation and maturation.

The tap root of a seedling tree originates from the radicle of a seed during germination. The radicle elongates, piercing the integuments, and grows into the soil. In the region of differentiation

Table 1. *Stomatal Density and Transpiration Rates of Some Fruit Tree Species*

Species	Stomatal Density[a]		Transpiration Rate[b]
	Upper	Lower	
Apple, *Malus domestica*	0	294	4.07
Black cherry, *Prunus serotina*	0	306	3.66
Sour cherry, *Prunus cerasus*	0	249	4.87
Peach, *Prunus persica*	0	225	4.85
Grape, *Vitis vinifera*	0	125	6.28
Mulberry, *Morus alba*	0	480	5.51/0.8
Pistachio, *Pistacia vera*	226	304	3.51/4121[c]
Pistachio, *Pistacia atlantica*	55	569	0.66/2.55[c]
Black walnut, *Juglans nigra*	0	461	4.89

[a] Stomates per square millimeter.
[b] Millimoles of water per square millimeter per second.
[c] Adaxial/abaxial sides of leaves (Lin et al., 1984).

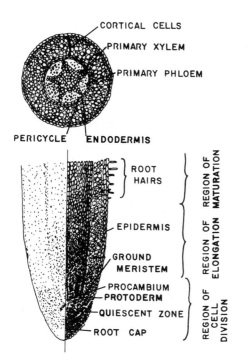

Figure 1.8
Cross and longitudinal sections of a root tip. The density of stippling on the lower sketch (left side) depicts the relative intensity of cell division. (From Jensen and Kavaljian, 1958)

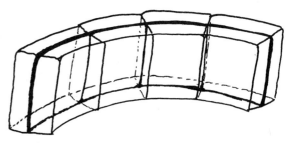

Figure 1.9
Four endodermal cells with their Casparian strip which is lignified and impermeable to water.

and maturation, zones of meristematic activity originate in the pericycle. These meristematic cells differentiate into lateral root initials (Fig. 1.10), which push through the cortical cells. They eventually protrude through the epidermis and begin functioning as roots.

Short-lived root hairs develop in many species from epidermal cells in the region of differentiation. They are sloughed off as the subepidermal cells become metacutinized and semipervious to water through the deposition of fatty substances, suberin, and lignin. Root hairs have not been observed on many species of *Prunus*, *Juglans*, or *Carya*.

Roots on stem cuttings arise from either masses of preformed initials, known as bur knots, in quince and apple, or from root initials formed by the resumption of meristematic activity in localized zones of the lateral cambium. These zones are usually located opposite a group of ray parenchyma cells. Once root initials are formed, their subsequent course of development is similar to that of roots which originate from the pericycle.

As cells in the region of elongation approach maximum size, they differentiate into specialized tissues: pith, central stele or vascular column, cortex, and epidermis. Cells in the stele, derived from the lateral cambium, differentiate into numerous vascular, storage, and structural elements: vessels, tracheids, fibers, and xylem parenchyma. Cells of the phloem differentiate into sieve tubes with companion cells, fibers, and phloem parenchyma. The cork cambium adds to the thickness of the bark. Thus, secondary growth of roots is similar to that of shoots.

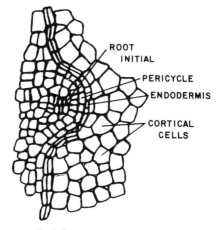

Figure 1.10
Enlarged section of a primary root through the pericycle illustrating the site of lateral root initiation.

The color of epidermal cells turns from white to brown as differentiation proceeds. Resistance to radial movement of some mineral elements increases with the formation of the endodermis and suberization of the walls of epidermal and cortical cells. In pear, the endodermal cells concurrently fill with tannins. Prominent lenticels develop from the periderm as secondary growth is initiated in *Prunus* and *Pyrus* roots. Lenticular cells are thin-walled and nonsuberized and are surrounded by large intercellular air spaces. These structures facilitate gaseous exchange between root tissues and the soil atmosphere.

Flower

MORPHOLOGY

Flowers are the reproductive structures of the angiosperms or Anthophyta. Although their shapes, sizes, coloration, and scents vary greatly, a typical flower consists of four whorls of modified leaves (Fig. 1.11). Acropetally, they are the sepals, petals, stamens, and carpel or pistil. Collectively, the sepals make up the calyx and the petals constitute the corolla. Together, the calyx and corolla are termed the perianth. The stamen and carpel are called androecium and gynoecium, respectively.

Pistil. A typical pistil has three parts: the stigma, upon whose surface the pollen germinates and penetrates the pistil; the style, which may be solid or a hollow canal; and the enlarged basal ovary, which develops into the pericarp of the fruit and houses the ovules or immature seeds.

Evolutionists believe that during the phylogeny of higher plants, the margins of modified seed-bearing leaves became appressed and united, enclosing the ovules and forming the locule (Fig. 1.12). Thus, the united ventral suture of the pistil is analogous to the leaf margins, and the opposite, adaxial, or dorsal side, is comparable to the midrib. During the ontogeny of a simple pistil, the pericarp closes as the ventral surfaces of the margins unite (concrescence) (Fig. 1.13; see also Figs. 5.10, 5.12).

Sepaloidy, a possible reversion of a pistil to an ancestral leaflike form (Fig. 1.14), occurs in strawberries. Botanically, the term is misleading, for these sepaloid structures arise from the receptacle, acropetal to the anthers, unlike the

Figure 1.11
Cutaway drawing of a peach flower showing four whorls of floral parts.

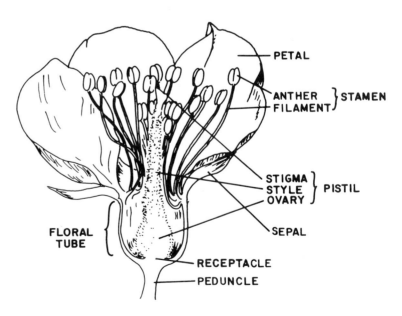

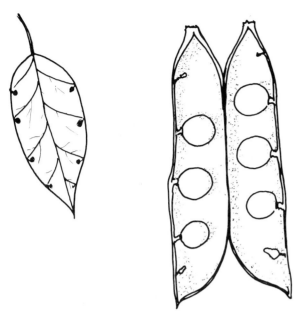

Figure 1.12
Hypothetical leaf (left) with naked seeds forming on the margins. Peapod (right) with ovules formed on funiculi arising alternately from ventral bundles. The dorsal bundle is analogous to the midrib of a leaf.

true sepals which are situated on the hypanthium or the floral tube.

A simple pistil consists of a single carpel, as in the drupe of a peach and the achene of a strawberry. A compound pistil has two or more united or fused carpels. Pistils of the currant, gooseberry, and grape (Fig. 1.15) have two fused carpels, and the persimmon has four. In some cultivars of persimmon, the carpels do not fuse during their ontogeny, a possible reversion to an ancestral form (Fig. 1.16). The pistil of the kiwifruit has more than 30 fused ovaries, but their styles are free (Fig. 1.17).

Pistils normally arise from the receptacle, but in abnormal cherry flowers they occur on filaments in place of anthers or adnate to the sides of the floral tube (Fig. 1.18). Could this be an evolutionary step toward the formation of fruit like the hips of the rose, in which pistils aggregate on the inner lining of the torus?

The stigma is a glandular tissue whose cells are rich in cytoplasm like those of nectaries. The epidermal cells develop into papillae or glandular hairs, as in cherry and other insect-pollinated species, or they remain nonpapillate. In wind-pollinated species, as in walnut, the long hairs give the bifurcate stigma a feathery appearance (Fig. 1.19). Hairs are often coated with a sticky stigmatic fluid, whereas the epidermal cells of the stigma are covered with cuticle. The tip of a germinating pollen tube secretes cutilase, an enzyme which dissolves the cuticle, allowing the tube to grow down between the stigmatic cells. The glandular stigmatoid tissue, also called the pollen-transmitting tissue or tract, extends from the stigma through the style to the placenta and nourishes the pollen tube.

The ovary develops into the edible fruit; in some species, such as apple and pear, a nonfloral organ called the accessory tissue makes up much of the edible part of the mature fruit. In walnut, the accessory tissue is the involucre or husk which consists of fused bracts (see Fruit Morphology later in this chapter).

Stamen. The stamen is a highly specialized, modified organ consisting of a very thin filament and an anther. The anther has two pairs of pollen sacs which dehisce longitudinally at anthesis and expose the pollen grains to various vectors (Fig. 1.20). In members of the Rosaceae, including stone fruits, pome fruits, strawberries, and brambles (bushberries), the androecia are spirally arranged. The anthers in the outermost whorl dehisce first, and dehiscence progresses acropetally as the filaments in the inner whorls elongate. The staggering of pollen dispersal over a period of several days increases the possibility of self- and cross-pollination.

Anthers seemingly produce gibberellins which cause filaments to elongate. Removal of anthers prevents further elongation of the filament, but this growth potential can be restored by applying gibberellic acid to the decapitated end of the filament (Greyson and Tepfer, 1967). Mutations of anthers and filaments, giving rise to petaloid structures, have resulted in "semi-

Figure 1.13
Whorls of anthers and the developing style of Redhaven peach pistil. Notice the ventral suture on the style. The outer whorls were removed. (From Warriner et al., 1985)

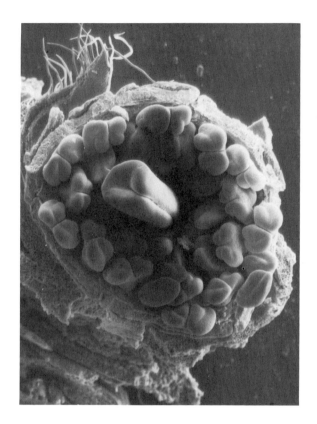

double" and "double" flowers. Many plants exhibiting these characteristics are prized as ornamentals.

Large numbers of androecia are borne on catkins in nut crops. The sessile anthers dehisce at different times along the spike, dispensing fine pollen grains that are transported by the wind. In chestnuts, basal catkins on a shoot bear only

Figure 1.14
Strawberry fruits in which sepal-like structures have developed in place of achenes.

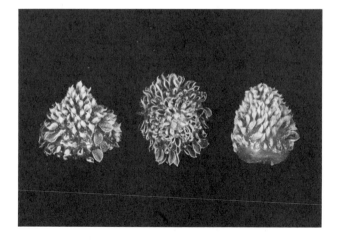

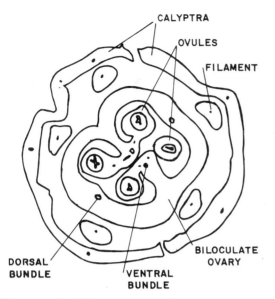

Figure 1.15
Cross section of a grape flower, *Vitis vinifera*, showing a bicarpellate pistil and calyptra.

Figure 1.17
Hypogynous flower of kiwifruit. Each style represents a fused carpel.

Figure 1.16
Persimmon fruits with varying degrees of fused carpels.

staminate flowers, whereas those nearer the terminal are bisexual. On the bisexual catkins, a few pistillate flowers are formed at the base, but the rest of the spike bears staminate flowers (Fig. 1.21). Occasionally, Persian walnut catkins manifest a bisexual condition.

Germinating pollen grains are the mature *male gametophytes*. They normally carry the haploid complement of chromosomes of the cultivar. Morphologically, pollen is classified as being binucleate or trinucleate, the binucleate condition being common in plants of the lower orders, including Juglandales and members of the Rosales. In both binucleate and trinucleate pollen, the vegetative nucleus becomes the pollen tube nucleus and trails just behind the elongating tip of the pollen tube as it penetrates the stigmatic surface and grows down the style. In trinucleate pollen the generative nucleus divides into two sperm nuclei within the dormant pollen grain, whereas in binucleate pollen this division occurs in the elongating pollen tube.

Figure 1.18
Abnormal cherry flower with pistils and petaloid structures borne on tips of filaments and the inner side of the floral tube.

Pollen has an elaborate external architecture (Fig. 1.22) consisting of a relatively thick, sturdy, chemically inert covering, the exine, and the underlying layer, the intine. The pollen tube emerges from one of several germ pores where the exine and intine are thin. The external appearance of pollen grains varies from species to species with respect to size, shape, and germ pore distribution.

Corolla. Petals that make up the whorl of the corolla possess the ability to synthesize a vast

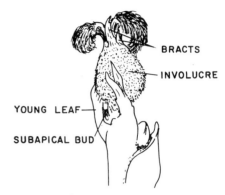

Figure 1.19
Walnut flower with bifurcate stigma.

array of specific pigments which attract pollinators such as bees, flies, moths, and even hummingbirds. Some species of tropical orchids and bees have coevolved so that the insects are indistinguishable from the parts of petals. Thus, bees are camouflaged and protected from predators while serving as pollen vectors.

Calyx. Sepals range in size from tiny lobes on an apple or pear flower to large green, fleshy structures in persimmons. In kiwifruit they are petal-like, whereas in the pomegranate the sepals are thick and leathery. In strawberry flowers the sepals and bracts (modified leaves intermediate between true leaves and the calyx) are indistinguishable.

CLASSIFICATION

Flowers are described and classified in many ways. When the floral parts, which are arranged in whorls or in spirals, are similar in shape and distributed equidistantly about the receptacle, the flower is said to be regular. If the parts are dissimilar in size and shape and do not radiate from the center, or if they are spaced unequally about the center, the flower is irregular.

Another distinguishing characteristic of flowers is the degree of coalescence or adnation of floral parts. When the parts of the perianth are not fused and each arises from the receptacle be-

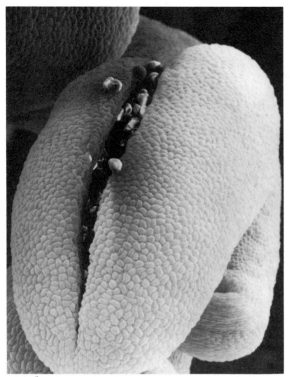

Figure 1.20
Anther of Redhaven peach rupturing along the suture, the stomium, consisting of thin-walled cells. (From Warriner et al., 1985)

low the pistil, as in kiwifruit and grape, the pistil is hypogynous or superior (Figs. 1.17, 1.23). When the basal portions of the calyx, corolla, and stamens are fused, as in stone fruits, the tubular part is called the hypanthium, torus, or floral tube. Horticulturists often refer to the floral tube as the shuck or jacket. When the hypanthium exists as a floral tube, the condition is said to be perigynous because the point of insertion or attachment of the floral parts is elevated and surrounds the pistil (Figs. 1.11, 1.24, 1.25). When floral parts arise above the ovary, as in pear and blueberry, the flower is epigynous, for the floral parts are above the gynoecium (Figs. 1.26, 1.27). An epigynous flower has an inferior ovary. The ovary of the pomegranate is only partially inferior (Fig. 1.28).

There are two schools of thought about how epigynous flowers originated. The *receptacular theory* holds that during the evolution of the se species, the ovary became enveloped by the fleshy receptacle. Anatomically, the receptacle is the stem end to which floral organs are attached. Hence, any manifestation of stem characteristics by the receptacle supports the receptacular theory. One of them is the occurrence of vegetative "pears" (Fig. 1.29) described by Kraus and Kraybill (1918). These pyriform structures borne on shoot terminals of Beurré d'Anjou trees lacked cores with their seed locules, indicating that the carpellary tissue was missing. The absence of other floral parts is not mentioned. Scaramuzzi (1953) observed pears with leaflike structures near the calyx. Since leaves originate from

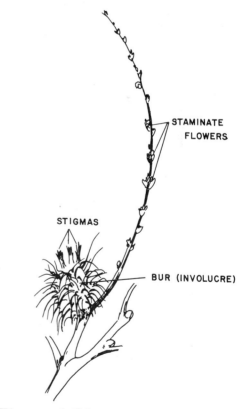

Figure 1.21
Terminal bisexual catkin of chestnut.

Flower

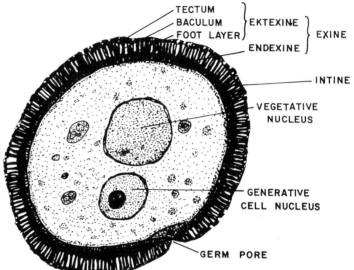

Figure 1.22
A binucleate pollen and its parts. The maze in the baculum fills with tapetal protein during sporogenesis.

stems, his observation supports the idea that the accessory tissue is of receptacular origin. Another supporting evidence is that a pear pedicel, which is of stem origin, often produces a soft juicy protuberance which tastes like the ripe fruit.

The *appendicular theory* contends that the accessory tissue evolved as a consequence of a fleshy hypanthium becoming adnate to the pericarp wall. Alternating vascular bundles in the accessory tissue of apple and pear lead to a sepal or petal, indicating that this part may have arisen from the torus. Some anatomists argue that if the accessory tissue were receptacular in origin, the floral bundles would separate at the distal end near the calyx lobes rather than from the end of the pedicel.

Bundle traces in the swollen portion of the tur-

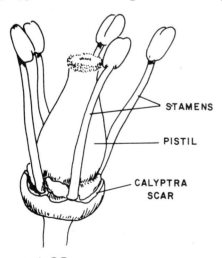

Figure 1.23
A hermaphroditic grape flower with a superior ovary. The calyptra has fallen.

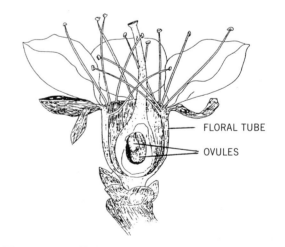

Figure 1.24
A perigynous apricot flower with part of the ovary cut away to show two developing ovules.

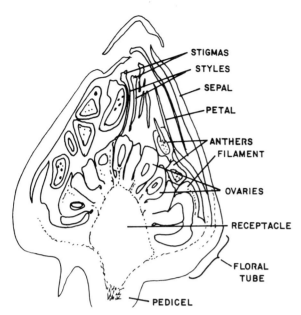

Figure 1.25
Longitudinal section of a perigynous raspberry flower bud showing multiple pistils. Each ovary develops into a drupelet.

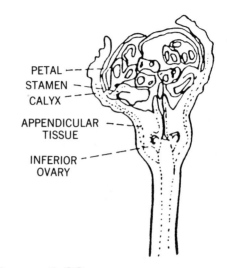

Figure 1.26
Longitudinal section of an epigynous pear flower bud.

ban squash indicate that the carpel is imbedded in the receptacle. Based on the vascular anatomy, the peel of the banana is also considered the receptacle and the edible portion is a tricarpellate ovary. Thus, inferior ovaries could have evolved through both pathways, depending on the species.

The flower is complete if all four whorls are present; it is incomplete if one or more of the whorls is missing. A flower is perfect or hermaphroditic if both male (androecium) and female (gynoecium) are present; it is imperfect if it lacks pistils or stamens. Hence, all imperfect flowers are incomplete, but incomplete flowers need not be imperfect. In categorizing some species for their floral characteristics, even a simple term as "imperfect" can be misleading. Some species in the genera *Actinidia*, *Diospyros*, *Vitis*, and *Fragaria* possess flowers that are functionally unisexual. Although both pistils and stamens are present, one or the other remains vestigial. Commercial strawberry cultivars have been selected for hermaphroditic flowers and fruit quality.

Normally, the sessile anthers in Persian walnut are borne on catkins or spikes, and pistillate flowers are borne terminally on short shoots or

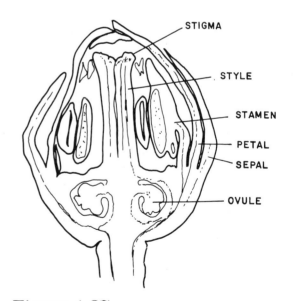

Figure 1.27
Longitudinal section of an epigynous highbush blueberry flower bud.

Flower 17

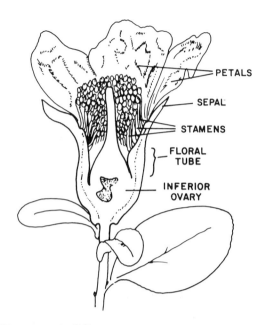

Figure 1.28
Semiepigynous pomegranate flower.

spurs. On rare occasions, catkins bearing pistillate flowers and pistillate flowers bearing stamens have been observed. Although these anomalies exist and are pointed out in this text, for the sake of simplicity, flowers with nonfunctional pistils or stamens are considered imperfect. Species are considered dioecious (*di*, two; *oikos*, house in Greek) if they bear these functionally male or female flowers on different plants. The plant is monoecious if pistils and stamens are borne on separate flowers but on the same plant.

CLASSIFICATION OF INFLORESCENCES

Inflorescences are structures or axes bearing two or more flowers. They are classified according to the pattern by which the flowers open and the manner in which the main axis branches to form lateral stalks. Pedicels are stalks that subtend individual flowers and arise from the main axis or peduncle. In the raceme (indeterminate), the first flowers to open are borne laterally while the terminal axis continues to elongate and produce more flowers. In the cyme (determinate), the first flower appears at the terminal of the main axis and younger ones appear below it (Fig. 1.30).

Racemose Types

1. **Spike.** Bears sessile (lacking a pedicel) or near sessile flowers, e.g., catkins of the walnut, chestnut, and filbert.

2. **Corymb.** Bears on a short peduncle several flowers having pedicels of nearly equal length, e.g., pear and cherry.

3. **Raceme.** Has an elongated peduncular axis with flowers on pedicels of equal length, e.g., red currants.

4. **Panicle.** Has a compound or multibranched inflorescence bearing loose racemose clusters of flowers, e.g., grape and pistachio.

Figure 1.29
Longitudinal view of a vegetative fruit of Beurré d'Anjou revealing the lack of ovarian tissues. (From Kraus and Kraybill, 1918)

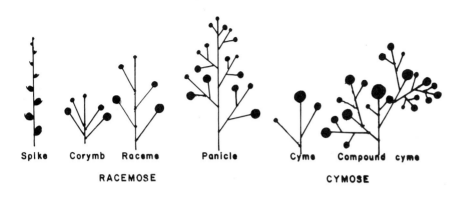

Figure 1.30 Diagrammatic representation of racemose and cymose inflorescences. The largest symbol on an inflorescence represents the oldest flower—the first to bloom.

Cymose Types

1. **Simple cyme.** Bears on a short peduncle flowers having pedicels of equal length, e.g., apple, kiwifruit, and male persimmon.
2. **Compound cyme.** Has a branched axis with flowers on pedicels of varying lengths, e.g., strawberry trusses, or of nearly equal lengths, e.g., raspberry and blackberry.

The inflorescence of the fig is unique; the individual fig flower is apetalous and imperfect. In the caprifig, the short-styled pistillate flowers line the inner wall (peduncle); the staminate flowers ring the ostiole or eye. Commercially important fig cultivars bear only long-styled pistillate flowers.

Fruit

MORPHOLOGY

Fruits are products of seed-bearing organs of higher plants, although some do not require fertilization of the egg and subsequent seed formation for the fruit to develop to maturity. Botanically, the pine cone that produces edible "pine nuts" or seeds and the seed-bearing structure of the gingko qualify as fruits. Horticulturists generally consider a fruit to be derived from the ovary, which may consist of a single carpel or a multicarpellate compound ovary, and/or any fleshy, edible tissue associated with it. The ovary wall or pericarp consists of three tissues: the epicarp or exocarp, mesocarp, and endocarp.

Epicarp or Exocarp. The epicarp or exocarp consists of the epidermal and subepidermal layers of fruits, which range in texture and appearance from glabrous (devoid of pubescence), e.g., nectarines; glaucous or pruinose (waxy bloom), e.g., plum; finely pubescent, e.g., apricot; or densely pubescent, e.g., peach, quince; bristly with branched coarse trichomes, e.g., kiwifruit; to a spiny covering, e.g., gooseberry.

Mesocarp. The mesocarp is the fleshy tissue between the epicarp and the endocarp, the innermost tissue of the pericarp. The mesocarp of an unripe drupe is fleshy, but, depending on the species, textural changes occur as the fruit approaches harvest. That of almond and pistachio becomes dry and leathery; that of coconut turns dry and fibrous. Mesocarps of stone fruits that soften as the fruits ripen differ widely in texture and juiciness. They may be firm or nonmelting, as in canning clingstone peach, or soft, as in a melting-type freestone peach. A comparable range in texture is found among cherry, plum, and apricot.

Endocarp. The morphology of the endocarp is an important key for classifying fruits. In true berries the texture of the endocarp is indistinguishable from the mesocarp. The endocarp of a pome fruit consists of thin layers of sclereids, giving it a papery texture. Endocarps of some blueberry (*Vaccinium*) cultivars are partially lignified. Endocarps of stone fruits vary from one species to another. Those of the peach and nectarine are thick and highly lignified and have a pitted surface. The apricot, cherry, and plum have flat, thin-walled endocarps with a coarse to smooth surface. The endocarps of almonds can be as hard as those of peach, or they can have a pithy outer covering and a highly lignified inner lining.

FRUIT TYPES

There are three types of fruits: simple, aggregate, and multiple. A simple fruit, such as peach and apple, develops from a single pistil. An aggregate fruit is derived from a single flower with many individual pistils. In the strawberry, for example, the individual pistils on the surface of the enlarged receptacle develop into achenes, the true fruits (Fig. 1.31). A raspberry is another example of an aggregate fruit in which numerous drupelets develop on the receptacle of a flower. In a multiple fruit, ovaries and floral parts of many individual flowers on an inflorescence become edible, for example, the mulberry, fig (Fig. 1.32), and pineapple.

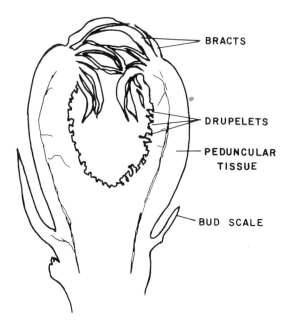

Figure 1.32
A multiple fruit of a young fig, a syconium. The true fruits are the drupelets, each surrounded by a fleshy sepal.

FRUIT CLASSIFICATION

The following key to fruit classification includes both pomological and nonpomological types to show that unrelated species may bear the same kind of fruit. Drupes, berries, nuts, and pomes, which make up most of the commercially important pomological crops, are described and illustrated in more detail than the non-pomological ones.

1. Fruits with dry pericarp at maturity.

    A. Indehiscent pericarp.

       a. Achene. Small, single-seeded, derived from a single ovary, e.g., the true fruit of the strawberry.

       b. Caryopsis or grain. Fruit of the grass family in which the pericarp is firmly united to the seed coat (integument).

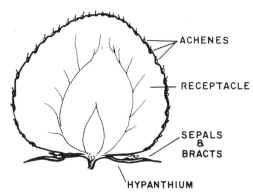

Figure 1.31
An aggregate fruit of the strawberry. The true fruits are the achenes appressed onto the surface of the receptacle.

c. Nut. Single-seeded and having a pericarp that ranges in texture from leathery, as in chestnut (*Castanea*), to highly lignified, as in filbert, syn. hazelnut and cobnut (*Corylus*). The pericarp can be partially or wholly surrounded by an involucre, as in filbert and walnut, respectively. The involucre consists of fused sepals, bracts, and bracteoles. In walnut (*Juglans*), the involucre or hull is adnate to the pericarp. The hull of the mature Persian walnut splits and separates from the inner endocarp, the shell; the hull of black walnut does not split or separate from the shell.

The morphology of the involucre ranges from a spiny bur in chestnut (Fig. 1.33), to a thin, papery shuck in filbert (Fig. 1.33), a smooth and fleshy shuck in pecan, hickory (*Carya*), and walnut (*Juglans*) (Fig. 1.33), or a woody cap in acorn (*Quercus*).

d. Samara. A winged achene, e.g., maple (*Acer*), elm (*Ulmus*), and wingnut (*Pterocarya*).

e. Schizocarp. A multicarpellate fruit of the carrot family (*Umbelliferae*); the carpels separate from each other, but the individual ovary does not dehisce along any suture.

B. Dry dehiscent fruits.

a. Capsule. Derived from a compound ovary, a fusion of two or more carpels that dehisce either lengthwise, e.g., iris (*Iris*) and tulip (*Tulipa*), or transversely, by splitting of caps or lids, e.g., plantain (*Plantago*) or by forming pores as in poppy (*Papaver*).

b. Follicle. Derived from a single carpel which usually dehisces along the ventral suture at maturity, e.g., magnolia (*Magnolia*), peony (*Paeonia*), and columbine (*Aquilegia*).

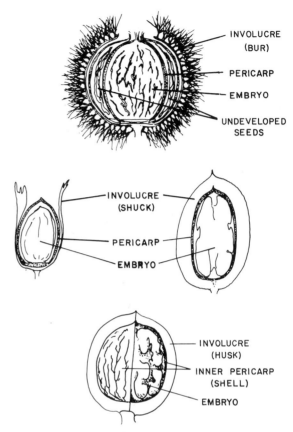

Figure 1.33
Nuts: chestnut (top), filbert (middle left), pecan (middle right), and walnut (bottom). Notice the variations in the morphology of the involucre.

c. Legume. Fruits of the pea family (*Leguminosae*) in which the mature ovary dehisces along the ventral and dorsal sutures.

d. Silique. Narrow, elongated, two-valved capsule, usually with many seeds, characteristic of fruits of the mustard family (*Cruciferae*). The fused carpels separate and dehisce, leaving a thin membrane to which a seed or seeds are attached.

2. Fruits in which the pericarp is entirely or partially fleshy or fibrous.

A. Indehiscent pericarp.

  a. Berry. Derived from a single or compound ovary and possessing one or more seeds. The mesocarp and endocarp are fleshy. The avocado (*Persea*) and the date palm (*Phoenix*) bear monocarpellate, single-seeded fruit. The berry of grape (*Vitis*) (Fig. 1.34) is a bicarpellate structure bearing a maximum of four seeds; that of persimmon (*Diospyros*) consists of four fused carpels, each with two seed locules (Fig. 1.35). In citrus (*Citrus*) and its close relative, kumquat (*Fortunella*), the fruit is a hesperidium. The modified pericarp is leathery and glandular, each segment representing a carpel with a membranous endocarp. The juice vesicle is comparable to internal hairs, originating and radiating from the inner lining of the endocarp (Fig. 1.36). The kiwifruit (*Actinidia*) has a compound ovary consisting of 30 to 45 carpels, with each locule containing from 10 to 30 seeds (Fig. 1.37). The banana (*Musa*) is an example of a monocotyledonous berry derived from an inferior ovary; the peel is receptacular in origin. The inferior ovaries of the cranberry (*Vaccinium*) (Fig. 1.38) are four-carpellate; those of blueberry are five-carpellate; and those of currant and gooseberry (*Ribes*) (Fig. 1.39) are bicarpellate. Because the endocarp is partially lignified in some

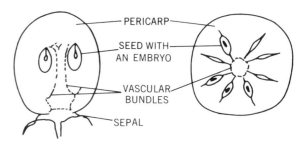

Figure 1.35
Longitudinal (left) and cross (right) sections of an oriental persimmon fruit. The berry consists of four fused carpels, each with two seed locules.

cultivars, morphologists often classify these two genera as pomes. The individual fruit of pineapple (*Ananas*) is also a monocotyledonous berry derived from epigynous flowers borne on a spike; the floral parts and bracts become tightly packed and succulent as they enlarge.

  b. Drupe. A fruit having one or more carpels in which the mesocarp is fleshy, fibrous, or leathery and the endocarp is lignified and stony. Most stone fruits, when ripe, possess a fleshy mesocarp. In some phenotypes the mesocarp separates readily from the endocarp or pit, a genetically

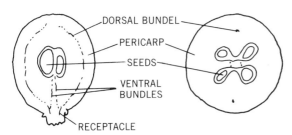

Figure 1.34
Longitudinal (left) and cross (right) sections of a bicarpellate grape berry.

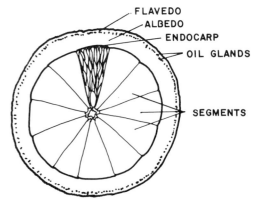

Figure 1.36
Cross section of an orange, a hesperidium. Each segment represents a carpel.

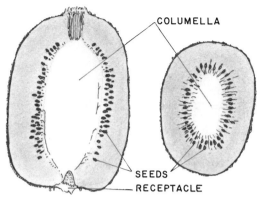

Figure 1.37
Longitudinal (left) and cross (right) sections of a mature kiwifruit. Each locule radiating from the columella represents a fused carpel.

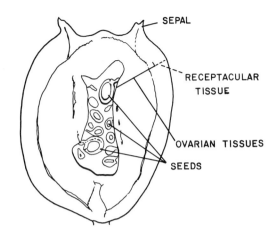

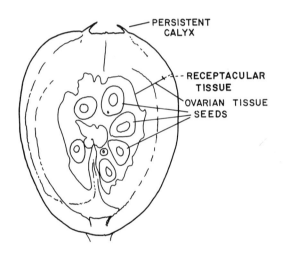

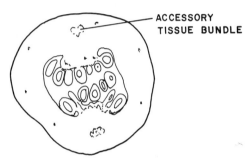

Figure 1.39
Longitudinal (upper) and cross (lower) sections of a young gooseberry fruit. The placentation is parietal.

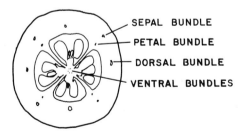

Figure 1.38
Longitudinal (upper) and cross (lower) sections of a young cranberry fruit. The placentation is axile.

dominant freestone characteristic; in clingstone cultivars, the mesocarp is tightly adherent to the endocarp (Fig. 1.40). Among the stone fruits, the almond is the exception; the thin leathery mesocarp does not enlarge during the last stage of fruit growth but dehisces at maturity. In some so-called paper shell cultivars, the endocarp cells contain little lignin so that the outer portion adheres to the mesocarp at maturity. The shriveling of the mesocarp forces the endocarp to split, exposing the embryo or kernel within. Pistachio is a unique drupe in which the mesocarp does not dehisce, but the endocarp within does (Fig. 1.41).

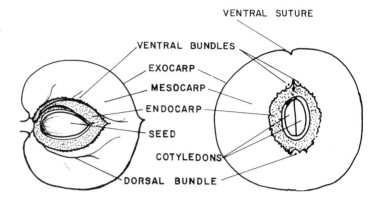

Figure 1.40
Longitudinal (left) and cross (right) sections of a ripe peach, a drupe.

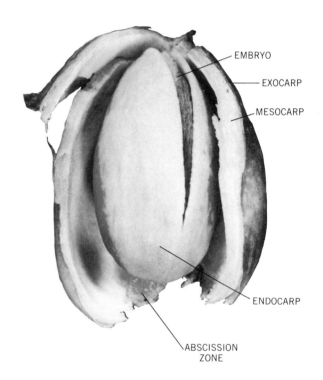

Figure 1.41
Drupe of a pistachio with the mesocarp partially removed to disclose a split endocarp and a fully developed seed. (Photograph courtesy of J. C. Crane)

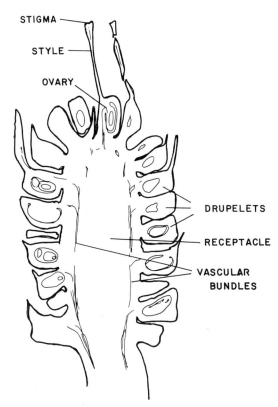

Figure 1.42
Longitudinal section of a young blackberry fruit, an aggregate fruit. The true fruits are the drupelets.

appendicular in origin and the ovarian tissues forming the core. In the apple (*Malus*), pear (*Pyrus*), and quince (*Cydonia* and *Chaenomeles*), the compound ovary has five locules, each representing a carpel (Fig. 1.43). The dorsal bundle of the carpel is just outside the locule, whereas the two ventral bundles that supply the seeds are located at the inner edges of the leathery endocarp. Bundles that supply the floral parts are located in the appendicular tissue. In some apple cultivars the ovarian tissue, or core, can

Individual fruitlets of raspberry, dewberry, and blackberry (Fig. 1.42) are drupelets. Those of raspberry are pubescent, whereas those of dewberry and blackberry are glabrous. Individual fruits of the mulberry and fig (*Moraceae*) are drupelets with accessory fleshy sepals. The fig inflorescence develops into a syconium, a large, urn-shaped fleshy fruit (Fig. 1.32). Upon ripening, the entire syconium, composed of a thick, fleshy peduncle bearing drupelets, becomes sweet and edible.

c. Pome. Derived from a multicarpellate inferior ovary, the accessory tissue being

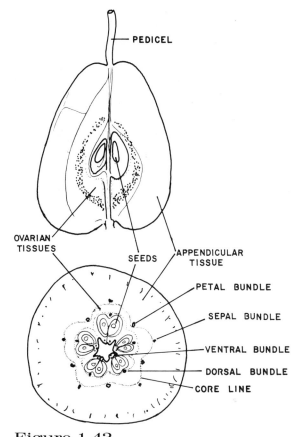

Figure 1.43
Longitudinal (upper) and cross (lower) sections of a European pear fruit, a pome.

Fruit

be delineated from the appendicular tissue by the difference in pigmentation.

PLACENTATION

The placenta is the ovule-bearing part of the carpel wall, and its location is related to how the margins of the carpel became united during the phylogeny or ontogeny of the fruit. In a simple pistil or apocarpous gynoecium which is described as being conduplicately folded, e.g., apricot, an ovule is borne on the inner side of the two ventral margins; the placentation is called marginal (see Fig. 5.12). When the margins of the folded carpels meet in the center of the fruit to form a bicarpellate or multicarpellate fruit, e.g., grape (Fig. 1.34), persimmon (Fig. 1.35), and cranberry (Fig. 1.38), the placentation is axile. In some species the partition between carpels disappears, resulting in a free central placentation. In species in which unfolded carpels are fused and ovules are formed at junctions of margins, e.g., gooseberry (Fig. 1.39) and currant, the placentation is parietal. When an ovule arises from the base of the ovary, e.g., walnut (Fig. 1.33), the placentation is basal.

References Cited

ESAU, K. 1958. *Plant Anatomy*, 2nd Ed. J. Wiley & Sons, Inc. New York.

FITZPATRICK, R. E. 1934. The ontogeny of the peach leaf. *Roy. Canad. Inst. Trans.* 20:73–76.

GREYSON, R. I., AND S. S. TEPFER. 1967. Emasculation effects on the stamen filament of *Nigella hispanica* and their partial reversal by gibberellic acid. *Amer. J. Bot.* 54:971–976.

JENSEN, W. A., AND L. G. KAVALJIAN. 1958. An analysis of cell morphology and the periodicity of division in the root tip of *Allium cepa*. *Amer. J. Bot.* 45:365–372.

KRAUS, E. J., AND H. R. KRAYBILL. 1918. Vegetation and reproduction with special reference to the tomato. Oreg. Agric. Coll. Exp. Sta. Bull. 149:1–90.

LIN, T-S, J. C. CRANE, K. RYUGO, V. S. POLITO, AND T. M. DEJONG. 1984. Comparative study of leaf morphology, photosynthesis, and leaf conductance in selected *Pistacia* species. *J. Amer. Soc. Hort. Sci.* 109:325–330.

MACDANIELS, L. H., AND F. F. COWART. 1944. The development and structure of the apple leaf. New York (Cornell) Agric. Exp. Sta. Mem. 258.

SCARAMUZZI, F. 1953. Ricerche sulla differenziazione delle gemme in alcune specie arboree da frutto. *J. Ital. Bot. Soc.* n.s. 60:1–101.

WARRINER, C. L., J. L. JOHNSON, AND M. W. SMITH. 1985. Comparison of the initiation and development of 'Redhaven' peach flowers in standard and meadow orchard trees. *J. Amer. Soc. Hort. Sci.* 110:371–378.

CHAPTER 2

Physiology and Functions of Buds and Stems

Control of Bud Development

DORMANCY AND REST OF BUDS

Most temperate zone plants, including deciduous fruit tree species, acquired certain gene controlled mechanisms and morphological structures as evolutionary adaptations to prevent buds and seeds from growing or to protect them during periods of inclement weather. Pomologists define dormancy and rest as two physiological conditions of the growing points that coordinate their growth with the onset of winter and that of spring. Dormancy is defined as a quiescent condition of shoot apices imposed by external conditions that are unfavorable for growth. Thus, in spite of being capable, buds fail to grow if the temperature is cold in the spring, and seeds do not germinate if the soil is too dry. Rest, on the contrary, is an internal condition that renders the apical meristem or the embryonic axis of a seed incapable of growing despite favorable environmental conditions (Chandler, 1957). Apparently, apices require time for endogenous changes to occur so that they can pass from the resting condition to the dormant one. Seed physiologists call this phenomenon after-ripening.

There is no agreement on the usage of these

terms. Plant physiologists substitute quiescence for dormancy and dormancy for rest (Salisbury and Ross, 1985). Westwood (1978) uses the terms *quiescence* and *dormancy* interchangeably and follows Chandler's definition for rest. Westwood defines the term *correlative inhibition* and cites the imposition of dormancy on lateral buds by the terminal bud of a shoot as an example (see Correlative Inhibition and Apical Dominance later in this chapter).

Lang and his colleagues (1985) have proposed the substitution of the terms *eco-dormancy* for dormancy, *endo-dormancy* for rest, and *ecto-* or *para-dormancy* for correlative inhibition. Pomologists are cooperating in this attempt to define and distinguish more precisely the different types of dormancy. This text will use Westwood's definitions of correlative inhibition, quiescence, dormancy, and rest.

Whereas the definitions of the status of buds are precise, the time when buds leave one stage and enter the next as the days become shorter and cooler in autumn is not precise because buds are not all the same age. This phenological period when buds undergo a gradual transition in the fall from a dormant stage into rest is called the onset of rest (Westwood, 1978). Under natural conditions, buds must be exposed to near-freezing temperatures to advance from the resting stage to a dormant one in the spring.

THE NECESSITY FOR REST AND CHILLING

The need for the bud to progress through a resting condition is an evolutionary adaptation, a safety mechanism, to assure that buds do not grow while inclement weather still prevails. The duration of exposure to cold required for resumption of normal shoot growth in the spring is called the rest period, and the amount of cold needed to satisfy the rest is called the chilling requirement.

The hours of chilling required before normal growth can occur vary from species to species and even within a given species (Table 2). Most of the old information on the chilling requirement was

Table 2. *Different Deciduous Fruit Species and the Duration Below 7° C (45° F) Needed to Satisfy Their Rest Requirements*

Species	Hours	Days
Almond	200–350	8–14
Apricot	700–1000	29–41
Apple	1200–1500	50–62
Cherries, sweet	1100–1300	46–54
Cherries, sour	1200	50
Fig	few hours	
Grapes, European	none	
Kiwifruit	450–700	19–29
Peach and nectarine	1000–1200	42–50
Pear, Asian	1200–1500	50–62
Pear, European	1200–1500	50–62
Persimmon, Japanese	less than 100	
Plums, European	700–1100	29–46
Plums, Japanese	700–1000	29–42
Walnuts, Persian	500–1500	21–62

derived empirically by comparing temperature data from weather stations with observations of fruit trees the following spring. If the bloom period was late and prolonged or if the trees exhibited symptoms of delayed foliation, inadequate chilling was suspected. Early pomologists integrated the degree-hours of chilling below 7° C from thermograph charts (Fig. 2.1) to estimate the chilling requirements of the different fruit tree species.

The technique for estimating rest requirements for different species has been refined. For most pome and stone fruit buds and seeds, 6° to 7° C (43° to 45° F) seems to be the optimal temperatures for satisfying the chilling requirement. Shoots collected periodically from trees growing in the orchard or in temperature-controlled growth chambers are placed in vases and forced into growth in a warm greenhouse. Buds on shoots collected early in the fall before exposure to cold do not grow; with prolonged chilling, the number of breaking buds increases and the growth rate of buds accelerates. The degree-hours required for 50 percent of the buds on these shoots to break are deemed to be the chilling requirement for the particular cultivar.

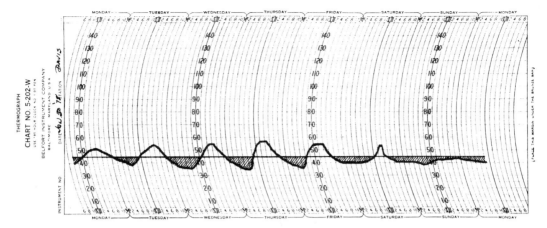

Figure 2.1
A thermograph chart showing daily temperature fluctuations during a week in January at a central California weather station. The hours below 45°F (hatched-line areas) were accumulated to derive the chilling requirements for different species in Table 2.

The end of rest is not predictable by a simple summation of the number of hours below 7° C because temperatures slightly above 7° C have a rest-breaking influence; temperatures below freezing are apparently ineffective. The Utah model (Richardson et al., 1974) for estimating the end of rest for peach trees takes into consideration the relative effectiveness of a specific temperature range. The model assigns a unit between 2.5° and 9.1° C and a 0.5 unit between 1.5° and 2.4° C and 9.2° and 12.4° C. The model assigns no unit for temperatures below 1.4° C and for the range between 12.5° and 15.9° C; it subtracts a 0.5 unit between 16° and 18° C, and a unit above 18° C.

Temperature summations alone do not give an accurate estimate of when the rest has been completely satisfied because the effects of other climatic factors such as rainfall, fog, and sunshine are not taken into consideration. Molisch (1921) demonstrated that budbreak could be advanced by bathing or sprinkling shoots of various species with water. He postulated that water-soluble growth inhibitors in buds were leached out by rain. Westwood and Bjornstad (1978) confirmed that rainfall shortened the rest period of apple and pear. In the Central Valley of California where fog may persist for days, budbreak appears normal even after a relatively mild winter. This behavior has been attributed to leaching of growth inhibitors, to evaporative chilling, and to low light intensity.

The role of light in breaking rest is unclear. Buds on potted Bartlett pear trees kept continuously in the dark broke sooner than those exposed to sunlight during the day but held at the same temperature. A question always arises whether solar radiation increases the temperature of the buds above that of the ambient air temperature, thereby prolonging the rest period. After rest is satisfied, any elevation of temperature should hasten budbreak.

Another complicating factor is the interaction between temperature and rootstock on the growth of the scion cultivar. Chandler (1960) demonstrated that shoots from apple scions grafted on chilled rootstocks grew much better than those on nonchilled ones. Westwood and Chestnut (1964) observed that Bartlett pear shoots from a Bartlett/*P. calleryana* combination grew more vigorously than those on a Bartlett/*P. communis*. Trees of both lots were chilled for a period of 850 hours, which is inadequate to satisfy the rest requirement of Bartlett. The difference in shoot vigor is attributed to the lower chilling requirement of the *P. calleryana* stock.

When Bennett and Skoog (1938) compared auxin contents of buds taken from two-year-old chilled and nonchilled cherry and pear trees, they found that the auxin level had increased in buds of chilled trees but not in buds of trees kept at room temperature. Injections into unchilled yearling apple, peach, and pear trees of auxin and yeast hydrolyzates containing substances promoting cytokinin activity induced budbreak. Bennett and Skoog concluded that the accumulation of an auxin precursor during the winter led to the breaking of the rest period.

The transmissibility of the stock effect to the scion and the implication that an auxin precursor acts as a rest-breaking agent are contradictory to the evidence that the effect of chilling is localized and not translocatable. That the cold factor is immobile has been demonstrated experimentally by exposing a branch to winter chilling outside a greenhouse while keeping the rest of the dormant plant inside. When the plant was taken outside in the spring, the chilled branch leafed out normally but the unchilled parts of the tree remained dormant (Fig. 2.2). Lavee (1973) has implicated the production of gibberellin and cytokinins and the depletion of abscisic acid as changes that lead to the completion of rest in the bud apices.

Chilling is known to cause other biochemical changes such as lowering the respiration rate of cells and activating certain enzymes. Enhanced amylase activity hastens starch hydrolysis so that more soluble carbohydrates are available for shoot growth. Contrarily, the exposure of unchilled tissues to elevated temperatures results in utilization of carbohydrates for respiration so that carbohydrates may become partially depleted. Tufts (see Chandler, 1957) found that flower buds on a chilled Lovell peach tree required only 16 days to reach the same stage of development that took 133 days for flowers on a tree kept in a warm greenhouse. Thus, chilling apparently increased the ability of flower parts to mobilize organic compounds. This ability may be attributed to increased cytokinin activity and/or to the increased permeability of cell membranes.

Figure 2.2
The effect of exposing branches (arrows) of a tree to winter chilling while keeping the remainder within the greenhouse. Notice that the unchilled portion has lost its leaves and the buds remain in the resting state.

OTHER ENVIRONMENTAL FACTORS GOVERNING DORMANCY AND REST

Rest in buds and cessation of shoot elongation probably evolved as defensive mechanisms against insufficient soil moisture, dry winds, heat stress during summer, and early frosts in autumn. In species such as the birch, cessation of shoot growth in the fall has been attributed to the shortening of day length. Shorter days are usually accompanied by lower temperatures, less intense light, and often depletion of soil moisture, all of which induce buds into the resting stage as a prelude to the onset of winter.

In contrast to chilling, high temperature may induce budbreak. Pear trees come into full bloom in the fall after a short heat spell; grape seeds soaked in warm water for 30 minutes will germinate and grow normally without chilling.

CORRELATIVE INHIBITION AND APICAL DOMINANCE

From ancient times, pruners have observed that removal of the terminal bud and notching (removing a semicircular piece of bark), scoring, or girdling branches above a lateral bud release it from dormancy. This suppression of growth of lateral buds by the terminal bud, known as *apical dominance*, is a form of correlative inhibition.

Other examples of correlative inhibition are (1) the presence of multilayered bud scales rich in abscisic acid which restricts bud expansion, but the restriction can be relieved by dilution or leaching of the hormone by rain; (2) the thick base of a grape petiole that suppresses bud enlargement, but its action is overcome by the leaf removal; (3) the thick endocarps that are rich in germination inhibitors, chemicals which are water-soluble; and (4) impervious integuments of seeds that prevent rapid diffusion of water and gases that can be made permeable by stratification, scarification, or removal to hasten seed germination.

Verner (1955) demonstrated that apical dominance is an auxin-imposed dormancy by showing that lateral buds of a chilled sweet cherry stem will grow if the apical bud is removed. But when he smeared the cut surface of a decapitated shoot with auxin suspended in lanolin paste, lateral buds below the cut did not grow. Verner therefore contended that auxin synthesized in the terminal bud is transported polarly, exposing subapical buds to supra-optimum levels of auxin that keep them dormant. As the auxin concentration becomes diluted during basipetal transport, lower buds on one- to two-year-old shoots overcome the inhibition and begin to grow. Experimental evidence supports the idea that polar transport of auxin is responsible for apical dominance: (1) notching or girdling a branch blocks downward transport of auxin (and other materials in the phloem), allowing buds below the wound to grow, and (2) treating dormant buds with triiodobenzoic acid (TIBA), an auxin-transport inhibitor, increases lateral branch formation, or "feathering," on nursery stock (Baldini et al., 1973).

Applications of benzyladenine, a synthetic cytokinin, release lateral buds on actively growing apple seedlings (Williams and Stahly, 1968). Brushing the hormone on peach branches during the winter was not effective in inducing budbreak (Weinberger, 1969), indicating that this hormone can overcome apical dominance but does not substitute for winter chilling to break the rest. The interaction between auxin and cytokinin in breaking rest is yet unclear.

The growth inhibition of an axillary bud by the subtending leaf is attributed to the production of growth-inhibiting substances and their subsequent export to the bud. Abscisic acid is presumed to be one such substance as it accumulates in bud scales. This idea is questionable because the removal of a grape leaf blade alone does not release the bud from its quiescent state. In this case the petiole must be stripped from the stem to induce the axillary bud to grow. Perhaps defoliation removes the source of a growth inhibitor and thus releases the bud from dormancy, but it also causes injury to the bark tissue adjacent to the bud. Injuries invariably induce evolution of ethylene gas, a wound hormone, which is known to cause budbreak. This dramatic response to defoliation has led grape growers in the Tropics to defoliate and severely prune vines to induce budbreak, thus allowing the harvest of three crops of grapes in two years.

The imposition of bud dormancy by bud scales is attributed to the accumulation of abscisic acid in the fall as the days become shorter. The release of buds from dormancy by removing bud scales is attributed to the elimination of growth inhibitors accumulated by the scales. Growth of a bud meristem following removal of bud scales has also been attributed to the evolution of a wound hormone.

If the winter has been cold enough to satisfy completely the chilling requirement of the buds and if the temperature rises rapidly so that growth can resume, buds emerge quickly. If, however, the temperature does not rise or rises slowly, buds will advance from the resting state but will remain dormant until temperatures become favorable for growth. When trees do not receive sufficient chilling, flower buds may abscise, and leaf buds will emerge slowly, giving rise to small, abnormal leaves. The last-named symptom is known as delayed foliation.

Bud, Shoot, and Spur Development

DETERMINATE AND INDETERMINATE GROWTH

A dormant terminal or lateral bud may contain all the leaf primordia and unextended internodes of a shoot. Extension growth involves only the elongation of internodes as a result of enlargement of preexisting cells followed by resumption of meristematic activity at the shoot apex. The initial flush of growth utilizes reserve carbohydrates, nitrogenous compounds, and minerals stored during the previous season (Davis and Sparks, 1974). As shoot elongation proceeds, new leaves that initially relied on the reserve food become independent and soon export photosynthates. The period in early summer when shoots elongate and leaves expand the most rapidly is called the grand period of growth.

Shoot growth in which a terminal leaf bud forms and elongates each year is said to be indeterminate or monopodial. Such growth is typically exhibited by the peach and almond. On other species such as the apple and pear, vigorous trees of some cultivars will produce long shoots with axillary flower buds and terminal leaf buds and occasionally terminal inflorescences. Hence, they exhibit both determinate and indeterminate shoot growth. On extremely vigorous apple and pear trees a lateral leaf bud on a cluster base or bourse (a thickened portion of a stem that subtends an inflorescence) does not merely elongate a few millimeters but often develops into a long shoot (see Fig. 11.1A). The growth of the grape shoot is truly indeterminate because the terminal bud does not set but continues to elongate, provided the temperature remains moderate and there is an adequate supply of mineral nutrients and soil moisture.

In a vigorous Persian walnut tree, the subapical bud pushes aside the terminal inflorescence and develops into an indeterminate vegetative shoot. The second flush of growth may elongate to 2 meters before ceasing. There are other species such as the apricot, filbert, and persimmon in which the terminal vegetative bud abscises; the subapical one resumes growth, provided the tree is vigorous. This cyclic shoot growth habit is called sympodial.

SPUR FORMATION

In spur-bearing species such as apple, pear, cherry, and plum, lateral buds that do not develop into branches the first season usually remain dormant. During the second year, some dormant buds may grow into long shoots while others develop into spurs, 3 to 10 millimeters in length, depending on their position on the shoot. In apples, basalmost nodes that lack buds are called "blind" nodes. Basal buds on the lower portion of long shoots tend to remain dormant, whereas the more distal ones often develop into spurs. As the branches grow older, spur development progresses downward. Spur formation on certain vigorous apple cultivars such as Granny Smith can be encouraged by heading back the shoot terminal (removing 25 to 30 percent) during the first dormant season. In the spring, buds near the cut will develop into extension shoots and the basal buds into spurs. Heading back of shoots is unnecessary on compact mutant-type apples because they readily form spurs.

In sweet cherry and some European plum cultivars, buds at the basal two to five nodes differentiate into flowers; there are no lateral leaf buds axillary to leaves as in peaches. Therefore, this portion of the shoot may bear flowers and fruits the following season but will be barren of leaves. Dormant buds near the shoot terminal of sweet cherry and some European plum cultivars are the first to form spurs or grow into lateral shoots; the development of spurs on shoots progresses basipetally, similar to that of apples and pears.

GROWTH IN GIRTH

The stem increases in diameter as a consequence of cell division in the vascular or lateral cambium and the cork cambium, or phellogen. The vascular cambium begins to divide and initiate daughter cells as buds swell prior to budbreak. Cell division begins at nodes just below a bud and progresses downwardly and laterally around the stem. In Bartlett pear, cambial activity begins at budbreak and ceases about four months later under conditions in California (Fig. 2.3; Evert, 1960, 1961). Early derivatives of the cambium differentiate into sieve tubes, companion cells, fibers, and phloem parenchyma cells. Differentiation continues in the phloem for eight weeks before vessel elements and tracheids begin to differentiate in the xylem. The two processes overlap for three weeks. Differentiation in the phloem then ceases, but that in the xylem continues for an additional six weeks until the first of August. The first phloem elements to appear early in the spring are those that formed late in the previous

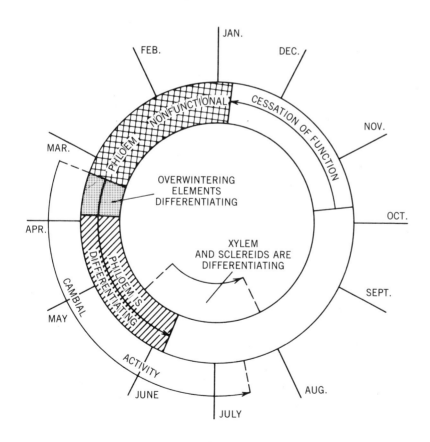

Figure 2.3
Diagram illustrating the seasonal changes that occur during one year of secondary growth in a branch of pear. (From Evert, 1961)

summer and remained undifferentiated through the fall and winter months. In other species the appearance of phloem elements precedes that of the xylem elements by only a week or two, unlike that of the Bartlett pear, indicating that species vary widely.

The amount of wood formed annually depends on environmental conditions. Under natural conditions, the width of the concentric annual rings consisting of spring and summer wood correlates well with the time and amount of summer and winter precipitations. In years of heavy rainfall, cambial activity continues longer into summer. Ecologists, foresters, and archaeologists take borings from trunks of old trees and timbers to measure widths of annual rings as clues to past rainfall patterns.

Under irrigated conditions, the increment of annual growth of fruit trees depends more on the crop load than on available soil moisture. A heavy crop, especially if the fruits ripen early and compete with shoot growth for current photosynthates and reserve food, curtails shoot elongation as well as growth in girth.

TENSION WOOD

Cross and longitudinal sections of upward curving branches reveal that annual rings are narrower on the upper side than on the lower side of the branch (Fig. 2.4). The upper portion containing tight, narrow annual rings is the tension wood, a form of reaction wood that is developed in response to the leaning of a trunk or bending of lateral branches. The formation of tension wood tends to reorient trunk and branches to the upright position. This unequal growth results from more auxin being translocated in the lower than in the upper side of limbs, thus establishing an auxin gradient. The elevated auxin concentration on the lower side promotes cell division and cell enlargement. Applications of auxin on the upper side of branches inhibit formation of tension wood, but administration of auxin-transport inhibitor, 2,3,5-triiodobenzoic acid, induces it (Mor-

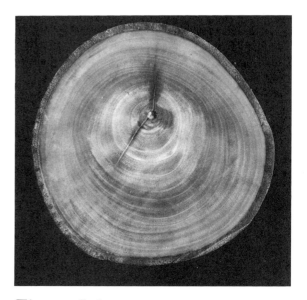

Figure 2.4
Cross section of a horizontally oriented walnut limb. The eccentric annual rings are wider on the lower side than on the upper side.

ey, 1973). When apple and peach branches were tied into tight arcs, the stress resulted in increased internal ethylene concentration (Leopold et al., 1972), but application of ethephon, an ethylene-generating compound, to eastern white pine enhanced formation of compression wood, a reaction wood peculiar to gymnosperms (Brown and Leopold, 1973). The interrelationship between auxin and ethylene in the formation of tension wood is not surprising because ethylene evolution is stimulated by application of synthetic auxin to tissue slices (Yang and Hoffman, 1984) and to intact fig fruits (Maxie and Crane, 1967).

One consequence of the thicker bark on the lower side is the formation of hangers (downward-growing branches), especially in peach trees (Fig. 2.5). These side branches begin as upright shoots the first year but gradually turn downward with age. This downward growth is attributed to (1) the overgrowth of the bark on the upper side of the hanger, and (2) the weight of fruits.

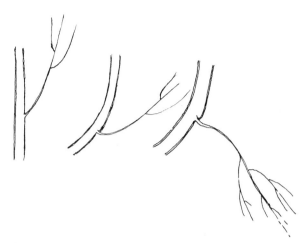

Figure 2.5
Formation of a hanger branch attributed to overgrowth of the bark on the upper side of the branch and to the weight of fruits.

SAPWOOD AND HEARTWOOD FORMATION

As trees age, the oldest cells in the xylem die and become nonfunctional. The vessels and tracheids no longer transport water and nutrients, and the ray parenchyma cells do not store starch. Phenolic substances, tannins, and other organic compounds in these old cells polymerize and become oxidized, staining the cell walls of the nonfunctional heartwood. The light-colored, functional wood adjacent to the bark is called sapwood.

Heartwood that forms early in the life of peach and apricot trees is susceptible to invasion by wood-rotting organisms. These organisms often gain entry through large pruning cuts. On the contrary, healthy apple and pear usually do not form heartwood, and the sapwood remains functional for several years. That the long-lived ray parenchyma cells in the sapwood of these species serve as storage tissue is manifested by the seasonal fluctuations in starch content; that is, starch increases during the summer and decreases in winter. Water transport, however, is restricted to the outermost, newly formed, mature vessels and tracheids of the xylem. Vessels are highly efficient in transporting water because they lack protoplasm. As vessel elements mature, their end walls dissolve and the protoplasmic contents become disorganized, so that they are drawn upward with the transpiration stream. This degeneration of nuclear material during the maturation of xylary elements may be one source of cytokinins found in xylem exudates.

Heartwood of some tree species such as the chestnut contains compounds that render the trees and the lumber made from them resistant to wood-rotting fungi. In species such as walnut, the heartwood with its deep rich color is highly prized for furniture or gunstocks. Heartwood of the tropical persimmon is the source of ebony wood. Religious carvings in medieval churches were made of pear wood because of its texture, workability, and light color. Perhaps in the future, our orchard trees will supply valuable wood for lumber as the natural stands are removed and not replenished.

Carbohydrate Storage

Xylem and phloem parenchyma (storage) cells in the above-ground portion of temperate zone tree species manifest at least two starch maxima and minima during their annual cycle (Hooker, 1920; Figs. 2.6, 2.7). The starch level reaches its lowest ebb at budbreak; it rises to the summer maximum as shoot growth slows down. Photosynthates, not converted to starch and stored in stems, are translocated downward and stored in roots. Rapidly ripening peaches (Ryugo and Davis, 1959) and figs create a large demand for carbohydrates over a relatively short period, causing starch previously stored in bearing branches to be remobilized and translocated to the fruits. Bearing apple and pear spurs and kiwifruit shoots do not manifest starch depletion as fruits enlarge because the fruits accumulate starch simultaneously. The starch content in pecan and

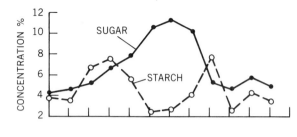

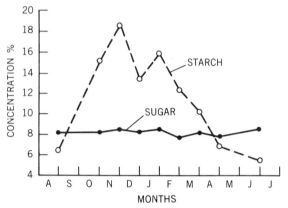

Figure 2.6
Seasonal changes in total sugar and starch contents in the pecan shoot (upper) and small roots (lower). (From Worley, 1979)

walnut branches does not fluctuate as the crop matures, presumably because the pericarp and embryo alternately assimilate dry matter over a relatively long period.

Starch is hydrolyzed to sugar in the fall as the days grow shorter and the nights become cooler. The conversion, completed by late fall to early winter, increases the osmotic concentration of the vacuoles. The high sugar concentration lowers the freezing point and prevents ice formation in the vacuole, thereby protecting trees from winter injury (see Protection Against Cold Injury, Ch. 7). A late winter peak in starch content occurs as starch is resynthesized from sugars. This reconversion of sugar to starch from midwinter to budbreak lasts about 90 days in pecan, 50 days in kiwifruit, and only 7 to 10 days in peach. Pecan roots do not exhibit a midwinter starch to sugar hydrolysis, possibly because the soil temperatures where they are grown are not sufficiently cold to induce this conversion.

The reconversion of sugar to starch exhibited by the pecan shoots and kiwifruit canes in late January is a common metabolic pattern for temperate zone deciduous trees and conifers. Almond, apricot, sweet cherry, peach, and European plum have widely different rest requirements, but analyses of their stems disclosed that

Figure 2.7
Seasonal fluctuations in total sugar and starch contents in kiwifruit branches. (From Grant, 1983)

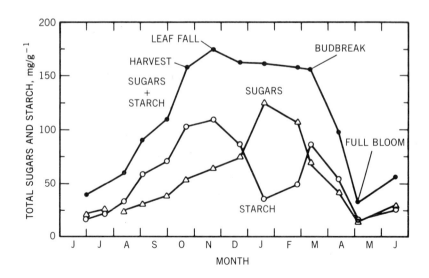

their midwinter starch minimum occurred at about the same time. Thus, the onset of reconversion of sugar to starch in midwinter does not seem to signal the end of rest. Starch grains in the amyloplasts may serve as part of the osmoregulatory function to prevent freeze damage during the winter by becoming hydrated (see Winter Injury and Acquired Cold Hardiness, Ch. 7). Starch rehydrolyzes to sugar rapidly as buds begin to swell; at budbreak little starch remains in stems.

Mutations, Bud Sports, and Chimeras

During rapid cell division in the shoot apex, the chromosomes of some cells frequently do not divide equally or become fragmented during mitosis, so that the two daughter cells are genetically different. These amitotic divisions occur at varying frequencies in somatic tissues such as the epidermis or the cortex and therefore go unnoticed. When these chromosomal aberrations occur in the apical cells of a bud and produce shoots and fruits noticeably unlike those of the parent stock, they are called bud sports.

When mutations occur so that the mutant and nonmutant characteristics are expressed phenotypically on the same organ, for example, the leaves on one side of a branch are variegated whereas other leaves are normal, the mixture is called a chimera. The Greek term *chimera* means "she-goat"; a chimera was a mythical flame-spouting female monster possessing a lion's head, the body of a goat, and a serpent's tail. In horticulture, the term applies to an organ such as a flower, fruit, or branch that consists of two or more unlike groups of cells (see Plates 1, 2, 3, 4). These aberrations are classified as sectorial, mericlinal, or periclinal according to their extent and depth (Fig. 2.8). In a sectorial chimera, the mutated stem or fruit may contain a pie-shaped sector that differs from the original organ. In a mericlinal chimera, a section of the epidermal and subepidermal tissues carries the mutant characters. In a periclinal chimera, a layer of mutant epidermal and subepidermal cells circumscribes the entire organ.

Mutations have given rise to many new cultivars, including many strains of Red Delicious apples and other commercially important fruit crops, such as plum, peach, nectarine, and European pear. The Delicious apple, the epidermis of which is red-striped on a green background, mutated to produce some solid red types (see Table 8, Ch. 11). The trees of these new mutant cultivars have produced isolated branches bearing fruits similar to the original Delicious, an indication that a reversion had taken place because the pigment-forming gene is unstable. These unstable mutant strains have been discarded by

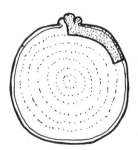 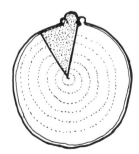 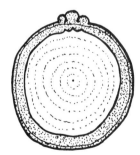

Figure 2.8
Cross sections of shoots depicting three kinds of chimera (stippled areas): mericlinal (left), sectorial (center), and periclinal (right).

growers and nursery personnel in favor of the more stable strains.

An apparent periclinal chimera occurred in the cortical layers of some apple cultivars. As a result, the inner cells of the cortex and the ovarian tissues of some apple cultivars remained diploid, whereas the outer cells and the epidermis became tetraploid and pigmented. A periclinal chimera in the epidermal cells of boysenberry gave rise to the cultivar Thornless Boysenberry. The mutant thornless character resides only in the epidermal cells from which thorns (prickles) arise. Therefore, cuttings from a Thornless Boysenberry mother plant will give rise to thornless daughter plants because the new shoots develop from buds that still maintain the thornless gene. Suckers arising from below ground, however, originate from roots as adventitious buds. These roots are not initiated from the epidermis of these cuttings but from subepidermal cells that carry the thorny gene. Hence, all cells of the adventitious buds carry the thorny character.

Mutant traits cannot be transmitted to offspring by hybridization unless the chimera includes tissues whose cells give rise to the embryo sac and/or the pollen mother cell. In the Thornless Boysenberry, only the epidermal cells mutated so that the chromosomes of the subepidermal cells that give rise to the embryo sac or pollen mother cell still carry the thorny character. Hence, self-pollination of this cultivar would not result in thornless individuals.

If the mutation involves tissues from which ovules and anthers arise, the egg and pollen grains may have the potential to transmit the mutant character to the next generation. Breeders have used this knowledge to incorporate the mutant trait into new cultivars.

Leaf variegation often originates as a periclinal chimera. If a vegetative bud originates from a juncture of a normally green tissue and a mutant tissue lacking chlorophyll, some leaves on this shoot will be variegated and others will be normal. This variegated pattern can be expressed by flowers and fruits on the same shoot. When a terminal bud acquires the albino characteristic (see Plate 2), the shoot that develops from it will eventually die because its leaves are not able to carry on photosynthesis, protein synthesis, and other vital processes.

On rare occasions, mericlinal or sectorial chimeras have originated near the base of grafted scions. Part of the shoot possesses characteristics unlike those of the stock or the scion cultivar. Such a shoot is called a graft chimera. On other occasions, entire shoots that are unlike those of the stock or the scion cultivar have arisen on a topworked tree. That these are true mutants has been questioned by some who consider them to be products of human error made during grafting. Recently, regeneration of shoot meristems from callus tissue *in vitro* has revealed that somatic mutations regularly occur during rapid cell division in callus tissues. This observation may explain graft chimeras and presumptive bud sports.

Fasciation is a mutation in which a flat, wide limb forms as though branches had fused together. As more branches join together, the grotesque structures resemble the antlers of a moose. Such abnormal growths have been noted in sweet cherry, pistachio, and Rome Beauty apple. Radiation of scionwood with gamma rays has induced fasciated branches. Natural cosmic radiation and some viruses are also believed to be responsible for the aberration.

References Cited

BALDINI, E., S. SANSAVINI, AND A. ZOCCA. 1973. Induction of feathers by growth regulators on maiden trees of apple and pear. *J. Hort. Sci.* 48:327–337.

BENNETT, J. P., AND F. SKOOG. 1938. Preliminary experiments on the relation of growth promoting substances to the rest period in fruit trees. *Plant Physiol.* 13:219–225.

BROWN, K. M., AND A. C. LEOPOLD. 1973. Ethylene and the regulation of growth in pine. *Can. J. For. Res.* 3:143–145.

CHANDLER, W. H. 1960. Some studies of rest in apple trees. *Proc. Amer. Soc. Hort. Sci.* 76:1–10.

DAVIS, J. T., AND D. SPARKS. 1974. Assimilation and translocation patterns of carbon-14 in shoots of fruiting pecan trees *Carya illinoiensis* Koch. *J. Amer. Soc. Hort. Sci.* 99:468–480.

EVERT, R. F. 1960. Phloem structure in *Pyrus communis* L. and its seasonal changes. Univ. of California, *Berkeley Publ. Bot.* 32:127–194.

EVERT, R. F. 1961. Some aspects of cambial development in *Pyrus communis*. *Amer. J. Bot.* 48:479–488.

GRANT, J. A. 1983. Influence of in-vine light environment on components of yield of kiwifruit (*Actinidia chinensis*, Planch). Master of Science thesis, University of California at Davis.

HOOKER, H. D., JR. 1920. Seasonal changes in the chemical composition of apple spurs. Mo. Agric. Exp. Sta. Bull. 40:1–51.

LANG, G. A., J. D. EARLY, N. J. ARROYAVE, R. L. DARNELL, G. C. MARTIN, AND G. W. STUTTE. 1985. Dormancy: toward a reduced, universal terminology. *HortScience* 20:809–812.

LAVEE, S. 1973. Dormancy and bud break in warm climates; considerations of growth-regulator involvement. *Acta. Hort.* 34:225–234.

LEOPOLD, A. C., K. M. BROWN, AND F. H. EMERSON. 1972. Ethylene in the wood of stressed trees. *HortScience* 7:175.

MAXIE, E. C., AND J. C. CRANE. 1967. Effect of ethylene on growth and maturation of the fig, *Ficus carica* L. fruit. *Proc. Amer. Soc. Hort. Sci.* 92:255–267.

MOLISCH, HANS. 1921. *Pflanzen-physiologie als Theorie der Gartnerei*, 4th Ed. Gustav Fischer. Jena.

MOREY, P. R. 1973. *How Trees Grow*. Arnold. London.

RICHARDSON, E. A., S. D. SEELEY, AND D. R. WALKER. 1974. A model for estimating the completion of rest of 'Redhaven' and 'Elberta' peach trees. *HortScience* 9:331–332.

RYUGO, K., AND L. D. DAVIS. 1959. The effect of the time of ripening on the starch content of bearing peach branches. *Proc. Amer. Soc. Hort. Sci.* 74: 130–133.

SALISBURY, F. B., AND C. W. ROSS. 1985. *Plant Physiology*, 3rd Ed. Wadsworth Publ. Co. Belmont, Calif.

VERNER, L. 1955. Hormone relations in the growth and training of apple trees. Idaho Agr. Exp. Sta. Res. Bull. 28.

WEINBERGER, J. H. 1969. The stimulation of dormant peach bud by a cytokinin. *HortScience* 4:125–126.

WESTWOOD, M. N. 1978. *Temperate Zone Pomology*. W. H. Freeman and Co. San Francisco.

WESTWOOD, M. N., AND H. O. BJORNSTAD. 1978. Winter rainfall reduces rest period of apple and pear. *J. Amer. Soc. Hort. Sci.* 103:142–144.

WESTWOOD, M. N., AND N. E. CHESTNUT. 1964. Rest period chilling requirement of 'Bartlett' pears as related to *Pyrus calleryana* and *P. communis* rootstocks. *Proc. Amer. Soc. Hort. Sci.* 84:82–87.

WILLIAMS, M. W., AND E. A. STAHLY. 1968. Effect of cytokinin on apple shoot development from axillary buds. *HortScience* 32:68–69.

WORLEY, R. E. 1979. Fall defoliation date and seasonal carbohydrate concentration of pecan wood tissue. *J. Amer. Soc. Hort. Sci.* 104:195–199.

YANG, S. F., AND N. W. HOFFMAN. 1984. Ethylene biosynthesis and its regulation in higher plants. *Ann. Rev. Plant Physiol.* 35:155–189.

CHAPTER 3

Physiology and Functions of Leaves

During the evolution of higher plants, leaves acquired specialized morphological and physiological characteristics to carry on photosynthesis and transpiration. Leaves also (1) biosynthesize compounds such as amino acids and proteins, florigen, pigments, growth-regulating hormones, phenolic substances that impart disease and insect resistance, and numerous others whose roles and functions are known and unknown; (2) store carbohydrates and nutrients, albeit temporarily, and (3) serve as photosensors for the phytochrome system.

Such functions as the production and accumulation of resins, gums, and latex do not seem particularly useful to plants, but these compounds have contributed greatly to our standard of living.

Photosynthesis

Photosynthesis is a process whereby atmospheric carbon dioxide is absorbed and reduced by green plant organs in the presence of light and water to form a simple carbohydrate. In the process, oxygen is released from water. The simplified chemical equation for photosynthesis is

$$6CO_2 + 6H_2O \xrightarrow[\text{chlorophyll}]{\text{light}} C_6H_{12}O_6 + 6O_2$$

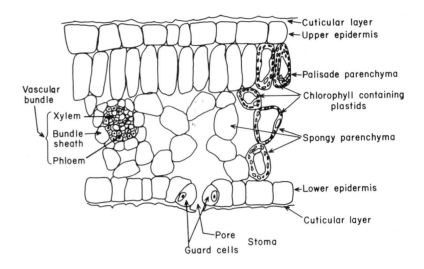

Figure 3.1
Cross section of a prune leaf. Chloroplasts are shown in a few mesophyll cells.

The primary organ adapted to carry out this process is the leaf because of its intricate morphology (Fig. 3.1), biosynthetic apparatus, and abundance of chlorophyll and accessory pigments. Bud scales, stems, fruits, and parts of flowers also possess chlorophyll, stomata, and lenticels for gaseous exchange, but their subepidermal tissues are not adapted for rapid diffusion of carbon dioxide and oxygen. Carbon dioxide assimilation by green, immature grape berries and calyces of persimmon has been measured, but the contribution of assimilates from these sources to total dry weight of fruits is relatively small (Kriedemann, 1968; Nakamura. 1967).

Chlorophylls a and b are the primary pigments that intercept light and ultimately transfer electrons necessary to combine hydrogen atoms from water with carbon dioxide, resulting in the evolution of oxygen. Based on their absorption spectra in an organic solvent (Fig. 3.2), chlorophylls were long thought to be the only pigments that participated in photosynthesis. But accessory pigments, carotenes, and other yet unknown pigments were found to be capable of absorbing light at wavelengths that were not absorbed by chlorophylls a and b (Fig. 3.3). The light energy absorbed by these accessory pigments is transmitted as high-energy electrons from pigment to pigment and finally to molecules of chlorophyll a. Since chlorophyll a, alone, can donate these light-activated electrons to electron acceptors, albino (nonchlorophyllic) leaves (see

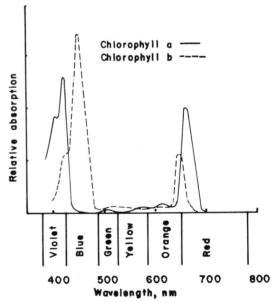

Figure 3.2
Absorption spectra of chlorophyll a and b.

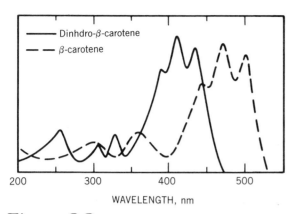

Figure 3.3
Absorption spectra of β-carotene and dihydro-β-carotene.

Plate 2) cannot fix carbon dioxide even though they possess yellow accessory pigments.

McCree (1972) calculated the relative quantum yield or the amount of carbon dioxide fixed per unit of absorption of quanta (photons) from the action spectrum, the energy per quantum, and the spectral absorptance of a leaf. He took leaves from 28 species grown in the field and growth chambers and exposed them to lights of different wavelengths. Simultaneously, their absorptance curves (Fig. 3.4A), which are the ratios of light absorbed by the leaves to the incident light and the amount of carbon dioxide assimilated at the same wavelengths, were determined. The relative quantum yield curves (Fig. 3.4B) reveal that carbon dioxide is fixed efficiently between 400 and 700 nanometers (nm). The fixation of carbon dioxide between the blue-violet and red lights is attributed to the presence of carotenes and related pigments. The energy per quantum is a func-

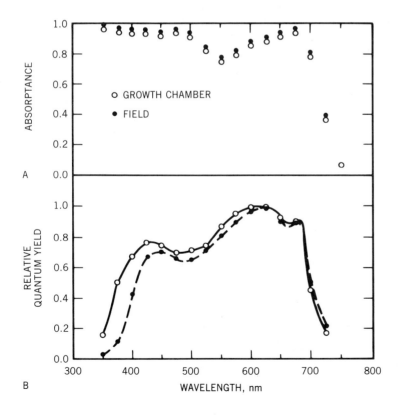

Figure 3.4
Mean absorptance values (upper) and relative quantum yield (lower) determined at different wavelengths for 28 species grown under field and growth chamber conditions. (Adapted from McCree, 1972)

Photosynthesis

tion of frequency; for example, radiations at 800 nm (infrared light) and 400 nm (violet) are 36 and 72 kilocalories per einstein, respectively.

Light intensity is measured in moles of photons or microeinsteins that impinge on a square meter of leaf area per second. (An einstein is a mole equivalent of photons or 6.02×10^{23} quanta.) The radiant energy is expressed as watts per square meter. In the literature, light intensity is referred to as the photosynthetic photon flux density (PPFD); previously, it was called the photosynthetically active radiation (PAR).

NET PHOTOSYNTHESIS AND NET ASSIMILATION RATE

Net photosynthetic rate, P_n, is measured in many ways, but one common method is to measure the amount of carbon dioxide absorbed by a unit leaf area per unit time. P_n is expressed as nanomoles of carbon dioxide assimilated per square meter of leaf area per second, or in some equivalent unit. P_n is used interchangeably with assimilation rate, A. The term *net assimilation rate* (NAR) was coined in England to express overall dry matter accumulation by entire plants per unit leaf area per unit time, usually as grams of dry matter per square meter of leaf area per 24-hour day or week. NAR values are lower than net photosynthesis rates because NAR takes into account the respiration of whole plants over a long time interval.

INFLUENCE OF LIGHT AND TEMPERATURE ON PHOTOSYNTHESIS

Assessing the total photosynthetic and transpiration capacities of fruit trees under field conditions is difficult for several reasons, but, first, because of their physical size. Second, not all leaves are equally exposed to light at any one time. Those on the tops of trees are continuously exposed to solar radiation most of the day, whereas those on lower scaffold branches intercept light intermittently as the sun moves across the sky or when winds move the limbs and leaves. Third, the mean leaf age and size are also constantly changing. During the grand period of growth in the spring, new leaves are rapidly unfolding while the full-sized basal ones are becoming older and more shaded as the foliar canopy thickens. Although very young leaves are highly efficient photosynthetically, much of their assimilate is utilized for their own growth and very little is exported to the shoot subtending them.

LIGHT SATURATION AND COMPENSATION POINTS

The light saturation point is the light intensity at which the photosynthetic rate reaches a maximum level at a given temperature, carbon dioxide concentration, and relative humidity (Fig. 3.5). Additional light beyond this intensity does not increase the carbon dioxide assimilation rate. The photosynthetic apparatus in most temperate zone deciduous fruit trees is saturated between 450 and 650 micromoles of photons per second per square meter of leaf, which are, respectively, 24 and 31 percent of full sunlight (about 2000 micromoles per second per square meter).

Depending on the light intensity, a single mature leaf will transmit only 10 percent of the incident light that falls on its surface; another 10 percent is reflected. Of the portion absorbed, some is dissipated to the surrounding environment as heat and a large fraction is used in the transpiration process, that is, vaporization of water. It is estimated that, on the average, only 0.5 to 3.5 percent of the incident solar energy is utilized for photosynthesis. Unless the canopy is opened up by pruning, leaves in the interior of the canopy are rarely exposed to light-saturating conditions, even for very short periods.

The compensation point is the point at which leaves neither gain nor lose dry matter, that is, the photosynthetic rate is equal to the rate of respiration (Fig. 3.5). Shaded, interior leaves of kiwifruit fix carbon dioxide at a rate barely above the compensation point, even when the sun is at zenith (Fig. 3.6).

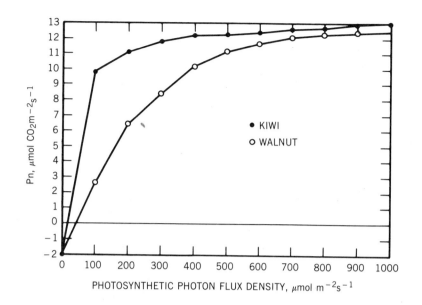

Figure 3.5
Relation between net photosynthesis (Pn) and photosynthetic photon flux density of kiwifruit and walnut leaves. At saturation point, Pn curves approach a constant rate; at compensation point, Pn equals 0. (From Grant and Ryugo, 1984; Tombesi et al., 1983)

Because the parameters that affect photosynthesis cannot easily be controlled under field conditions, compensation and saturation points for plants are better estimated under closely controlled laboratory or greenhouse conditions. The advantages of measuring photosynthesis on potted plants in a growth chamber under a closely monitored system are, first, that a single environmental factor can be altered stepwise while others are kept constant. For example, the light intensity can be raised in small increments while keeping the ambient temperature constant. Similarly, the temperature can be varied while keeping light intensity constant. Second, the test

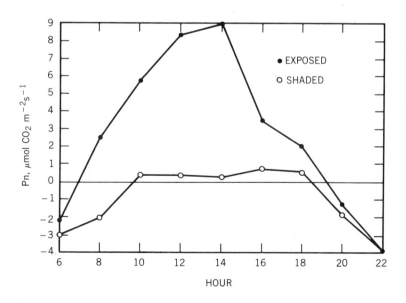

Figure 3.6
Daily trend of net photosynthesis (Pn) of Hayward kiwifruit leaves in exposed and shaded portions of the canopy. (From Grant and Ryugo, 1984)

Photosynthesis

plants can be undergoing differential treatments with growth regulators or fertilizers concurrently.

Under such steady state conditions, light saturation and compensation points can be more precisely determined with minimal variations compared to readings obtained in the field. The optimal temperature range for Pn was from 18° to 25° C for almond, kiwifruit, peach, and walnut, provided light intensity was maintained at 900 micromoles of photons per square meter per second (DeJong, 1983). The mean values of Pn obtained from leaves of detached walnut shoots and potted kiwifruit vines under such controlled conditions closely approximated those collected under field conditions (Tombesi et al., 1983; DeJong et al., 1984).

Transpiration

Transpiration is the process by which plants lose water. Stomatal transpiration, or the loss of moisture from the leaves via the stomata, accounts for over 90 percent of the water absorbed by roots. The remaining water is lost through cuticular transpiration, lenticels, and guttation.

WATER POTENTIAL

Plant physiologists use the term *water potential*, denoted by the Greek letter psi (ψ), to express the status of water pressure in soil–plant–atmosphere relationship. Water moves along a water potential gradient from a region of high ψ to that of low ψ. The metric unit of water potential is the bar, which is equivalent to 0.1 megapascal, or 0.987 atmospheric pressure, or 750.12 millimeters of mercury, or 10 dynes per square centimeter. The water potential of a soil solution is about -0.1 bar; that of the water in the xylem during the day may approach -12 bars. A tension of 12 bars is estimated to be sufficient to pull a column of water in the xylem to a height of 30 meters.

The osmotic potential, pressure potential, and matric potential are three factors that determine the water potential of cells at a given temperature. The lowering of water potential as a result of imbibition or adsorption of water by cellulosic micelles or clay particles is called matric potential, ψm. It is not considered in the water potential equation of an osmotic system:

$$\text{Psi } (\psi) = \text{osmotic potential } (\psi \pi) + \text{pressure potential } (\psi p)$$

Osmotic potential and pressure potential were formerly called osmotic pressure and turgor or wall pressure, respectively.

The opening and closing of the stomata are controlled by the osmotic concentration of the guard cells in relation to the surrounding epidermal cells. The mechanism by which solutes move in and out of the guard cells is yet unclear. The rapid widening of the apertures has been associated with the hydrolysis of starch, whereas the closure of stomata because of water stress has been attributed to the increase in abscisic acid and decrease in cytokinins (Kuiper, 1972). The influx and efflux of potassium ions have been correlated with the opening and closure of stomata (Humble and Hsiao, 1970; Humble and Raschke, 1971). Guard cells respond quickly to light; those of walnut and kiwifruit leaves open within 15 minutes of sunrise when the light intensity, PPFD, is less than 50 micromoles of photons per second per square meter.

The process of transpiration not only moderates leaf temperature by evaporative cooling, but also serves to bring dissolved salts and organic substances synthesized in the roots or arising from disorganization of maturing xylem vessels and tracheids to the shoot extremities. Since both water loss and carbon dioxide uptake occur through the stomata, there is a good correlation between stomatal conductance, g, the rate at which a volume of water vapor moves out of leaves, and photosynthetic rate (Fig. 3.7). Stomatal conductance is determined with a porometer and is expressed as centimeters per second. Stomatal resistance, r, is the reciprocal of

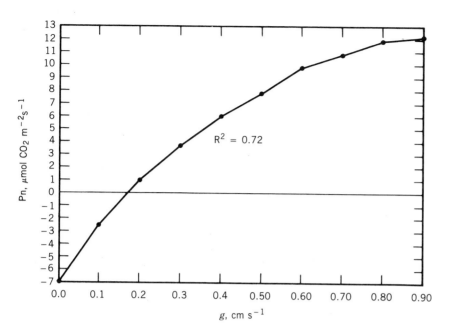

Figure 3.7
Curvilinear relationship between net photosynthesis (Pn) and leaf conductance (g) in kiwifruit leaves. (From Grant and Ryugo, 1984)

conductance, $1/g$, and measures the resistance of the stomatal aperture to movement of water. Transpiration rate, T, is also measured by means of a porometer and is expressed as millimoles of water lost per square meter of leaf area per second.

RELATIONSHIP BETWEEN Pn AND T

The term *water use efficiency* (WUE) is used to describe the relative efficiency of plants in assimilating carbon dioxide per unit weight of water transpired. WUE is determined for different species or cultivars grown under certain environmental conditions by plotting Pn versus T. The slope of the derived curve, dPn/dT, gives an estimate of WUE; in Figure 3.8 a plant exhibiting a steep curve a, uses water more efficiently than the one with slope b.

DAILY TRENDS OF Pn AND g

The daily trends of Pn of fully expanded leaves of apple (Fig. 3.9), kiwifruit, (Fig. 3.6), pistachio, and walnut (Fig. 3.10) taken on different dates and at different places vary slightly. These variations may be reflections of differences in the response of species to the environment and/or the conditions under which the measurements were

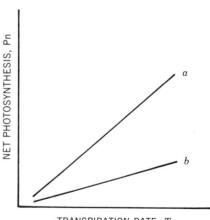

Figure 3.8
Relation between net photosynthesis (Pn) and transpiration rate (T). The slope indicates water use efficiency (WUE), so that a plant with slope a is more efficient than that with slope b.

Transpiration

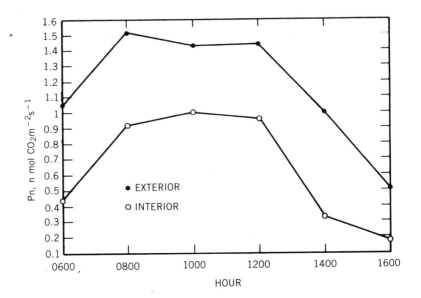

Figure 3.9
Daily trend of net photosynthesis (Pn) of Bancroft apple leaves located on the exterior and interior of the canopy. (From Mika and Antoszewski, 1972)

taken. The daily trends of light intensity closely parallel the amount of incident light that impinges on the leaf surface.

Pn and g values for leaves reveal that stomata open within 15 minutes of daybreak. This early opening reflects the sensitivity of guard cells to light. Hence, Pn can surpass the compensation point at very low light intensities. Pn reaches a near-maximum at about 0800 hours and slowly increases to a maximum level near midday. The gradual increase in Pn during the late morning hours is attributed to the saturation of the photo-

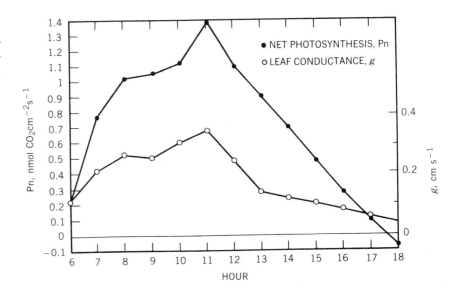

Figure 3.10
Daily fluctuations in net photosynthesis (Pn) and leaf conductance (g) of walnut leaves. (From Tombesi et al., 1983)

3 / Physiology and Functions of Leaves

synthetic apparatus. Pn decreases in the afternoon, owing to increased water stress and to feedback inhibition of photosynthesis.

The lower light saturation point of kiwifruit leaves, compared to that of walnut (Fig. 3.5), probably reflects an evolutionary adaptation of plants having a liana-type growth habit. In the wild, kiwifruit vines are natural tree climbers which must spend their early years on the dimly lit forest floor while seeking other plants that would physically support them.

ENDOGENOUS AND ENVIRONMENTAL FACTORS THAT INFLUENCE Pn, *g*, AND *T*

Leaf Age. Young leaves emerging in the spring rely on stored food in the adjacent stem, scaffold branch, trunk, and roots for initial growth. When they are about one-half to three-fourths full size, they become self-sufficient and export part of their photosynthetic products to younger leaves, more distal on the stem. By midsummer, they export much of their current photosynthates to branches, trunk, and roots. Pn of apple leaves attains the maximum value 30 to 45 days after full leaf expansion (Fig. 3.11) and then declines.

Chlorophyll and Nitrogen Content. An important participant in the photosynthetic reaction is the green pigment chlorophyll. A correlation between Pn and the chlorophyll or nitrogen content of leaves has been difficult to establish because under field conditions Pn fluctuates daily, whereas chlorophyll and nitrogen contents of leaves remain relatively constant during the day. However, when experiments were conducted on five *Prunus* species and Golden Glory peach which were nitrogen-sufficient and nitrogen-deficient, a high correlation existed between Pn measurement and nitrogen content (DeJong. 1982, 1983; DeJong et al., 1984; Fig. 3.12). A similar experiment with kiwifruit revealed a proportionality between Pn and leaf chlorophyll content (Fig. 3.13A). Nitrogen-deficient kiwifruit leaves with about 60 percent less chlorophyll assimilated carbon dioxide about 40 percent more slowly than nitrogen-sufficient leaves. Starch content of nitrogen-deficient leaves was greater than that in leaves supplied with adequate amounts of nitrogen (Fig. 3.13B). Normally, one would expect plants with greater Pn to contain more starch. Under conditions in which the nitrogen supply is inadequate, the process of amination seemingly becomes limiting, increasing sugar levels. As a

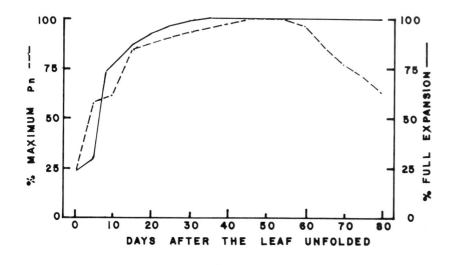

Figure 3.11
Relation beween net photosynthesis (Pn) and leaf size and age. (From Sams and Flore, 1982)

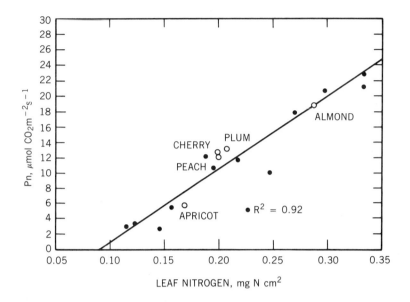

Figure 3.12
Relation between maximum net photosynthesis and leaf nitrogen content within a single peach cultivar, Golden Glory (●), and among five *Prunus* species (○). (Adapted from DeJong, 1982, 1983)

consequence, starch is deposited in the chloroplasts.

The correlation between Pn and nitrogen might be expected because (1) a chlorophyll molecule contains four atoms of nitrogen, and (2) the photosynthetic process involves many enzymes that are proteinaceous and require nitrogen for their synthesis. The predominant photosynthetic enzyme in green leaves is ribulose bisphosphate carboxylase. If nitrogen is deficient, synthesis of certain anabolic proteins and enzymes and cytokinins ceases, and the synthesis of enzymes that promote senescence, for example, chlorophyllase, is initiated.

Presence of a Crop. Chandler and Heinicke (1925, 1926) studied dry matter residues accumulated by bearing and nonbearing Oldenburg apple and grapevines for several seasons. At the end of the experiment, they found that bearing plants had fewer leaves and were smaller than nonbearing plants. But when the weight of the dry matter removed annually by the crops was included, the total dry matter produced by bearing trees and vines was greater than that of nonbearing ones. Hence, they concluded that the presence of a crop improves photosynthetic efficiency.

Figs and peaches import large quantities of photosynthates during their ripening process. Starch, previously stored in the parenchyma cells of the bark and wood, is hydrolyzed to supply the crop with carbohydrates (Crosby, 1954; Ryugo and Davis, 1959). The rapid increase in starch content in branches immediately after harvest suggests that, although leaves are photosynthesizing at a maximum rate, the assimilation rate is still not adequate to meet the large demands created by the competition among the crop, shoot and root growth, and storage cells.

The presence of a crop on the tree not only enhances the photosynthetic efficiency of leaves (Avery, 1975) but allows the tree to transpire more water because the stomata are open for a longer period. Comparison between bearing and nonbearing apple trees grown under two nitrogen levels revealed that bearing trees utilized more water than nonbearing trees. Trees given higher nitrogen dosages used more water than those under low nitrogen regimens (Lenz, 1985) because they had larger, greener leaves.

Bearing trees require more water than non-

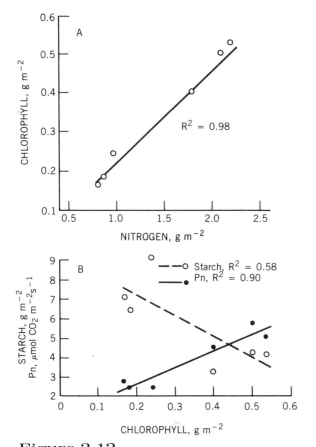

Figure 3.13
Relations between chlorophyll and nitrogen contents on an area basis (upper) and between net photosynthesis (Pn) and starch content versus chlorophyll contents of kiwifruit leaves (lower). (Adapted from DeJong et al., 1984).

bearing trees, even though the fruiting ones may have a smaller leaf area. The amount of water retained by the crop and tree, however, is a small fraction of that transpired by leaves during the season. For example, a peach fruit consists of 85 percent water. Thus, a crop of 22.7 metric tons per 0.404 hectare (25 tons/acre) contains 19.3 metric tons (21.3 tons) of water. To produce such a heavy crop in the San Joaquin Valley of California, growers irrigate the average peach orchard on loamy soil with a volume of water equal to 15.6 centimeters per 0.404 hectare (6 acre-inches) at least seven times during the growing season. This is equivalent to 4.41 million liters (1.14 million gallons) or 4410 metric tons (4,860 tons) of water per 0.404 hectare (acre). Thus, the crop uses only 19.3/4410 × 100, or less than 0.5 percent of the irrigation water.

Light Quality. Plants grown in glass or plastic houses have lower Pn values than comparable plants grown outdoors. Not only do glass and plastic change light quality, but their coating of whitewash during the summer months also decreases light intensity. Atmospheric pollutants, smog, and dust particles scatter and alter wavelengths of light, especially ultraviolet rays, so that light quality changes. Dust particles that settle on leaves reduce light intensity and absorb heat that could adversely affect net photosynthesis.

Light Intensity. Some species cannot tolerate intense light or the resultant heat created by it. Leaves of *Camellia japonica* exposed to direct solar radiation tend to fade and sunburn. Hartley walnut leaves exposed to sunlight have greater specific leaf weight (SLW) and contain more chlorophyll per unit area than those in the shade (Fig. 3.14). Specific leaf weight is the dry weight of leaves expressed on a per unit area basis, usually as milligrams per square decimeter or centimeter.

The heavier SLW of exposed leaves is attributed to the greater Pn and T than those of shaded leaves. Cells of exposed leaves have thicker walls and a greater number of proteins. The high transpiration rate of exposed leaves allows greater accumulation of nutrients, including nitrates, than in shade leaves. The nitrate ions entering grape, kiwifruit, and walnut leaves are reduced to ammonia more rapidly in light than in the shade. The resulting ammonia participates in the amination process whereby amino acids and other nitrogenous compounds are formed.

Kiwifruit leaves, contrary to leaves of fruit

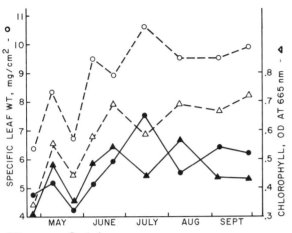

Figure 3.14
Seasonal changes in specific leaf weights and chlorophyll contents of exposed (open symbols) and shaded (solid symbols) of Hartley walnut leaves. (From Ryugo et al., 1980).

tree species, grow larger in the shade and contain more chlorophyll, but their specific leaf weights are lighter than those of leaves grown in the sun. This may be typical of plants that evolved as tree climbers under diffuse light conditions.

Soil and Plant Water Potentials. Xylem water potential, ψ, is restored to nearly zero by daybreak if trees are well supplied with water and their leaves are fully turgid. The exception would be a species that exudes xylem sap under root pressure at certain times of the year. Xylem water potential may reach 1 to 2 bars (see Root Pressure, Ch. 4). As the morning light strikes the leaves, stomata open and transpiration begins. By midmorning the transpiration rate soon exceeds the rate of water absorption by roots. The column of water in the transpiration stream comes under tension, thus becoming increasingly negative. This results in stomatal closure and often leads to incipient wilting and/or leaf curvature, which reduces the leaf area facing the sun; Pn and g decline rapidly. If soil moisture is adequate, plants normally recover from daily incipient wilting during the night when stomata are closed. By dawn, leaves become fully turgid so that Pn and T may proceed at rapid rates. Xylem water potential approaches but rarely attains zero, even though water absorption by roots exceeds transpiration loss during the night because of (1) the resistance of xylem cells to water movement, (2) the distance that water must travel, and (3) the short recovery interval between sunset and dawn. If irrigation water is withheld and the soil is allowed to reach *permanent wilting percentage* (PWP) before the orchard is irrigated again, leaves may require several days before Pn recovers to the prestressed levels. At PWP the soil water potential is nearly −1.5 megapascals (−15 bars), at which point roots are unable to extract moisture from the soils and plants remain wilted until irrigated.

Wind. When the air is very still, water vapor diffusing from the stomata accumulates as a thin layer on the leaf surface. This accumulation increases the stomatal resistance to the movement of vapor from the stomata and decreases the vapor pressure deficit between the mesophyll and ambient atmosphere. The impedance of stomatal transpiration by the formation of a vapor layer is called boundary effect. When the wind blows and leaves move, this boundary layer is destroyed so that the vapor deficit increases and the transpiration rate rises.

A light breeze also exposes leaves to intermittent light that has been shown to be beneficial in enhancing Pn as compared to continuous light. Moderate to strong winds cause rapid transpiration and leaf damage. Excess transpiration causes stomatal closure, whereas leaf damage disrupts transport of photosynthates, thus impeding their export.

Flooding. Flooding for an extended period by irrigation or a combination of irrigation and heavy rainfall also has a negative effect on the assimilation rate. Excess water, especially in poorly drained soils, often leads to anaerobic conditions, which inhibit water absorption by roots. The value of ψ in the xylem becomes increasingly more negative as though there were a water deficit in the soil.

Ambient Temperature. Leaf temperature usually reaches a maximum in the midafternoon. Since the rate of respiration increases at a steeper gradient than that of photosynthesis, the net assimilation rate diminishes. Increased photorespiration may be responsible for part of this diminution. Supraoptimal temperatures of 30° to 35° C occurring almost daily in the summer in the western and southwestern United States have adverse effects on Pn. At temperatures over 40° C, the photosynthetic apparatus in some of the more heat-sensitive species is damaged irreversibly, causing symptoms of sunburning to appear.

Humidity. In arid and semiarid regions of the world where fruit trees are grown, the humidity (the water vapor concentration of the atmosphere) can become so low that there is a large difference in water vapor pressure between the intercellular air space in the mesophyll and the external environment. This initially tends to hasten the rate of transpiration but eventually leads to a water deficit within the plant, causing stomatal closure.

CULTURAL MEANS OF MAINTAINING OPTIMAL RATES OF Pn AND T

Knowing how the various external and internal parameters can influence photosynthesis and transpiration, what can one do to optimize these factors to maximize orchard yields? The amount of solar radiation and the length of the photoperiod during the growing season are fixed conditions for a given geographical locality. But soil fertility and moisture content, temperature, wind velocity, and the relative humidity of the atmosphere are environmental factors that can be manipulated by cultural practices. For example, net photosynthesis can be increased by changing the canopy configuration by dormant and summer pruning and thus exposing leaves to optimal light intensity and duration (see Canopy Configuration later in this chapter).

Preventing and Correcting Chlorosis. Chlorosis is a diseased condition in which leaves are unable to synthesize chlorophyll so that they become a pale yellowish green. If the chlorotic condition can be associated with a lack of certain mineral elements by chemical analysis, the amount of chlorophyll in leaves can be manipulated through careful monitoring of the levels of mineral nutrients in the leaf blade and/or petiole. Chlorosis is brought on indirectly by excess lime that increases the pH, thereby rendering iron, copper, and zinc insoluble, or the condition is produced directly by a deficiency of certain essential elements. Chlorosis can be corrected by (1) adding soil amendments, (2) spraying the trees with appropriate nutrient solutions, (3) applying the nutrient to the soil, or (4) selecting rootstocks that are tolerant of high pH or are more efficient in extracting minute quantities of essential elements from the soil. (see Fertilizers, Ch. 8).

Moderation of Temperature. In the Mediterranean areas or in the Sonoran life zones of the southwestern United States where relative humidity becomes very low at times, the transpiration rate can exceed the rate of water uptake by roots relatively early in the day. Concurrently, the ambient daytime air temperature about the leaves often surpasses 40° C.

Intermittent overhead sprinkling has been used successfully to ameliorate the detrimental effects of high temperatures and low humidity and to reduce stress. Sprinkling also washes off dust particles, which tend to absorb heat and reduce light intensity. Sprinkling should be done soon after sunrise so that water can evaporate quickly; otherwise, species that are sensitive to mildew organisms will soon be infected. The use of sod cover crops reduces ambient temperature because the plants intercept solar radiation and prevent the soil from absorbing heat from the direct rays of the sun. Transpiration by the cover crop also provides a cooling effect (see Sod Culture, Ch. 7).

Creating Windbreaks. Winds prevent the formation of a vapor boundary layer, hasten transpiration rates, and force early stomatal closure, resulting in a decrease in the photosynthetic efficiency of plants. Winds are responsible for shoot breakage, bruising of fruits, and tattering of leaves, which reduce the yield potential of trees. Hence, where there is a constant prevailing wind as in the Rio Negro Valley of Argentina, windbreaks are essential. Although Lombardy poplars harbor many pests, they are commonly used in South America because their tall, upright growth effectively impedes the wind velocity. Kiwifruit plantings have been encircled and covered with perforated plastic sheets to reduce wind damage.

A windbreak should not be a solid wall of vegetation, fencing, or sheeting; frequent gaps in the wall allow some wind to pass through and decrease the amount of turbulence behind the windbreak. In areas where frost is a hazard, a solid windbreak blocks air drainage, causing an accumulation of cold air which could increase damage.

Selecting Suitable Scion–Stock Combinations. Although increased photosynthetic rates might be achieved by altering environmental conditions, the greatest gains can be obtained by selecting cultivar and rootstock combinations best suited for the particular orchard site. Thus, by using dwarfing or size-controlling rootstocks to reduce tree size, the grower will have photosynthetically efficient trees in which the dry matter partitioning coefficient favors the crop rather than the vegetative parts of a tree. These scion–stock combinations, coupled with the use of growth retardants to enhance flowering, promote early cropping, facilitate fruit thinning, harvesting, and pruning, and increase the potential yield per unit of land area.

In highly fruitful walnut cultivars, for example, Chico and Vina, the use of black walnut rootstocks produces an assimilate partitioning coefficient that favors the crop at the expense of vegetative growth. Such cultivars should be topworked on vigorous rootstocks and pruned annually to maintain tree vigor which, in turn, will prevent dwarfing and promote better nut size and quality.

Other genetic characteristics that increase cropping efficiency are leaf morphology and sun-tracking ability. Leaves of some spur-type, compact apple cultivars and pistachio species have two or more layers of palisade parenchyma cells, and/or longer cells than those of standard cultivars. When the mesophyll cells are tightly packed together, the ratio of the cell wall area to leaf surface increases. Since carbon dioxide is absorbed through the walls of mesophyll cells, photosynthetic efficiency per unit of leaf area increases with greater cell density. The leaves of these compact phenotypes also contain more chlorophyll per unit leaf area than do the standard varieties.

Another plant trait that improves P_n is the ability to track the sun, enabling the plant to distribute leaves evenly throughout the tree canopy. For example, peach leaf blades remain relatively horizontal owing to the twisting of the petioles, even though branches sag as the crop gets heavier. When kiwifruit shoots are tied down to reduce breakage, the petioles become twisted, turning the lamina upward within hours. This twisting, a phototropic response, is believed to be auxin-mediated.

That the isolateral leaves of *Pistacia vera* do not track the sun is an evolutionary trait to reduce transpiration and avoid direct solar radiation. This trait, which is common to xerophytic plants, allows considerable sunlight to penetrate the canopy onto the orchard floor.

Canopy Configuration, Leaf Area Index, and Canopy Area Index. Changing canopy configuration through different pruning and trellising systems (see Ch. 9) increases foliar exposure to sunlight and improves photosynthetic efficiency. The yield of dry matter per unit leaf area or per plant may interest a plant physiologist, but it means little to a grower because farm income is based on yield per unit land surface. Watson (1947) proposed the term *leaf area index* (LAI), which is the ratio of leaf area of a plant to the unit land area the plant occupies. Since LAI is a measure of how effectively the land surface is covered by the leaves of the crops, yields of different species could be compared on a unit land area basis. Hence, an LAI value of 4 means that there are 4 square meters of leaf area per square meter of land surface occupied by the plant. If LAI is less than 4, measurements show that the plants are small and/or the planting distance so wide that sunlight is falling onto the orchard floor. Conversely, if LAI is greater than 7, the plants are tall and/or crowded so that the lower leaves of the canopy are carrying on photosynthesis at suboptimal rates.

LAI varies with cultivars, plant age and shape, and spacing. Peach trees trained to a vase shape and spaced 6.7 meters apart (22′ × 22′) had 102,000 leaves with a leaf area of 210 square meters. Each tree covered an area of 44.9 square meters, so the LAI was estimated to be 4.7:1. In a deciduous tree community, LAI lies between 3:1 and 6:1. The orchardist should strive to attain these LAI values to optimize Pn by shaping the foliar canopy.

Foliar canopy management and tree and vine modeling are simplified by treating the canopy area like a sheet rather than a mosaic of leaves. Therefore, the term *canopy area index* (CAI) will be used in lieu of LAI. To do so assumes that canopies have the same thickness independent of the training system, and that they equally intercept all solar radiation. The relationship between LAI and CAI is derived from the vase-shaped peach tree just mentioned. If the vase shape is considered as an inverted cone with a base diameter of 3.35 meters and a height of 4.85 meters, the inner surface of the conical canopy is 62 square meters, and the ground surface area covered by the cone is 45 square meters. The ratio of the canopy surface to land area is approximately 1.4:1. If LAI of 4.7 is divided by CAI of 1.4, the resulting quotient is 3.4. This is construed to mean that the leaf area of the peach tree is about three and a half times that of a canopy treated as though it were a sheet.

CAI for different training systems varies according to planting distances, tree height, and canopy configuration. A vineyard in Argentina, or a kiwifruit plantation in New Zealand, or an Asian pear orchard in Japan in which the plants are trained to a flat pergola or the tanazukuri (shelf construction) system has a CAI value of 1.0, for the canopy area exposed to solar radiation is equal to the land surface it covers. But if the same plants are trained to a vertical fruiting wall, a hedgerow, measuring 4.5 meters tall (no branches within a meter from the soil), 3.5 meters long, and 1 meter thick, with an alleyway of 2.5 meters between the fruiting walls, CAI will be 2.0. Thus, orienting the foliar surface from the horizontal to the vertical position doubles the potential photosynthetic area.

Theoretically, if plants are trained to a tall spindle or cone shape, leaving minimal space between trees for over-the-tree spraying and harvesting equipment to traverse, CAI exceeds 3.0. Although it is possible to increase CAI with these canopy configurations, the efficiency of conducting other cultural operations, for example fruit thinning, spraying for pests, and harvesting of the crop, should not be sacrificed. Increased tree density and taller trees soon lead to shading of the lower and inner canopy so that, although CAI (and LAI) will increase, the functional leaf area will decrease. Thus, there seems to be an upper limit of CAI values at which yields will begin to decrease.

CAI may not be substituted for LAI in discussing certain physiological parameters. For example, brachytic dwarf peach trees have a large LAI because they may have as many leaves as the normal peach genotypes, but the trees occupy a much smaller land area. Because their internodes are short, the leaves are superimposed one on top of another. Under such growing conditions, most of the mature leaves are functioning at near-compensation point during the daylight hours. Therefore, with a large LAI, the overall photosynthetic efficiency of the plant may reach a point of diminishing return. Excess shading under the thick canopy causes shoot dieback and a reduction in fruit quality and size. The term *foliar density*, which is expressed as leaf area per unit volume of canopy, has been recently introduced to compare the relative looseness or compactness of canopies.

Other Photochemical and Related Leaf Functions

SOURCE OF FLOWERING STIMULUS

Synthesizing and exporting the floral stimulus or florigen to buds are two important functions of leaves (Guttridge, 1969; Lang, 1952). The stimulus that is translocated in the phloem requires the action of living cells. If a petiole is chilled or its cells are killed by steaming, movement of the floral stimulus from the leaf to the bud in its axil is impeded or arrested. When leaves from a plant induced to flower by a favorable photoperiod are grafted onto a noninduced plant, the noninduced plant flowers, indicating that the stimulus is graft-transmissible.

NITRATE REDUCTION AND AMINO ACID SYNTHESIS

The reduction of nitrate ions to nitrite and subsequently to ammonia normally occurs in roots. This two-step process is mediated by the enzymes nitrate and nitrite reductases. The reaction is the first step in amination of carbohydrates to form amino acids. If roots are exposed to abnormally high concentrations of nitrate ions, some nitrate ions will escape being reduced and enter the transpiration stream. They are then translocated to leaves where their reduction takes place. The enzyme has been induced experimentally in leaves of potted plants by elevating the nitrate concentration in the nutrient solution.

In actively photosynthesizing walnut leaves, reduction rates are such that the ratio of nitrate to ammonium ions is about 1.5:1. In shaded leaves, the ratio is 8:1, indicating that with little reductase activity, nitrate ions accumulate. Nitrate reductase activity is relatively high in walnut, grape, and apricot leaves; the enzyme activity is barely detectable in sweet cherry leaves and is yet to be detected in peach leaves (Perez and Kliewer, 1978).

BIOSYNTHESIS OF TERPENOIDS, INCLUDING HORMONES

Chloroplasts contain enzymes that metabolize mevalonic acid to terpenoid compounds. Decarboxylation of mevalonic acid gives rise to isopentenyl pyrophosphate and its isomer, dimethylallyl pyrophosphate. These 5-carbon isomers join "head-to-tail" to form monoterpene, the starting point for many diverse families of chemicals, including gibberellic acid precursors, carotene, and phytol, the "tail" of the chlorophyll molecule. Isopentenyl alcohol is an integral part of zeatin, a cytokinin. Other compounds include steroids, abscisic acid, rubber, and gutta. Turpentine is a crude distillate of terpenes and resins usually collected from pine as an exudate, known as pine pitch. The resin, collected and distilled from pistachio exudate, is a source of turpentine for high-quality varnish.

SYNTHESIS OF PHENOLIC SUBSTANCES AND TANNINS

There are innumerable phenolic compounds whose roles are yet unclear in plant metabolism. Among them are numerous phenolic glycosides occurring in various tissues of fruit plants. Arbutin (*p*-hydroxyphenyl-β-glucopyranoside) and phloridzin (phloretin-β-glucoside) exist in large quantities in pear and apple, respectively. They are found in the phloem of the stem with sucrose and sorbitol and are seemingly metabolized as they enter the fruit tissue; there is some evidence that hydrolysis occurs in the pedicel. Hence, they may be serving as carriers of phenolic substrates and glucose in the transport system.

Amygdalin (mandelonitrile-β-gentiobiose) and its precursor, prunasin (mandelonitrile-β-glucoside), are found in the leaves and seeds of *Prunus* species. These bitter constituents, which yield hydrocyanic acid upon hydrolysis, are stored in the cotyledons of most stone fruit species and are utilized as reserve food during germination. They are thought to be translocated in the phloem with other water-soluble photosynthates produced by leaves. Cotyledons of almond, apricot, and peach that have sweet kernels lack amygdalin.

Leaves are also rich in tannins and phenolic compounds. The term *tannin* is derived from *tan* the Latin word for oak. Ancient people used extracts of oak leaves and bark for converting animal skins into leather. Many tannins are hydroxyphenols which form cross-linkages between proteins and large macromolecules in the skin to make the resulting leather stable and pliable. Chestnut tannins are highly prized because they make the leather pliable and impart a fine finish. This capacity of tannin to bind with proteins inactivates many enzymes that would otherwise cause proteins to become putrid. The same binding reaction occurs on our tongue, giving us a puckery sensation when we eat food or drink liquids containing tannins.

Some of these phenolic compounds are naturally present in plants, but others are synthesized in response to wounding by invasions of insects, fungi, or bacteria. These compounds, known as "phytoalexins," ward off or reduce infection by sealing off the organisms. They play an important role in the disease-resistance mechanism of plants. Infection by a mycoplasma or a virus induces abnormal secretions at the graft union that plug the phloem transport system, resulting in incompatibility (see Graft Incompatibility, Ch. 10).

Juglone, a naphthoquinone, is a toxic compound found in the leaves and hulls of walnuts. As this compound is released by decomposing leaves, the germination and growth of many herbaceous species under walnut trees are inhibited.

FOOD AND MINERAL NUTRIENT STORAGE

A leaf serves as a temporary and long-term storage organ. Under optimum light, temperature, and moisture conditions, the photosynthetic rate is faster than the rate of export of photosynthates. As sugars accumulate in leaves, starch is synthesized and stored in the chloroplasts during the day. After sunset, starch is hydrolyzed to glucose, which is respired or converted to such

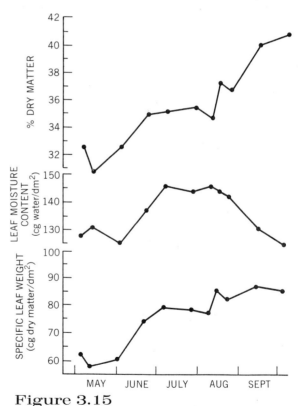

Figure 3.15
Seasonal changes in dry weight and moisture content of French prune leaves concurrent to the increase in specific leaf weight. (From Hansen et al., 1982)

translocatable compounds as sucrose and sorbitol and exported to other parts of the tree. By morning little or no starch remains in leaves.

Just as the specific leaf weight of exposed walnut leaves increases with age and light intensity (Fig. 3.14), that of apple (Oland, 1963) and prune (Fig. 3.15) increases as the season progresses. The increase in dry matter is attributed to thicker cell walls, assimilation of more nitrogenous compounds, such as chlorophyll and enzymes, thickening of the cuticular layers, and a small accumulation of mineral constituents. With the concurrent increase in leaf age and chlorophyll content, the capacity to store starch in the chloroplasts of prune leaves increases from 2 percent at the end of July to nearly 10 percent by September. With the increase of dry matter and storage capacity for starch, the moisture content of prune leaves decreases proportionately (Fig. 3.15). SLW of sun-exposed leaves is consistently heavier than that of shade leaves, indicating that leaves are capable of integrating the effects of the environmental conditions that influence transpiration and photosynthesis. Hence, a comparison of SLW values derived from leaves collected from different parts of a tree offers an estimate of (1) how crowded the trees are in an orchard and (2) the density of the foliar canopy.

As leaves begin to senesce and chlorophyll degradation ensues, breakdown products stemming from the catabolism of nitrogenous and carbohydrate compounds and mineral elements are exported to the stem. This process is known as the backflow phenomenon. Oland (1963) found that during this three- to four-week period of remobilization prior to leaf abscission, SLW and nitrogen content of apple leaves decreased 16 and 65 percent, respectively (Fig. 3.16); the potassium content of yellowing leaves decreased 21 percent. Much of the substances remobilized by the stem during the leaf senescing period is utilized again for growth of new shoots the following spring.

The preceding discussion emphasized the importance of training and pruning trees and vines judiciously in order to assure maximum efficiency of the foliar canopy in trapping solar radiation. In view of the export of nutrients from senescing

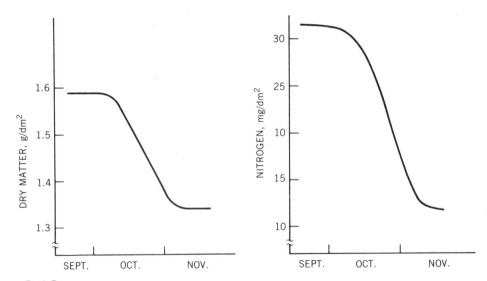

Figure 3.16
Decrease in dry matter (left) and nitrogen content (right) per square decimeter of senescing apple leaf area. (From Oland, 1963)

leaves to the adjacent spurs and shoots, dormant pruning should be delayed until leaf fall to conserve the mobile nutrients as part of the reserve food store. Furthermore, pruning prior to leaf fall may delay the hardening of shoots and buds, making them susceptible to early fall frost.

References Cited

AVERY, D. J. 1975. Effects of fruits on photosynthetic efficiency. In *Climate and the Orchard*. Ed. H. C. Pereira. Commonwealth Agricultural Bureaux, Farnham Royal Slough, England.

CHANDLER, W. H., AND A. J. HEINICKE. 1925. Some effects of fruiting on the growth of grape vines. *Proc. Amer. Soc. Hort. Sci.* 22:74–80.

CHANDLER, W. H., AND A. J. HEINICKE. 1926. The effect of fruiting on the growth of Oldenburg apple trees. *Proc. Amer. Soc. Hort. Sci.* 23:36–46.

CROSBY, E. A. 1954. Seasonal fluctuations of carbohydrates in the branches and fruit associated with growth regulator accelerated development of the Calimyrna fig. Doctoral thesis, University of California at Davis.

DEJONG, T. M. 1982. Leaf nitrogen content and CO_2 assimilation capacity in peach. *J. Amer. Soc. Hort. Sci.* 107:955–959.

DEJONG, T. M. 1983. CO_2 assimilation characteristics of five *Prunus* tree species. *J. Amer. Soc. Hort. Sci.* 108:303–310.

DEJONG, T. M., A. TOMBESI, AND K. RYUGO. 1984. Photosynthetic efficiency of kiwi (*Actinidia chinensis*, Planch.) in response to nitrogen deficiency. *Photosynthetica* 18:139–145.

GRANT, J. A., AND K. RYUGO. 1984. Influence of within-canopy shading on net photosynthetic rate, stomatal conductance, and chlorophyll content of kiwifruit leaves (*Actinidia chinensis* Planch.). *HortScience* 19:834–836.

GUTTRIDGE, C. G. 1969. Fragaria. In *The Induction of Flowering*. Ed. L. T. Evans. Cornell University Press. Ithaca, N.Y.

HANSEN, P., K. RYUGO, D. E. RAMOS, AND L. FITCH 1982. Influence of cropping on Ca, K, Mg, and carbohydrate status of 'French' prune trees grown on potassium limited soils. *J. Amer. Soc. Hort. Sci.* 107:511–515.

HUMBLE, G. D., AND T. C. HSIAO. 1970. Light-dependent influx and efflux of potassium of guard cells during stomatal opening and closing. *Plant Physiol.* 46:483–487.

HUMBLE, G. D., AND K. RASCHKE. 1971. Stomatal opening quantitatively related to potassium transport (evidence from electron probe analysis). *Plant Physiol.* 48:447–453.

KRIEDEMANN, P. E. 1968. Observations on gas exchange in the developing Sultana berry. *Austr. J. Biol. Sci.* 21:907–916.

KUIPER, P. J. C. 1972. Water transport across membranes. *Ann. Rev. Plant Physiol.* 23:157–172.

LANG, A. 1952. Physiology of flowering. *Ann. Rev. Plant Physiol.* 3:265–306.

LENZ, F. 1985. Water consumption of apple tree for high production. *Compact Fruit Tree* 18:21–26. East Lansing, Mich.

MCCREE, K. J. 1972. The action spectrum, absorptance and quantum yield of photosynthesis in crop plants. *Agric. Meteorol.* 9:191–216.

MIKA. A., AND R. ANTOSZEWSKI. 1972. Effect of leaf position and tree shape on the rate of photosynthesis in apple tree. *Photosynthetica* 6:381–386.

NAKAMURA, M. 1967. Physiological and ecological studies on the calyx of the Japanese persimmon fruit. *Res. Bull.* #23. Gifu University. Gifu, Japan.

OLAND, K. 1963. Changes in the content of dry matter and major nutrient elements of apple foliage during senescence and abscission. *Physiol. Plant.* 16:682–694.

PEREZ, J. R., AND W. M. KLIEWER. 1978. Nitrate reduction in leaves of grapevines and other fruit trees. *J. Amer. Soc. Hort. Sci.* 103:246–250.

RYUGO, K., AND L. D. DAVIS. 1959. The effect of the time of ripening on the starch content of bearing peach branches. *Proc. Amer. Soc. Hort. Sci.* 74:130–133.

RYUGO, K., B. MARANGONI, AND D. E. RAMOS. 1980. Light intensity and fruiting effects on carbohydrate contents, spur development, and return bloom of 'Hartley' walnut. *J. Amer. Soc. Hort. Sci.* 105:223–227.

SAMS, C. E., AND J. A. FLORE. 1982. The influence of age, position, and environmental variables on net photosynthetic rate of sour cherry leaves. *J. Amer. Soc. Hort. Sci.* 107:339–344.

TOMBESI, A., T. M. DEJONG, AND K. RYUGO. 1983. Net CO_2 assimilation characteristics of walnut leaves under field and laboratory conditions. *J. Amer. Soc. Hort. Sci.* 108:558–561.

WATSON, D. J. 1947. Comparative physiological studies on the growth of field crops. I. Variation in net assimilation rate and leaf area between species and varieties, and within and between years. *Ann. Bot.*, n.s. 11:41–76.

CHAPTER 4

Physiology and Functions of Roots

Root extension and exploration of soils are governed by (1) edaphic conditions such as soil density, depth, fertility, and temperature and height of the water table during the winter months; (2) cultural practices such as pruning and fruit thinning; and (3) the inherent growth characteristics of the species.

Roots grow both in length and in diameter—in length by cell division and enlargement of primary tissues, and in diameter as a result of secondary meristematic activity in the lateral and cork cambia. Roots penetrate the soil mass because cells in the region of elongation, extending a few millimeters behind the root cap, enlarge and push the yet undifferentiated meristematic apex ahead (see Fig. 1.8) Roots bend when their tips encounter obstacles. Cells of the root cap are eroded, but they are quickly replaced by those originating from the apical meristem.

The endodermis and pericycle, being primary tissues, become isolated, eventually degraded, and sloughed off with the outer cortical and epidermal cells as secondary growth is initiated by divisions in the lateral cambium. The capacity to transport and store food and mineral elements increases as secondary xylem and phloem tissues are laid down and roots grow in diameter.

How well roots develop depends a great deal on soil texture, temperature, and native fertility. Fine roots grow and extract moisture throughout much of the soil mass in sandy loam soils because water is held less tenaciously by large sand particles than by the fine clay particles. Roots do not grow extensively in extremely sandy soils in spite of the large air volume and good water drainage (porosity) because these soils have a small water-holding capacity and are usually infertile. Such orchard soils require frequent irrigation and fertilization.

Root growth slows down as water is absorbed and soil moisture content diminishes. Eventually, the moisture content will decrease to a point where roots will no longer be able to extract water. The soil moisture is at its permanent wilting point (see Soil–Plant–Water Relationships, Ch. 7). Fruit tree roots also do not effectively explore extremely heavy clay soils because soil particles are too fine and tightly compacted. Because pore space is small in these soils, oxygen is likely to become limiting and reduce growth.

Growth Periodicity

The seasonal increment of root growth depends on the severity of dormant pruning which regulates not only the amount of shoot growth but also the crop load for the following summer. Shoot growth and crop load, in turn, influence the amount of photosynthates that is translocated to roots to be stored as carbohydrate reserves and/or used for root extension. The alternation in seasonal periodicity of root growth is not consistently observed in young, vigorous, nonbearing trees, presumably because there is a plentiful supply of photosynthates.

Root growth is slow during the depth of winter but accelerates in late winter, two to six weeks before budbreak, provided the soil temperature is above 4° C (Head, 1967; Fig. 4.1). This resumption of growth coincides with the period when walnut, kiwifruit, and grape begin to exhibit positive root pressure. Pruning of these species at this time results in xylem exudation.

As shoot growth commences in the spring, root growth slows but resumes again as shoot growth decelerates in late summer. Less root growth was observed on pruned Worcester apple on MM.104 rootstock than on unpruned trees. Pruning results in a smaller canopy, even though shoots on a pruned tree grow three to four weeks longer than do shoots on an unpruned tree (Head, 1968). The periodicity of root and shoot growth is reflected in the accumulation of photosynthates. When Hansen (1967b) administered radioactive carbon dioxide to leaves of one-year-old apple trees in early summer while shoots were still elongating rapidly, he found that most of the radioactivity remained in the aerial portion of the trees. In another set of trees treated in October when shoot growth had ceased, he recovered more radioactivity from roots than from shoots.

Root growth on bearing four-year-old Golden Delicious apple trees was negligible; the poor growth was attributed to the interception of photosynthates by fruits because of their superior sink strength and favorable competitive position (Hansen, 1967a, 1971). For the same reason, the roots of heavily bearing French prune trees accumulated starch to a lesser extent than lightly cropped trees (Hansen et al., 1982). With prune trees, the lack of carbohydrates in roots not only curtails growth but also limits respiratory energy required for active uptake and retention of cations against a concentration gradient. Thus, heavy cropping eventually leads to potassium deficiency, especially in soils with a relatively large magnesium–calcium ratio (see Fertilizers, Ch. 8). These observations illustrate how two cultural practices—thinning of fruits and pruning—may have a carryover effect that regulates growth of the tree the following year.

Functions

ABSORPTION AND CONDUCTION OF WATER AND MINERAL NUTRIENTS

Young, white roots in the regions of elongation and maturation absorb most of the water and min-

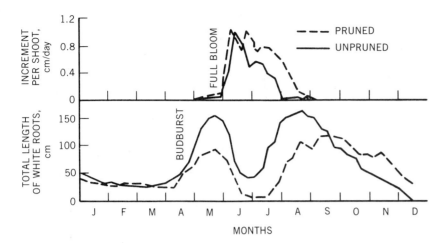

Figure 4.1
Seasonal alternation in the length of white roots (lower) and shoot growth (upper) for unpruned and pruned Worcester apple on MM.104 rootstock. (From Head, 1968)

eral nutrients utilized by the tree. These non-suberized roots expend much respiratory energy, absorbing and retaining nutrients within epidermal cells against a concentration gradient. Hence, the concentration of nutrients such as potassium is higher within vacuoles of root hairs than in the external soil solution. Evidence that respiratory energy is required for uptake and accumulation of nutrients is provided by exposing roots to respiratory poisons. Protoplasts of poisoned root cells lose their ability to accumulate or retain salts against a concentration gradient. Therefore, a leakage of previously accumulated nutrients occurs.

Minute amounts of minerals are absorbed by an ion exchange process by direct contact between fine roots and clay particles. Hydrogen ions secreted at the root surface are exchanged for potassium, magnesium, or calcium ions that are adsorbed on the surface of the clay particles. In turn, these cations can be exchanged for others. Similarly, phosphate ions may be absorbed through an anion exchange mechanism. The degree to which cations and anions are exchanged depends on the exchange capacity of the particular soil.

Once these elements are absorbed, they are transported through the cortical cells into the interior of the roots by two pathways: symplastic and apoplastic. Symplastic movement occurs through the cytoplasm, and apoplastic movement through the cell walls. Ions can move through the apoplast to the endodermis, but at the Casparian strip they must pass through the cytoplasm of the endodermis in order to enter the central stele. Once nutrients and root metabolites reach the vessels and tracheids, they move passively with the transpiration stream, a continuum of water from the root hairs to the mesophyll cells of the leaves.

As water is absorbed, the soil about the root hairs becomes relatively dry. The rate at which roots are able to extract soil moisture is a function of the rapidity with which water can move through the soil and replenish what was absorbed (see Irrigation Systems, Ch. 7).

STORAGE AND METABOLISM OF RESERVE CARBOHYDRATES

Bearing and nonbearing four-year-old Golden Delicious apple trees produce about the same total amount of dry matter. But because the crop removed about half of the assimilates, the weight of the perennial part of bearing trees was only half that of nonbearing trees. The dry weight of roots from nonbearing trees was nearly three times that of roots from bearing trees. Much of the dry matter in the roots was remobilized and utilized for new growth the following spring (Hansen, 1971).

When pecan seedlings were allowed to fix radioactive carbon dioxide during one summer and were grafted with nonradioactive scions during the dormant period, new shoots arising from the scions the following spring became radioactive (Lockwood and Sparks, 1978). The labeled carbohydrate which was stored in the roots as starch and sugars (Worley, 1979) was utilized for shoot growth. Sugars and other osmoregulatory solutes in cells not only provide substrates for respiration and growth but also afford protection against winter injury by lowering the freezing point of the sap. Soil temperature in the root zone under California conditions rarely approaches the freezing point. Thus, hydrolysis of starch may not occur in the roots to the same extent that it does in the above-ground parts or in roots of trees growing in areas where the soil freezes.

Although a relatively large quantity of photosynthates is stored as reserve carbohydrates, portions are respired or assimilated into permanent structures, for example, the cell wall. Part of the stored materials is transformed to various nitrogenous compounds, especially amino acids. The amino group originates from enzymatic reduction of nitrate ions by nitrate and nitrite reductases in the epidermal and subepidermal cells. Amino acids are formed by (1) reductive amination of an organic acid such as α-keto-glutaric acid yielding glutamic acid in the presence of a coenzyme, nicotinamide adenine dinucleotide (NAD), or its reduced form nicotinamide adenine dinucleotide phosphate (NADP); or (2) transamination, in which the amino group is transferred from an existing amino acid to another organic acid. The amino acids serve as building blocks for the synthesis of proteins as well as substrates for other nitrogenous compounds, for example, indoleacetic acid, nicotinic acid, purines, and alkaloids. Some of these water-soluble substances, which are not utilized *in situ*, are translocated upward in the xylem and phloem when growth resumes in the spring. Some nitrate ions which are not reduced in roots are translocated to the leaves where they are reduced by the same enzyme systems and utilized. Along with the transformed metabolites, newly synthesized gibberellins and cytokinins are exported from the roots to shoots (Carr et al., 1964; Carr and Reid, 1968).

ANCHORAGE AND SUPPORT

Nothing can be more disheartening to an orchardist than to discover a full-bearing tree lying on its side after an irrigation or a windstorm. Most fruit trees and rootstocks started from cuttings or nurse-root grafts have shallow, horizontally growing roots (Fig. 4.2). Prune trees grafted on Marianna 2624 rootstocks have blown over when in full leaf, especially when the orchard had just been irrigated before a strong wind. The weight of the crop and the resistance of the foliage to the wind cause the trees to be uprooted.

Some rootstocks, when propagated as seedlings or cuttings, develop roots predominantly on

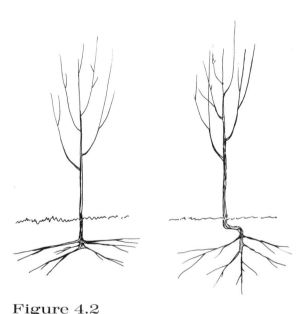

Figure 4.2
A shallow-rooted tree typical of one propagated from a cutting (left) and a doglegged one (right); both provide poor anchorage.

one side, resulting in a tree with a crooked "dogleg" form (Fig. 4.2). The horizontal portion of such roots twists easily, causing the tree to lean or topple over. Apple trees grafted on the dwarfing rootstock, Malling 9 (M.9), often suffer from this problem. Therefore, in some localities it is illegal to sell these "doglegged" trees as nursery stock.

Some dwarfing stocks develop a willowy trunk, especially if it is high-grafted so that the tree bends or topples when the top becomes excessively large and heavy. Such occurrences can be avoided by supporting the tree with a post or a trellis.

ROOT PRESSURE AND XYLEM EXUDATION

Several species, for instance, walnut, grape, and kiwifruit, exhibit the phenomenon of root pressure or xylem exudation in which copious amounts of sap are exuded from pruning cuts and wounds during late winter and early spring. Mature grapevines and kiwifruit vines exude as much as 4 liters of sap per day for several days, especially if the branch ends are recut to prevent the xylem from becoming plugged. In walnuts the volume of exudate from pruning wounds is greater on days when the air temperature fluctuates than when the temperature remains fairly constant. How this change in air temperature from night to day activates the mechanism of root pressure is yet unknown. In California, soil temperature at a depth of 50 centimeters remains nearly constant whatever the fluctuations in air temperature on even the coldest days.

The flow of exudates in walnuts begins in mid-January and continues until mid-April, a period of about 12 weeks (Fig. 4.3). At the height of the exudation period, walnut exudates contain as much as 28 milligrams of dry matter per milliliter, including growth regulators, phenolic substances, and many nitrogenous, ninhydrin-positive compounds (Prataviera, 1982). Some of these substances are produced at the apices of elongating roots. Others are presumably degra-

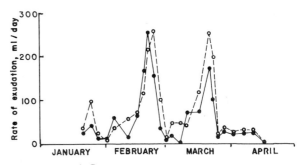

Figure 4.3
Seasonal fluctuation in rates of xylem exudation from two walnut trees. (From Prataviera, 1982)

dation products from the protoplasts of maturing vessels and tracheids. As these cells differentiate and mature, the end walls of vessels dissolve and disappear. Their cytoplasmic contents, rich in nitrogen, break down so that they are carried upward with the transpiration stream, only to be reutilized by younger tissues in the shoot.

Species that are prone to xylem exudation should be pruned early in winter while the plants are in the rest period to reduce loss of these nitrogenous compounds and carbohydrates. Grapevines pruned in early winter before bleeding occurred have higher yields than vines pruned in late winter and allowed to bleed profusely.

Root exudates make top-grafting of these species difficult, especially if the exudation occurs adjacent to the grafted scion. The exudates bathe the cut surface of the scion and apparently prevent growth-promoting substances from accumulating at the graft union, thus inhibiting callus formation. A toxic compound, juglone, in the walnut sap prevents the formation of a union between the stock and scion.

Root Distribution

SPECIES CHARACTERISTICS

Not all seedlings necessarily give rise to well-anchored trees because species differ widely in their root size and distribution patterns. All stone fruit species have large roots radiating from the crown, but peach and plum roots tend to grow horizontally, whereas those of almond and apricot grow laterally and vertically (Fig. 4.4). In areas where the soil is fertile and deep, almond seedlings produce several large roots growing nearly vertically below the crown to depths of 3 to 4 meters.

The density of small, fine roots is an inherent characteristic of the species; the number of fine roots increases progressively from almond to apricot to peach to myrobalan plum.

Three species are used for sweet cherry rootstocks: *P. avium*, the sweet cherry, commonly referred to as mazzard; *P. mahaleb*, the mahaleb cherry; and *Prunus cerasus*, the sour cherry. Sweet cherry seedlings produce large, fibrous, shallow roots. Mahaleb seedlings develop a deep system of unbranched, nearly vertical roots (Fig. 4.5). They are especially adaptable to well-drained gravelly soils but are very vulnerable to excess soil moisture or "wet feet." The sour cherry rootstocks, Stockton Morello and Vladimir, are clonally propagated by cuttings or from suckers, and, therefore, they develop shallow roots (Fig. 4.5). They tolerate relatively heavy, wet soils and have considerable resistance to infection from the soil fungus, *Phythophthora*.

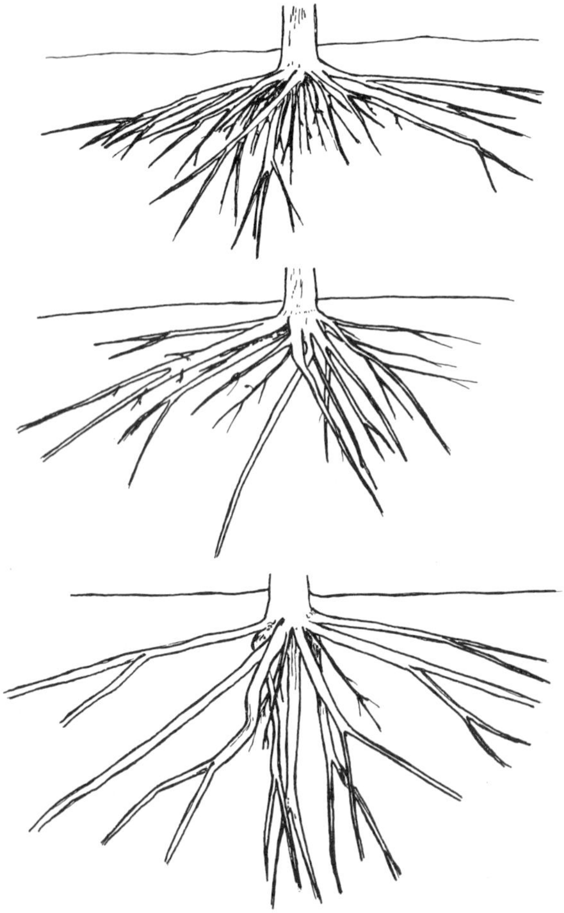

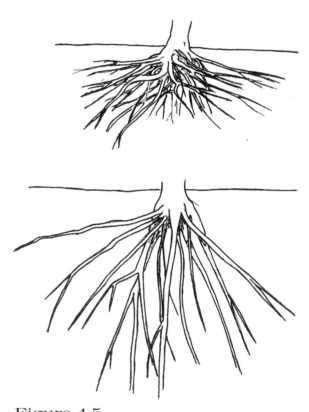

Figure 4.4
Root systems of three *Prunus* species illustrating the depth of rooting: peach (upper), apricot (middle), and almond (lower) seedlings.

Figure 4.5
Root systems of two sweet cherry rootstocks: Stockton Morello started as a cutting (upper) and a mahaleb seedling (lower).

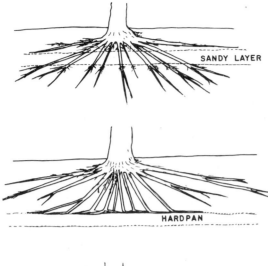
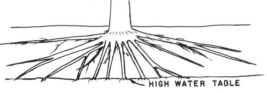

Figure 4.6
Influence of root environment on growth. Roots growing through a sandy stratum (upper) lacking fibrous roots; deflection of roots unable to penetrate a hardpan (middle); and limitation of growth by a water table (lower).

Root distribution differs with certain specific scion–stock combinations. For example, the growth habits of bench-grafted French crabapple understocks varied with the scion apple cultivar (Tukey and Brase, 1933).

Nursery trees that naturally develop lateral roots grow better after they are transplanted in orchards. A walnut seedling develops a large, swollen cotyledonary node from which a large tap root emerges. To induce lateral roots on the tap root, nursery personnel routinely remove the root tip (pinch) of a seedling when it is dug from a seedbed in preparation for transplanting into a nursery row. Otherwise, yearling black walnut stocks are undercut in the nursery row with a U-shaped digging tool to encourage lateral root formation. Any wounding of roots by pinching or cutting predisposes the cut surface to an infection by the crown gall bacteria.

ENVIRONMENTAL EFFECTS

Although the distribution pattern of roots is a heritable characteristic of a species, it may be altered by soil texture, a high water table, or competition with roots of adjacent trees and/or with cover crops. In orchards planted on overflow lands adjacent to rivers, peach roots develop

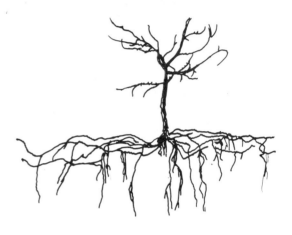
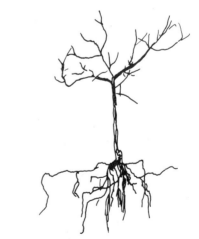

Figure 4.7
Root systems of five-year old Golden Delicious/M.9 planted at wide (upper) and narrow (lower) tree spacings. (Adapted from Atkinson et al., 1976).

Root Distribution 67

many fine fibrous roots in the more fertile and moist silt layers, whereas roots growing through the sandy stratum are smooth and devoid of fibrous roots (Fig. 4.6). In areas where the water table rises within a meter or two of the soil surface during winter months, roots are killed back annually to the point of submersion (Fig. 4.6). Trees grown in such an area periodically exhibit symptoms of nutritional deficiencies, such as of potassium (see Fertilizers, Ch. 8) because destruction of roots reduces the absorbing ability and the high water condition leaches the essential nutrients. Roots will not grow into a hardpan, a compacted layer of soil, or into a plow sole, a layer of earth at the bottom of a furrow, compacted by repeated plowing to the same depth, or into a stratum of pure sand (Fig. 4.6).

Crowding of trees has an effect on root distribution. The roots of the dwarfing apple rootstock, M.9, grow horizontally if uncrowded but predominantly downward in high-density plantings (Atkinson et al., 1976; Fig. 4.7).

References Cited

ATKINSON, D., D. NAYLOR, AND G. A. COLDRICK. 1976. The effect of tree spacing on the apple root system. *Hort. Res.* 16:89–105.

CARR, K. J., AND D. M. REID. 1968. The physiological significance of the synthesis of hormones in roots and of their export to the shoot system. In *Biochemistry and Physiology of Plant Growth Substance*. Ed. F. Wightman and G. Setterfield. Runge Press, Ltd. Ottawa, Canada.

CARR, K. J., D. M. REID, AND K. G. M. SKENE. 1964. The supply of gibberellins from the root to the shoot. *Planta* 63:382–392.

HANSEN, P. 1967a. ^{14}C-Studies on apple trees. I. The effect of the fruit on the translocation and distribution of photosynthates. *Physiol. Plant.* 20:382–391.

HANSEN, P. 1967b. ^{14}C-Studies on apple trees. III. The influence of season on storage and mobilization of labelled compounds. *Physiol. Plant.* 20:1103–1111.

HANSEN, P. 1971. The effect of cropping on the distribution of growth in apple trees. *Tidsskr. Planteave* 75:119–127.

HANSEN, P., K. RYUGO, D. E. RAMOS, AND L. FITCH. 1982. Influence of cropping on Ca, K, Mg, and carbohydrate status of 'French' prune trees grown on potassium limited soils. *J. Amer. Soc. Hort. Sci.* 107:511–515.

HEAD, G. C. 1967. Effects of seasonal changes in shoot growth on the amount of unsuberized root on apple and plum trees. *J. Hort. Sci.* 42:169–180.

HEAD, G. C. 1968. Seasonal changes in the diameter of secondarily thickened roots of fruit trees in relation to growth of other parts of the tree. *J. Hort. Sci.* 43:275–282.

LOCKWOOD, D. W., AND D. SPARKS. 1978. Translocation of ^{14}C in 'Stuart' pecan in the spring following assimilation of ^{14}CO$_2$ during the previous growing season. *J. Amer. Soc. Hort Sci.* 103:38–45.

PRATAVIERA, A. G. 1982. Exudation characteristics and composition of walnut xylem sap and its possible role in graft failure. M.S. thesis, University of California at Davis.

TUKEY, H. B., AND K. D. BRASE. 1933. Influence of the scion and of an intermediate stem-piece upon the character and development of roots of young apple trees. N.Y. Agric. Exp. Sta. Tech. Bull. 218.

WORLEY, R. E. 1979. Fall defoliation date and seasonal carbohydrate concentration of pecan wood tissue. *J. Amer. Soc. Hort. Sci.* 104:195–199.

CHAPTER 5

The Flowering Process and Seed Formation

Juvenility

CHARACTERISTICS OF JUVENILE PLANTS

Fruit trees grown from seeds undergo developmental phase changes from juvenility through transition to maturity. During the juvenile period trees do not produce flowers. The gradual production of flowers marks the onset of the adult or mature phase. As tree size and the number of shoots bearing flowers increase, the annual increment of shoot length lessens, especially if the flowers develop into fruits. The stage when few flowers are being formed is termed the transition phase by some researchers. This transition from juvenility to maturity is gradual because the lower parts retain their juvenile characteristics while the upper extremities of the tree begin to bear flowers. Whereas juvenility is usually concerned with ungrafted seedling plants, morphological and physiological evidence reveals that all three developmental phase changes exist in mature, bearing trees of some species, whether or not they have been grafted.

Flowering is only one outward expression or characteristic that differentiates a juvenile from the mature part of a tree. Other morphological manifestations of juvenility are leaf shape and size, thornlike shoots, and ease of inducing cuttings to root.

Leaves of juvenile pecan trees are simple and pubescent, whereas those of mature plants are compound and less hairy. Shoots of juvenile apple, citrus, pear, and plum trees are thorny. Morphologically, these thorns are short, pointed shoots with aborted terminal buds. As shoot extension continues with tree age, distal buds develop into lateral shoots and spurs instead of spines.

Stem cuttings from these juvenile parts of trees initiate roots more readily than do cuttings taken from mature parts. Some species such as the quince possess preformed root initials on mature wood so that their cuttings root easily.

If scions are taken from a spiny, juvenile shoot and are grafted onto scaffold limbs of mature bearing trees, subsequent growth from these scions will enter the mature stage and begin to bear flowers sooner than shoots left behind on the mother stock tree. Conversely, mature stems may revert to the juvenile phase when they are used as scions for grafting or as cuttings. Scions from mature citrus trees may produce spiny shoots when grafted onto seedlings; lemons are notorious in this respect. If scions taken from bearing pear and plum trees are grafted onto seedling rootstocks, the basal portions of their new shoots tend to be spiny. Quince and marianna plum cuttings from mature stems form spiny shoots.

The portion of the trunk adjacent to the roots remains juvenile; some physiologists attribute this phenomenon to the production of root hormones which are translocated to the top. Changing levels of these hormones in the apical meristems may be responsible for the transition from a juvenile to a mature stage. For example, juvenile and mature shoot apices have different types of ribonucleic acid (RNA). Since the genetic code of a particular clone is constant, this change may be a reflection of gene repression and expression brought about by hormones.

Gibberellins are translocated from the roots to shoots (Carr et al., 1964). As the tree grows larger and more branches are formed, not only does the distance from the roots to shoot terminals become greater, but the number of shoot terminals also increases. Thus, it is speculated that these growth-regulating substances originating in the roots become more dilute, to the point that they are ineffective as floral inhibitors and promoters of spines.

Evidence to support this hormone concept is demonstrated by the effect of gibberellic acid sprays on bearing almond, apricot, cherry, and plum branches. A single application of 200 parts per million made during the early part of the growing season, when flowers are being initiated, induces spurs to revert to spines and completely inhibits flower bud formation for the following season.

LENGTH AND HERITABILITY OF THE JUVENILE PERIOD

The length of the juvenile period differs among and within species, depending on the genotype of the clone. It is a heritable characteristic which apple breeders believe is transmitted through the cytoplasm of the egg, an example of cytoplasmic inheritance. That is, if a hybrid is made between a mother plant having a short juvenile period and a pollen parent with a long one, the progenies will more than likely have short juvenile periods like the maternal parent.

Wellington (1924) observed that seedling progenies derived from parent apple cultivars with long juvenile periods were also slow in coming into bearing. Scions taken from these seedling progenies and grafted on rootstocks required several years before they initiated flowers. Grafting shortens the juvenile period, perhaps by causing a temporary girdling at the graft union, but it does not overcome the inherent juvenile characteristic of seedlings. Seedlings that come into flowering at an early age are said to be precocious, and the phenomenon is termed "precocity."

Zimmerman (1971, 1972) found that plants

grown under continuous light were larger and had shorter juvenile periods than those grown under natural photoperiods. The juvenile period was also shortened by inducing plants into dormancy by withholding water and then forcing them into growth for several cycles. After a number of growth cycles, plants grew more and flowered sooner than those that had not been stressed. Based on these experiments, Zimmerman postulated that trees must attain a critical height, or possess a minimum number of nodes, or undergo a number of growth cycles before flowering is initiated.

Nucellar or apomictic seedlings normally progress through the juvenile phase just as zygotic seedlings do. Occasionally, however, citrus breeders have observed flowers on newly germinated seedlings grown under a strict temperature regimen.

The juvenile period is of particular interest to fruit breeders and propagators but is of little concern to fruit growers. The length of the juvenile period ranges from two to three years for almond and peach and as long as seven to ten years for apple and pear seedlings. Plant propagators prefer to retain juvenile characters in rootstock materials so that own-rooted cuttings can be obtained easily and inexpensively. They accomplish this objective by maintaining the stock trees in mounds or stool beds (see Layerage under Propagation, Ch. 10) and by dehorning trees (severely pruning) to force suckers and sprouts. Fruit breeders prefer to shorten the juvenile period of hybrid progenies in order to evaluate them for fruit quality as soon as possible. By shortening the generation time, they can accumulate genetic information faster and hasten the pace of the breeding program.

The length of the juvenile period does not affect fruit growers because all commercially grown trees are propagated by grafting or by cuttings using mature scion wood. Depending on the species, young trees obtained from nurseries do not produce flowers for two to five years after planting; heavy pruning during the formative period is partly responsible for this delay. Mature, bearing trees that are topworked over to new cultivars similarly require a few years before flowering. Visser (1964) refers to this period after planting or grafting during which the plants do not flower as the vegetative phase, a temporary stage of shoot development.

Trees should be kept in the vegetative phase during the formative period when their structural frameworks are being developed. Experimentally, the vegetative phase has been bypassed with fruitful walnut, stone fruit, and grape cultivars because flower parts or inflorescence primordia are already present by winter on certain parts of their shoots. Scions taken from these regions will proceed to flower if they are grafted onto appropriate rootstocks.

Flower Bud Formation

FOUR PHYSIOLOGICAL AND MORPHOLOGICAL STAGES

Morphologically, the conversion of a vegetative apex to a floral one from inception to anthesis is a relatively gradual progression. Lang (1952) separated the process of flower formation into four distinct stages: (1) floral initiation, the differentiation of floral primordia; (2) floral organization, the differentiation of the individual flower parts; (3) floral maturation, consisting of several, often concurrent, processes—growth of floral parts, differentiation of the sporogenous tissues, meiosis, pollen and embryo sac development; and (4) anthesis. Floral initiation is the stage during which a vegetative bud is biochemically stimulated and changed to a reproductive bud. There is no outward morphological manifestation that this stage has occurred.

The first visual signal that the change from the vegetative to reproductive apex has taken place is the flattening of the apical dome (Fig. 5.1). This is the beginning of the floral differentiation or organization period when the primordia of individual floral parts become morphologically apparent at the apex which becomes the receptacle.

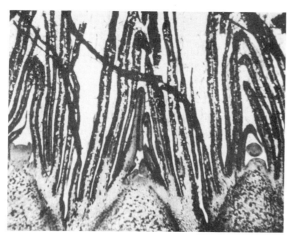

Figure 5.1
Longitudinal section of three peach buds at a node. The middle bud is vegetative; the dome of the right bud is beginning to flatten, and that on the left has formed sepal primordia. (Adapted from Tufts and Morrow, 1925)

The use of the terms *floral initiation* and *floral induction* synonymously with *floral differentiation* has led to confusion among plant scientists. In this textbook, Lang's definition of floral initiation is used synonymously with floral induction to facilitate the discussion of flowering.

INVESTIGATIONS ON FLOWER INITIATION

History of the Nutritional Concept

Long before pomology was established as a science, horticulturists and fruit growers gathered information regarding the responses of fruit trees to pruning and thinning. They focused especially on the flowering and fruiting processes because the livelihood of these people depended on the production of fruits and nuts. Horticulturists were aware that certain cultivars of apple, pear, and other fruit species would produce a crop one year and not the next, a phenomenon known as alternate or biennial bearing. The year the tree is bearing a crop is called the "on-year," and the year when the tree is barren is termed the "off-year" (Fig. 5.2). In cultivars that tend not to alternate, about half of the spurs on a tree differentiate flowers and produce a crop every other year. The annual yield of the tree remains relatively constant. Thus, the presence of fruit on a spur appears to have a strong influence on flower formation on that spur. Horticulturists also knew that trees fertilized heavily with nitrogen develop lush, vigorous growth, whereas those defoliated early in the spring by insects or strong winds form few or no flowers.

Observations that vigorous trees tend to produce few flowers led Kraus and Kraybill (1918) to investigate the role of nitrogen in flower formation and set. Although their data revealed that high nitrogen and low carbohydrate contents in tomato led to poor fruit set, the nutrient status did not deter the plants from differentiating flowers. Their work stimulated interest in the carbohydrate:nitrogen (C:N) relationship in various fruit tree species and led to extensive fertilizer and pruning trials in many parts of the world. Gourley and Howlett (1947) summarized these findings and developed a model for apple trees (Fig. 5.3), similar to that of Kraus and Kraybill for tomato. Apple trees in Class I are carbohydrate-deficient and weakly vegetative, in spite of adequate nitrogen, and do not form flowers; those in Class II are slightly carbohydrate-deficient and moderately vigorous from heavy nitrogen fertilization. Trees in this category also do not flower. Trees in Class III have adequate carbohydrate and nitrogen supplies, produce abundant flowers, and set crops; those in Class IV, lacking nitrogen, produce few flowers that rarely set. Trees can be shifted from one class to another by adjusting the severity of pruning and monitoring the amount of fertilizers applied.

History of the Hormonal Concept

Julius Sachs (1888), a German scientist known as the father of modern plant physiology, conceived

Figure 5.2
A sugar prune tree that was partially deflorated at full bloom (left). The same tree photographed the following spring (right). The part that flowered and fruited one year is barren the next. (Night photographs courtesy of L. D. Davis and Pomology Department, University of California)

the term *flower-forming substance* (*Blutenbildenstoff*), a substance he presumed to be synthesized in leaves. Chailakhyan (1968) changed the name to *florigen*, which means flower genesis in Greek. This theoretical flowering hormone has been neither isolated in a pure form nor identified; thus, some physiologists still prefer to use the nondescript term *floral stimulus*. Here, the terms *florigen* and *floral stimulus* will be used interchangeably for want of a better term to describe the substance or substances that activate the flowering process.

The pathway to the investigation of flower initiation diverged into (1) the interaction between nutritional and cultural practices by horticulturists, and (2) the pursuit of florigen, primarily by plant physiologists.

The Nutritional and Nutrient Diversion Concept

Based on his observation that nonbearing apple spurs contained high starch content compared to bearing apple spurs during the period of flower initiation, Hooker (1920) concluded that carbohydrate deficiency inhibited flower formation. Baxter (1970) proposed a theory on the flowering process based on the promotive effect of phosphate and nitrogen (Fritzsche et al., 1964; Grasmanis and Leeper, 1967; Kobel, 1954), emphasizing the interplay of genes, light energy, mineral elements, hormones, and inhibitors. Sachs and Hackett (1983) formulated the idea of nutrient diversion which causes the shoot apex to initiate floral primordia. They conceived that

Figure 5.3
Relation between the nitrogen and carbohydrate composition in the apple and its response to various cultural treatments with respect to flowering and fruiting. (From Gourley and Howlett, 1947, in *Modern Fruit Production*)

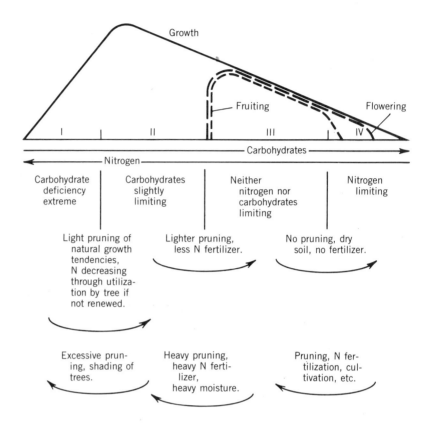

a complex of chemicals is diverted to the vegetative shoot apex under certain conditions. The chemicals educe specific localized genes to initiate morphogenesis, converting the apex to a floral one.

The Florigen Concept

Garner and Allard (1920) accidentally discovered that short days and long nights play a key role in the flowering of tobacco. They found that Maryland Mammoth tobacco plants that grew vegetatively in summer bloomed when grown in greenhouses during the winter. The duration of day and night periods as a unit was called photoperiod, and the biology of day–night lengths on the flowering response was termed photoperiodism. They categorized plants that flowered when days were 12 hours or less as short-day plants (SDP); those that flowered when days were over 12 hours as long-day plants (LDP); and those that flowered regardless of day length as day-neutral plants (DNP). Subsequently, Borthwick and Parker (1953) discovered that, for short-day plants, a long dark period was equally or more important than the short light duration. They also found that, with short-day plants, interrupting the long dark period in the middle with one hour of low light intensity (20 to 40 footcandles) reversed the short-day effect and inhibited flowering.

Experimental evidence with herbaceous species indicates that florigen, the elusive substance, is (1) translocatable through living tissue, presumably the phloem, (2) graft-transmissible from one species to another, and (3) accumulated above a graft union (Chailakhyan, 1968; Lang, 1952). The floral stimulus is translocated from a mother strawberry plant growing under short-day conditions to a daughter plant growing under a noninductive regimen (Hartmann, 1947), but it does not move if the daughter plant is as large as the mother plant (Guttridge, 1969). The stimulus was imparted in the daughter plant if it was defoliated or put in the dark, indicating that florigen moves with the photosynthate gradient. By covering spinach buds or leaves, Knott (1934) demonstrated that leaves rather than buds were the source or site of synthesis of the stimulus. There is some evidence, however, that if compound buds of grapes are exposed to intense light, the inflorescence within them will initiate more florets. Furthermore, more nodes within the compressed buds will produce inflorescences. Grape buds may perceive the stimulus directly or contribute some factor that enhances the signal coming from leaves.

Little or no information is available about photoperiodism in fruit trees because of their large size and the high cost of maintaining growth chambers large enough to hold a sufficient number of them. Their growth and flowering characteristics suggest, however, that fruit trees are not responsive to photoperiodic influences (see Time of Leaf Expansion later in this chapter).

TIME OF FLORAL INITIATION AND SOME PREREQUISITES

On the assumption that leaves are the receptor of the floral stimulus, pomologists investigated the basis of alternate bearing by defoliating and defruiting different spurs and branches at short intervals. Species or cultivars that alternated markedly from one year to the next were most commonly used. Apple spurs, defoliated within 40 to 60 days after full bloom, flowered the following spring; spurs that were defoliated after a lapse of 60 days did not (Magness, 1927). It appears that about 50 days are required for the flowering stimulus to reach a threshold level in the apical vegetative bud. After a bud on a spur attains this level of florigen, it is stimulated to flower. Defoliating the spur after this critical period has no influence on the progression of the flowering process except that the quality of the flower is poor. The lack of vigor is ascribed to the paucity of photosynthates because defoliated spurs fail to accumulate starch (Harvey and Murneek, 1921).

Davis (1957) deflorated and defruited Sugar prune trees at ten-day intervals from full bloom for 80 days. He found that the longer the fruits were left on the tree, the poorer was the return bloom the following season (Fig. 5.4). Delaying fruit removal for 30 days inhibited flower initiation on some spurs, but with a delay of 50 days after full bloom there were still 5 percent of the spurs that had return bloom. This variability of bearing spurs to form flower buds for the next season may be attributed to the light exposure that their location on the tree affords them or to the number of fruits they bore.

In order to be effective in preventing return bloom, defoliation must be done during the flower initiation period. After flower parts become evident, the treatment becomes ineffective (Gourley and Howlett, 1947). The antagonistic effect of fruits and the promotive influence of leaves on flowering partially explain why spurs bear in alternate years.

Roles of Endogenous Hormones. Florigen is presumably translocated from leaves to buds, inducing the buds to initiate flowers. But if a fruit is present on the spur, florigen is diverted into the developing fruit along with other photosynthates and is thus prevented from reaching the bud. Conversely, there is evidence that the fruit exports some of its own metabolites to the bud, which prevents the bud from initiating flowers. Luckwill (1970) removed developing flowers and fruits of Emneth Early apple at weekly intervals and bioassayed seed extracts for hormonal levels. (A bioassay is the use of biological materials as a means of estimating minute quantities of growth-regulating substances.) When the return bloom on these spurs was evaluated the following spring, spurs from which fruits were removed eight weeks after full bloom bore few or no flowers (Fig. 5.5). These data agree with those obtained from Sugar prune. The onset of the flower bud inhibition period corresponded to the

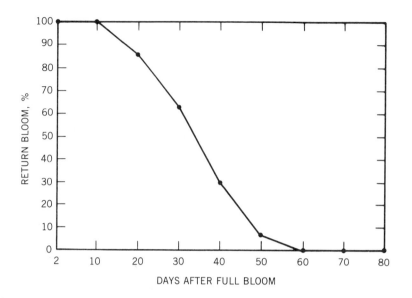

Figure 5.4
The effect of the time of removal of Sugar prune flowers and fruitlets on flower initiation for the following season. (Courtesy of L. D. Davis)

time when the gibberellin content of seeds began to increase.

Analyses of bearing and nonbearing spurs of Cox's Orange Pippin apple for growth-promoting substances revealed that levels of auxinlike compounds were high between the fifth and tenth weeks after full bloom when floral stimulus was presumably being imparted to spur buds. The level decreased to that of the control on the eleventh week and then increased. On nonbearing spurs, the level of these compounds remained nearly constant (Fig. 5.6). The presence of seeded fruit was correlated with an accumulation of a growth-promoting substance in spurs during the critical period when flower initiation was taking place. The amount of gibberellin that diffused

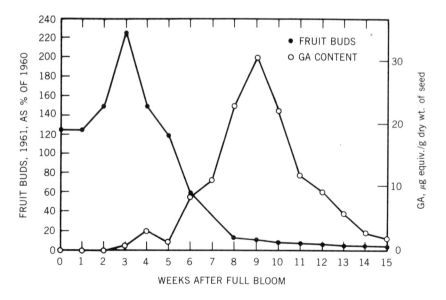

Figure 5.5
The effect of fruit removal at different times after full bloom on fruit bud formation on spurs of Emneth Early apple, and the corresponding concentration of gibberellins in the seeds. (From Luckwill, 1970)

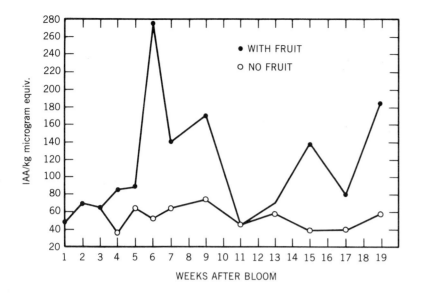

Figure 5.6
Concentration of growth-promoting substances in fruiting and defruited bourses (spurs) of Cox's Orange Pippen, as measured in the oat mesocotyl test. (From Luckwill, 1970)

from fruits of Laxton's Superb, an alternate bearing cultivar, was greater than that from Cox's Orange Pippin, an annual bearer; but the level of auxins in Laxton's Superb was lower than that of Cox's Orange Pippin (Hoad, 1979).

Influence of Exogenously Applied Growth Regulators. Grochowska (1968) excised seeds from apples still attached to spurs and replaced them with cotton swabs soaked with indoleacetic acid (IAA), naphthaleneacetic acid (NAA), and gibberellic acid (GA). Spurs treated with 20 parts per million (ppm) IAA and 500 ppm GA had return blooms equal to that of the water control; spurs treated with 10 ppm NAA had more flowers, whereas spurs treated with 1000 ppm GA possessed a third fewer flowers the following spring. Perhaps the higher GA application did not completely suppress flower formation because fruit growth was enhanced, indicating that the hormone was metabolized *in situ* and not exported to any extent. Such was seemingly the case when radioactive carbon dioxide was administered to young GA-treated grape leaves. Leaf size was increased and little radioactivity was exported (Manivel, 1973), suggesting that GA caused the leaves to act simultaneously as both source and sink. The supposition that limited export of GA from deseeded apples occurred is based on findings that spray applications of only 20 to 40 ppm GA during the flower initiation period decreased flower bud formation in cherry, peach, plum, apple, and strawberry; at 100 to 250 ppm inhibition was complete. (Strangely enough, application of this hormone to some conifer species promotes cone formation.)

Hoad (1979) confirmed that auxins favored flower initiation on Cox's Orange Pippin. Auxins may have caused ethylene evolution (Maxie and Crane, 1967) that altered the permeability of cell membranes and resulted in greater diffusion of other growth-promoting substances to the spurs. Hence, the auxin effect may not be direct.

Applications of zeatin, a natural cytokinin, and benzyladenine, a synthetic one, to the cut ends of petioles of debladed leaves on Egremont Russet apple spurs overcame the effect of defoliation on flower initiation. The percentage of return bloom on these cytokinin-treated spurs was higher than that on water-treated spurs (Hoad, 1979; Ramirez, 1979). Grape tendrils treated with zeatin and benzyladenine also initiated florets *in vitro* and *in vivo* (Mullins, 1966; Srinivasan and Mullins, 1978, 1979). Zeatin-treated mother strawberry plants did not flower but produced runner plants that bloomed under long-day conditions (Ryugo, 1985; Fig. 5.7), indicating that the flowering was stimulated by zeatin and that this stimulus was translocated via stolons. These observations suggest that flowering depends on a fine balance of hormones, whatever their roles.

Interaction Between Seeds and Leaves. Chan and Cain (1967) evaluated the role of seeds in flower initiation by hand-pollinating flowers of Spencer Seedless apple, an apetalous cultivar. This cultivar normally sets seedless fruits because its flowers do not attract bees. At harvest the number of seeds per fruit and per spur was recorded. The following spring, examination of these same spurs revealed that over 95 percent of the spurs bearing parthenocarpic fruits had return bloom. But on spurs that bore fruits totaling between 1 and 15 seeds, less than 20 percent of the spurs initiated flowers the following spring. When the number of seeds per spur exceeded 15, the return bloom was about 1 to 2 percent (Fig. 5.8).

Huet (1972) adjusted the number of leaves on bearing, seeded Bartlett pear spurs and on nonbearing spurs and found that increasing the number of leaves per spur diminished the inhibitory effect of seeds on the return bloom (Fig. 5.9).

Girdling. Gourley and Howlett (1927) reported that girdled apple trees produced more flowers the following spring than nongirdled trees, suggesting that the floral stimulus is translocated in the phloem and accumulated above the girdle. Huet (1972) found that girdled pear branches with the same number of leaves as nongirdled branches had greater return bloom, thus supporting the findings of Gourley and Howlett.

Time of Leaf Expansion. The concept that flower initiation occurs within six to eight weeks after anthesis was formulated from observations made on spur-bearing species, especially apple, pear, and Sugar prune. In these species, the spurs elongate several millimeters and produce limited numbers of leaves so that the leaf area becomes fixed early in the season. Bartlett pear bears most of its flowers on spurs, but it also produces flower buds laterally and terminally on shoots. This is not the case with stone fruits such as almond, apricot, peach, and plum, in which flower buds are formed in axils of leaves along the current season's shoot. Distal flower buds, borne on tips of these long shoots, are initiated as late as 16 weeks after full bloom. This indicates that basal flower buds are initiated by midspring when the days are short; the distal flower buds are stimulated to flower in late July when days are long. Thus, flower bud initiation at a given node seemingly depends on the time when the leaf at that node becomes fully expanded and begins to export florigen to the bud in its axil and not on any photoperiodic stimuli.

With the highly fruitful walnut cultivar Chico, removing very young leaves from the shoot terminal down to the eighth node did not deter buds in their axils from differentiating flower primordia (Ryugo and Ramos, 1979), nor did removing a few mature peach leaves inhibit buds in their axils

Figure 5.7
Zeatin-treated Tioga strawberry plant (rear) with flowering daughter plants. The mother plant was sprayed three times in March before runners were formed. (Photograph taken in July, Proc. 5th Intnl. Sym. Growth Reg. in Fruit Production. Rimini, Italy, September 1985.)

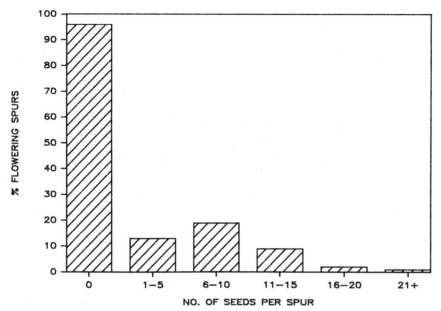

Figure 5.8
Influence of the number of seeds per spur of Spencer Seedless apples on return bloom. (Adapted from Chan and Cain, 1967)

from differentiating flowers. Apparently, florigen is abundant in these cultivars and is translocated about freely so that even buds at defoliated nodes are readily induced to initiate flowers.

Vernalization. Vernalization, a term coined Trofim Denisovich Lysenko, a Russian scientist, means making a plant behave as though spring has arrived by artificially chilling seeds or plants to induce flower bud initiation (Murneek and Whyte, 1948). Most fruit trees do not require vernalization for floral initiation; some olive cultivars and kiwifruit vines are exceptions. Olive trees exposed to a diurnal temperature fluctuation be-

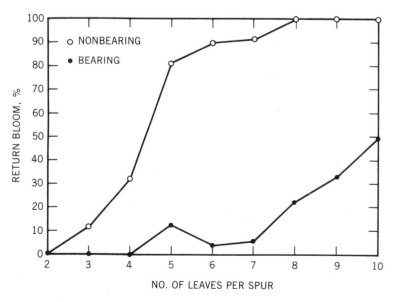

Figure 5.9
The related effects on flower bud formation of the number of leaves on spurs bearing seeded Bartlett pears and of those on nonbearing spurs. (Adapted from Huet, 1964, 1972)

Flower Bud Formation

tween 10° and 13° for 1500 to 2000 hours during winter months produce the maximum number of inflorescences. When olive trees were defoliated, winter chilling had no effect on flower formation on the panicles, indicating that some factor in the leaves contributes to the induction process (Hackett and Hartmann, 1964). In kiwifruit vines, canes exposed to winter chilling for longer periods produced mixed buds with more fruitful nodes per shoot and more flowers per inflorescence (see Flowering in Kiwifruit Vines later in this chapter).

FLOWERING IN STRAWBERRIES

Strawberry cultivars are classified as either facultative short-day types or day-neutral types. Facultative short-day plants will flower after several exposures to short-day cycles of eight to ten hours. They are facultative because they will also flower under long-day conditions provided that the mean temperature does not exceed 17° C (Darrow, 1936). Thus, short-day cultivars planted in early winter will produce flowers and fruits from March to November along the central and southern California coast where the summer temperatures are cool. The same cultivars flower and fruit for only four to five weeks when planted in the Central Valley of California because the plants become vegetative and produce many stolons as the days become longer and warmer. A planting is maintained for only one year because the amount of chilling received during the second winter is not predictable. If the second winter should be cold and the chilling requirement is completely satisfied, the crowns will flower for a short period, then grow very vigorously, and produce many runners. Should plants enter this vegetative phase, they are slow to revert to the reproductive phase after the days become short and favorable for flowering. On the contrary, if the winter is mild, plants will display symptoms of delayed foliation, having small leaves with short petioles. In either case, yield and fruit quality are reduced. Therefore, strawberries are not grown commercially along the northern California coast where the temperature is cold even during the summer. At these sites current cultivars tend to remain dormant during the winter and grow vigorously during the cool summer.

All day-neutral or everbearing types initiate flowers under both short and long photoperiods. They have different vegetative characteristics under long-day conditions, however; one type forms stolons or runners, whereas the other forms lateral crown primordia.

Inflorescence initials are formed at the apex of the strawberry crown, but as the subterminal bud elongates and leafs out, the flower bud is pushed aside and becomes laterally oriented. A minimum of two to four leaves is necessary before the next flower truss is initiated at the apex under favorable light and temperature conditions. Vegetative runner buds or lateral crown buds are initiated between nodes bearing flower buds.

Stolons that originate near the base of crowns are modified stems having two nodes. The first node does not form a daughter plant but occasionally produces a lateral branch; the second node forms a bud that develops into a rosetted shoot. If soil moisture conditions are favorable, the daughter plant initiates roots, becomes independent, and soon gives rise to stolons, creating a chain of runner plants.

The size and growth habit of strawberry plants make a good model system for investigating transport of florigen and other compounds between the mother and daughter plants, a donor–receptor relationship. Hartmann (1947) demonstrated that the floral stimulus from a mother plant growing under a short-day regimen migrated to a daughter plant growing under long-day noninductive conditions through a stolon connecting them. Guttridge (1969) showed that when daughter plants, growing under similar conditions, attained a size nearly equal to that of the mother plant, the flowering stimulus was not translocated. Mother plants under this donor–receptor relationship required nine plastochrons to flower. The first flower bud was in the axil of the ninth leaf initial before the next one was initiated at the apex. When the attached daughter plant in the receptor position was defoliated, only two plastochrons were needed to initiate another flower bud at the apex of the mother plant; the defoliated daughter plant also flowered. Guttridge (1959) attributed the delay in the initiation of floral buds in the mother plant to the translocation of a promoter of vegetative growth and of an inhibitor of floral initiation by the daughter plant growing under a long-day condition. This happens in spite of the distance between the apex and the point of attachment of the stolon to the mother plant.

FLOWERING IN GRAPEVINES

European grapes have two kinds of bearing habits: (1) those such as Zinfandel in which the basal second to the fifth buds on one-year-old canes are fruitful, and (2) others such as Thompson Seedless in which fruitful nodes are located beyond the eighth or ninth node from the base. Hence, the different pruning systems of grapes are predicated on these bearing habits (see Ch. 9).

The reproductive shoot arising from a compound bud bears a primary cluster opposite a leaf from the basal second to the fifth node. The florets on a primary cluster differentiate on the rachis primordium at budbreak or shortly thereafter. Depending on the cultivar, secondary clusters are borne on lateral branches emerging from the main shoot late in the season. The inflorescence primordia for the secondary clusters are initiated and differentiated in the same season while the main shoot is still elongating. The inflorescences range from those having many florets on all peduncular branches of a rachis to those with few florets on what are essentially tendrils. Florets have been induced to form on tendrils *in vivo* and *in vitro* with cytokinins (Srinivasan and Mullins, 1978, 1979). Hence, tendrils are considered to be morphologically homologous with the rachis.

The rachis primordia are initiated in the summer prior to flowering, but differentiation of florets on the rachis takes place at budbreak or shortly thereafter. The number of florets that differentiate depends on how well the elongating shoot was exposed to light the previous season; shoots that develop in shady portions of the canopy produce few florets. Leaf removal is often practiced to promote better aeration and reduce fungal infections, but floret formation is unaffected because shoots are repositioned concurrently to optimize photosynthesis.

That flower differentiation occurs in the spring, rather than during the previous summer as with most fruit trees, was demonstrated by Sugiura and his colleagues (1976). They were able to induce inflorescence primordia in place of tendrils on Muscat of Alexandria shoots, 15 to 40 centimeters long, by repeatedly treating developing buds and shoots with the growth retardant, cycocel. Spraying Zinfandel shoots shortly after budbreak with gibberellic acid, which inhibits flower initiation, had little or no influence on the number of florets on the primary cluster, but it significantly reduced the number of florets on the secondary cluster (Grover, 1980). This finding supports the concept that florets on secondary clusters are initiated and differentiated in the same season.

The results of double-pruning are other evidence that floral differentiation in grapes may occur relatively late. In areas where late spring frosts are common occurrences, viticulturists try to minimize frost damage by leaving four- to six-bud spurs rather than the customary two- to three-bud spurs. If a young developing shoot is damaged by a late frost, it is removed by shortening the spur immediately to force the basal dormant bud to grow. The new shoot requires another seven to ten days to reach the same developmental stage as the damaged shoot that was removed. The clusters on the second flush of growth will be smaller than those on the initial flush but larger than the frost-damaged clusters that were removed by pruning. Because the days become warmer, delaying budbreak by double-pruning affords a seven- to ten-day grace period and lessens the possibility of frost damage.

The crop on a double-pruned vine is usually lighter than that on a single-pruned vine because the basal buds that are forced into growth by double-pruning possess smaller clusters with fewer florets. This reduction in the number of florets per cluster as a result of double-pruning has been attributed to (1) the utilization of flower-forming substances and reserve food by the initially elongating shoots, leaving a much depleted supply to the inflorescence primordia in the basal dormant bud; (2) the loss of reserve food through xylem exudation that occurs during this period; or (3) both.

FLOWERING IN KIWIFRUIT VINES

Mature kiwifruit are dormant-pruned to short spurs or long canes in order to force vigorous shoots in the spring. These shoots bear inflorescences of one to three flowers in the axil of each leaf located between the second and seventh nodes. When dormant canes were collected periodically and brought into the laboratory to study flower initiation, the number of flowers per inflorescence was found to increase proportionally to the length of time the canes were exposed to winter chilling. With the same length of chilling, some cultivars produced more flowers than others (Brundell, 1975). However, the number of flowering nodes per cane depended on whether or not the canes were exposed to adequate light during the previous summer (Grant, 1983).

That the flowering stimulus is reversible is best exhibited by the kiwifruit bud. When long, dormant Hayward canes, previously exposed to winter chilling, were brought into the laboratory, cut into two-bud pieces, and placed in water to sprout, 70 to 90 percent of the developing shoots were fruitful. Hence, flower initiation in a large majority of dormant buds must have occurred by midwinter. Consequently, nearly all buds that break in the spring and elongate into shoots bear inflorescences. Yet, if any of these flowering shoots are removed manually or are broken by winds after they are more than 8 centimeters long, the adjacent dormant buds that are predictably fruitful in winter lose their potential to flower and develop into vegetative shoots.

Similarly, when chilled scions with potentially fruitful buds were grafted onto dormant rootstocks, the emerging shoots bore flowers, whereas comparable scions, grafted later in the spring onto growing stocks or onto branches of mature vines bearing fruitful shoots, developed vegetative shoots. It is postulated that the development of the first flush of growth gradually deprives the adjacent dormant buds of the flowering stimulus and/or nutrients that are prerequisite to flowering. Or the developing shoot produces an inhibitor, perhaps, a hormone that diffuses basipetally to adjacent dormant buds, eventually overcoming the promotive effect of the flowering stimulus.

The practice of tying flowering shoots to prevent their breakage not only increases yield but also keeps vigorous vegetative shoots from forming.

CULTURAL PRACTICES THAT INFLUENCE FLOWER BUD FORMATION

Certain cultural treatments such as girdling, using dwarfing rootstocks, and applying growth-retarding chemicals reduce tree size and favor the formation of flower buds. In contrast, invigorating treatments such as overthinning the crop, heavy dormant pruning, and applying excessive amounts of nitrogenous fertilizers seem to delay flower formation (Visser, 1964).

Girdling or Scoring. Girdling or ringing is a process of removing with a knife a 2- to 5-millimeter band of bark. The bark of the trunk is cut to the wood around the tree. Girdling or scoring temporarily restricts movement of photosynthates from leaves to roots and results in an accumulation of carbohydrates and hormones above the girdle. The accumulation in the growing shoots tends to stimulate floral initiation. Roots must rely on stored reserve food for respiratory energy and growth until the girdle heals over.

Girdles should be made in the spring or early summer when the lateral cambium is actively dividing and the bark is easily removed. The treatment is most effective during this period because flower bud initiation is in progress. The wound heals quickly because callus forms readily, minimizing damage to the tree. The narrow gap closes within seven to ten days, especially if the girdle is covered with grafting wax to prevent desiccation.

Rootstock–Scion Combination. Highly floriferous cultivars grafted onto dwarfing rootstocks usually come into flowering at an earlier age than those grafted on standard seedlings. Precocity of dwarfed trees is attributed to (1) the ability of the scion cultivar, especially the spur or compact types, to initiate flowers even under adverse conditions, (2) the girdling effect at the graft union, and (3) limited root growth. The overall result of such a scion–rootstock combination is that both top and roots do not grow vigorously. Hence, assimilates are directed toward the accumulation of floral stimulus, fruit development, and the formation of spurs rather than for long vegetative shoots.

Scion–stock combinations can be induced to precocity by minimizing the amount of dormant pruning and by practicing summer pruning. The ability to flower is not necessarily restricted by vigor; yearling plants of some kiwifruit and laterally fruitful walnut cultivars can grow 3 meters and still initiate flowers.

Growth Retardants. Growth-retarding chemicals such as cycocel and daminozide were originally developed to control plant size, but they were soon found to promote flower bud formation. Bloom density on daminozide-treated apple and sweet cherry trees was greater the following spring than that on untreated trees. These growth-retarding chemicals are believed to act by interrupting gibberellin synthesis. Different chemicals block the synthetic pathways at different sites. When daminozide and radioactive mevalonic acid, an early substrate for gibberellin biosynthesis, were injected into the nucellus–endosperm tissues of almond ovules, the ratios of gibberellin precursors stemming from mevalonic acid were found to be altered as compared to those in control tissues that were injected with radioactive mevalonic acid alone. This indicates that daminozide interrupts diterpene metabolism by causing levels of some gibberellin precursors to be elevated while depressing others (Ryugo, 1976). When a mixture of GA and a growth retardant was sprayed on plants, shoot growth was normal, indicating that the effect of the growth retardant was overridden. Evidently, exogenous GA enters the biosynthetic pathway beyond the point of blockage and supplies the tissues with whatever is necessary for normal development.

The proposal that the growth retardant interrupts GA synthesis was tested by determining the amounts of GA that were diffused and could be extracted from treated and control plants. GA from excised shoot apices and fruit was allowed to diffuse into agar or water; after a period of diffusion, the organs were macerated and extracted. The extractives were analyzed for residual hormonal levels. These methods revealed that less diffusible and extractable amounts of gibberellin-like substances were detectable in daminozide-treated tips of sweet cherry shoots than in those of untreated apices (Ryugo and Sansavini, 1972). Similarly, smaller quantities of diffusible gibberellins were found in daminozide-treated apple shoot tips than in nontreated ones (Hoad and Monselise, 1976; Ramirez and Hoad, 1978). The treated shoots contained more abscisic acid and less auxin. This indicates that the growth retardant not only interrupts GA synthesis but also upsets the metabolic pathways of other hormones.

Pruning versus Nonpruning. Dormant-pruning a young tree reduces the size of the aerial portion of the tree but leaves the roots intact. The resulting imbalance between the shoot and root enhances the vigor of the top the following spring because the food stored in the roots and trunk during the previous season is used by fewer buds.

As pruned trees increase in size, shoots elongate less and require less dormant pruning. Trees enter into Class III of the carbohydrate:nitrogen relationship (Fig. 5.3) and begin to initiate flower buds and set fruits. Thus, pruning delays the onset of flower formation, but because the trees are properly trained and maintained, they tend to bear good-quality fruits regularly.

Unpruned trees tend to be precocious as compared to pruned trees because unpruned trees lose their vigor early and flowers that form are not removed. If, however, trees are left unpruned for an extended period, their canopies thicken and limit light penetration. Under such shady conditions, most leaves carry on photosynthesis just above the compensation point, which results in very poor flower bud formation.

Nonpruning favors flower bud formation during the early stages of tree development, but it eventually leads to barrenness. Fruits on unpruned trees are usually of poor quality and impossible to harvest (see Training Systems, Ch. 9).

Photoperiodically sensitive plants such as Cosmos and some strawberry cultivars may be stimulated to flower at light intensities of only 5 to 10 nanomoles of photons per second per square meter. Owing to excessive top growth or crowded field conditions, buds located in shady parts of fruit trees initiate few flower; those that do form are weak.

Summer Pruning. Early summer pruning is debilitating for a mature tree, although it is an excellent practice for training young trees (see Ch. 9). Summer pruning of mature trees of early-ripening cultivars is practiced in conjunction with the use of growth retardants and size-controlling rootstocks in order to (1) maintain a tree within its allotted volume, (2) foster flower bud formation, and (3) improve the coloration of maturing fruits.

Opening up the canopy by thinning out vigorous shoots and heading back others exposes leaves within the canopy to better light and enhances the formation of flower buds. However, the treatment has led to vegetative regrowth in the fall that exposes the tree to potential winter injury. Summer pruning has occasionally resulted in early budbreak in the spring, subjecting flowers to frost damage.

Root Pruning. Root pruning is an old horticultural practice that (1) reduces the total stored food supply available to buds and (2) removes roots apices that are sources of certain hormones. Thus, each bud on the tree has a smaller share of available food, water, and hormones so that developing shoots grow less than do those on trees in which roots are left intact. The reduction in top growth favors flower bud formation.

Excessive Nitrogenous Fertilizers. Application of excessive nitrogenous fertilizers invigorates trees and results in long, willowy shoots with large, tender leaves. This condition leads to a rapid filling of the canopy early in the season and shading of the interior leaves, a situation that is unfavorable for flower bud initiation.

Branch Positioning. Bending branches toward the horizontal is also an old horticultural practice that inhibits shoot elongation and promotes flowering. French horticulturists developed various systems of espaliering fruit and ornamental trees to control growth and to promote early bearing. When a limb is bent away from the vertical orientation, the terminal bud loses its dominance. Analysis of these terminal buds revealed that their gibberellic acid content steadily decreased for several weeks when limbs were bent from the vertical position and that shoot growth was reduced at the same time (Ryugo and Sansavini, 1972). This reduction in gibberellin content, which correlates with enhancement of flower bud formation, supports the idea that gibberellins are antagonistic to the flowering process.

Limb positioning may enhance flower formation in two other ways: (1) by exposing leaves to light which is richer in red in the mornings, cytokinin synthesis is favored (Hewett and Wareing, 1973); and (2) basipetal translocation of auxins, GA, and carbohydrates is hindered by the formation of tension wood.

Knowledge pertaining to the inverse relationship between vegetative growth and the potential to flower in fruit trees is the basis for many of our cultural practices. If trees are very vigorous and set few fruits, the following procedures may correct the situation in a few years: (1) prune lightly to encourage many buds to grow, (2) reduce or omit nitrogen applications for a season or two, (3) do not thin the crop or thin sparingly, (4) plant a cover crop that will compete with the trees for nutrients and water, and (5) apply any or all of the manipulations such as limb positioning, summer pruning, and girdling just mentioned.

Strongly alternate bearing cultivars should not be planted. Existing trees with this tendency may be topworked over to cultivars that do not bear biennially. Cropping can be stabilized in cultivars that have a slight tendency to alternate by the following: (1) prune the trees moderately after an "off" year, and (2) thin flowers, or (3) thin fruitlets within three to five weeks after full bloom in the ensuing "on" year. These treatments (1) promote return bloom by adjusting the leaf to fruit ratio, (2) avoid the effect of seeds on flower bud formation, and (3) improve the light environment of the tree for the following season.

SUMMARY OF TREATMENTS INHIBITORY TO FLOWER INITIATION

During the floral initiation stage, completion of the process may be prevented in some fruit species by (1) lowering light intensity through shading or overcrowding, (2) allowing defoliation by insects or diseases, (3) applying excessively large amounts of nitrogenous fertilizers, (4) spraying with gibberellic acid, (5) double-pruning developing shoots or breaking limbs of kiwifruit vines, or (6) interrupting the night cycle of short-day strawberry cultivars with a short interval of light.

It is not yet known whether, to activate flower differentiation, a bud must accumulate florigen to a threshold level with or without a supply of carbohydrates. Phytochrome, a conjugated proteinaceous blue pigment which exists in two interconvertible forms, phytochrome-far-red (Pfr) and phytochrome-red (Pr), plays a yet unknown role in flower initiation of these photoperiodically sensitive short-day (long-night) plants.

Flower Differentiation

Tufts and Morrow (1925) examined buds of several *Prunus* species, the Bartlett pear, and Gravenstein apple and found that those stimulated to flower manifested morphological changes from early June to mid-September, depending on species and cultural practices. The shoot apex of a bud destined to flower elongates and flattens into a mesa-shaped dome (Fig. 5.10). In members of the Rosaceae family, the dome develops two distinct zones, the tunica or mantle and the corpus or body. As the dome continues to flatten, sepal primordia differentiate at the outer edge. This occurs in early July for Elberta peach grown in California. By middle to late August, petal primordia appear and enlarge. The stamens with anthers appear in early September and the pistil shortly afterward. While the outer organs are enlarging slowly during late fall, the pistil primordium enlarges and the style becomes apparent; additional rows of stamens appear. The ovarian cavity becomes noticeable in September, and by January sporogenous tissues in the anthers become apparent (Fig. 5.11). In February, the pollen mother sac gives rise to tetrads of future pollen. The flower bud is now ready to bloom.

The pistil, during its ontogeny, arises first as a horseshoe- or omega-shaped protuberance in the center of the receptacle. As the sides of this protuberance rise to form a mound, the open margins of the omega-shaped structure come together to form the ventral suture (Fig. 5.12). As the style with its stigma elongates, two ovule primordia appear, one on each side of the ventral suture. The ventral suture eventually seals and com-

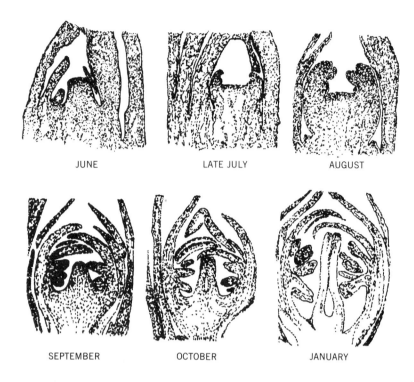

Figure 5.10
Developemental sequence of an Elberta peach flower bud. (Adapted from Tufts and Morrow, 1925)

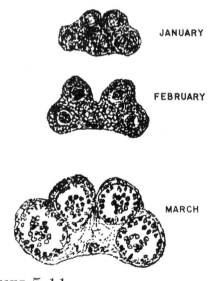

Figure 5.11
Microsporogenesis in the Elberta peach anther. Sporogenous tissue forming from the tapetum (January); pollen mother cells (February); and pollen grains ready for dispersal (early March). (Adapted from Tufts and Morrow, 1925)

pletely encloses the ovarian cavity or the seed locule.

The first sign of differentiation of pistillate flowers in Persian walnut is the appearance of sepals or tepals. The time of their appearance varies from late summer to late winter depending on the cultivar. The remaining floral parts appear in the spring prior to bloom (Lin et al., 1977). The course of floral differentiation depends on whether the cultivar is protogynous or protandrous (Polito and Li, 1985). Walnut catkins differentiate in axils of leaves in the spring, one year before they shed their pollen. Occasionally, catkins with several perfect flowers have been observed on young vigorous trees.

In the grape, rachis differentiation occurs the year before flowering, but the time when sepals appear is debatable (Pratt, 1971). Most research indicates that florets are differentiated in the spring at budbreak or shortly thereafter, but other studies suggest that sepal primordia are visible in the fall. Flower differentiation in kiwi-

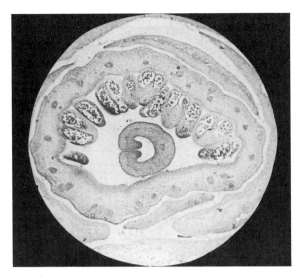

Figure 5.12
Cross section of an apricot flower bud taken in late winter. Ovule primordia are initiated from the ventral margins of the ovary. (Slide by E. B. Morrow)

fruit canes occurs at budbreak (Polito and Grant, 1984; Watanabe and Takahashi, 1984).

Buds can be induced to initiate flowers by various cultural manipulations such as spraying with growth retardants and cytokinins and girdling and positioning of limbs; buds can be inhibited from forming flowers by the administration of gibberellins during the critical period. Thus, buds may be stimulated to flower, but this stimulus can be reversed during the floral induction period, causing the bud to revert to the vegetative state. After floral parts begin to appear on the shoot apex, however, the progression to flowering is irreversible, though its rate and flower quality may be altered. A dormant, potentially fruitful kiwifruit bud loses its ability to flower by merely removing the adjacent elongating shoot.

Droughts or withholding irrigation and extreme heat during the summer when stone fruit flower parts are being differentiated often result in doubling of sweet cherry, apricot, and peach pistils. A heat spell following a period of relatively cool weather or defoliation caused by an ethylene-generating compound or by a mite infestation will hasten floral differentiation and often cause a second bloom in late summer.

Abscission of Buds, Flowers, and Fruits

BASES FOR FLOWER BUD ABSCISSION

Buds may differentiate floral parts, but whether these buds progress to anthesis and the resulting fruits persist to maturity depends on endogenous and environmental conditions. Not only flower buds but also flowers and immature and mature fruits abscise or separate from the tree in distinct waves. The amount of organs remaining after each wave of abscission is commonly referred to as having set. Hence, horticulturists refer to the proportion of fruitlets that persist after petal fall as fruit set. The same term is used to describe those that remain on the tree after "June" drop. Insofar as climatic and genetic factors govern abscission and, ultimately, yield, they should be considered during site and cultivar selection.

Winter Injury. When winter temperatures drop below $-21°$ C ($-5.8°$ F), the flower buds of many species are killed. Damage is more severe if the drop in temperature is rapid. Outwardly, buds show no symptoms of injury, but the meristematic floral parts become discolored and necrotic. These buds will fail to develop and will eventually drop.

Lack of Winter Chilling and Prebloom Abscission. The lack of winter chilling is a common cause of flower bud abscission in stone fruits, especially apricots. It occurs prior to bloom as buds begin to swell. Cultivars vary even within species with respect to their chilling requirements. Hence, those requiring long chilling periods drop more buds after a relatively warm winter than do those requiring short chilling periods.

Influence of Rain. Occasionally, after a wet but marginally cold winter, flower bud abscission does not occur in the spring as anticipated. Such an unexpected event is attributed to leaching of abscisic acid and other growth-inhibiting compounds in bud scales and in the meristematic zones of the bud. Molisch (1921) and Westwood and Bjornstad (1978) found that moistening the flower buds shortens their rest period and advances the time of flowering.

Extreme Summer Heat. Abscission of flower and vegetative buds in certain almond cultivars takes place in late summer when the temperature hovers around 40° C for several days. This disorder is called noninfectious bud failure. Bud abscission is more severe on branches facing the southwest where they are exposed to the highest solar radiation. Trees propagated with scions taken from these branches manifest the same symptoms early in their orchard life. But the disorder is not transmitted to the rootstock or reciprocally from the rootstock to the top through the graft union. Nevertheless, it can be sexually transmitted from parents having the symptoms to seedling progenies.

Competition among Developing Organs. In the pistachio cultivar Kerman, inflorescence buds on the current season's growth abscise when embryos on the one-year-old stem portion on the same branch (on-year) begin to accumulate dry matter in July. Bud abscission does not occur on nonbearing branches (off-year) because the leaves are able to supply the necessary photosynthates and hormones for inflorescence bud development. This influence of the crop on flower bud abscission does not extend to adjacent branches.

In pecan, *Carya illinoensis*, a heavy crop of nuts results in poorly developed pistillate flowers. In the spring, these flowers or the entire inflorescences on the elongating shoot terminals abscise. This abscission is attributed to the inability of (1) the shoot apex to form normal flowers and (2) the leaves to replenish the carbohydrate reserve in storage cells between harvest, when it was depleted, and leaf drop in autumn. Early defoliation of trees in the fall caused by insects or windstorms produces the same results as overcropping. The inadequate food supply leads not only to poor pistillate flower development and abscission but also to the formation of fewer catkins per bud (Worley, 1979).

Some cultivars of pecan exhibit another unusual type of abscission that is attributable to internal competition. A compound bud in pecan contains a preformed shoot consisting of basal staminate catkin initials and distal leaf and pistillate flower primordia (Wood and Payne, 1983). As the catkins begin to expand in the spring, the terminal part of the mixed bud elongates 1 to 3 centimeters and then abscises, eliminating the potential leaves and flowers. The abscission of the lateral shoot terminal would not occur if the main leader had been headed back to the compound bud during the previous dormant season. Shoot abscission is thus attributed to a type of apical dominance. The apical bud that breaks first in the spring appears to have greater food-mobilizing potential than the basal lateral buds on the one-year-old wood.

A similar abscission phenomenon occurs in walnut, *Juglans regia*, especially in the protandrous cultivar Serr, which may produce as many as 10 catkins per spur and 35,000 catkins per tree when grown under crowded conditions. The average mature catkin accumulates 350 milligrams of dry matter, of which 3 percent is nitrogen. The amassing of organic compounds by such a large number of developing catkins during a short period apparently deprives the terminal mixed buds of nutrients. Consequently, some buds abscise; others persist and elongate, but as many as 60 to 95 percent of the immature pistillate flowers on the terminals of these shoots abscise. The abscission of buds and flowers is attributed to the low nitrogen and cytokinin contents of the bud prior to budbreak. This postulation is supported by the reduction of pistil-

Figure 5.13
Serr walnut shoot with two bearing spurs from which six catkins abscised (pointer) and a nonbearing spur with intact catkins.

late flower abscission by spraying branches with benzyladenine, a synthetic cytokinin.

Abscission of immature catkins (Fig. 5.13) takes place in August and continues through September as embryos begin to fill with oil and other storage and structural materials. This phenomenon is comparable to inflorescence bud abscission in pistachio, a reflection of a strong internal competition for nutrients.

Anthesis and Zygote Formation

ANTHESIS

Anthesis, the opening of flowers, heralds the onset of spring and changes an orchard of twiggy trees into a sea of pink or white flowers, a beautiful sight to behold. The bloom is short-lived because most flowers wither soon after anthesis.

The expansion of petals and stamens requires considerable cell enlargement in each organ. This process utilizes respiratory energy and results in the evolution of ethylene gas. Filaments elongate, setting the stage for the anthers to dehisce and disperse the enclosed pollen (see Fig. 1.25). The style simultaneously straightens out, and the stigma becomes receptive to pollen.

POLLINATION

The transfer of pollen grains from anther to stigma is termed pollination. Cross-pollination is the transfer of pollen between flowers of two different cultivars of the same or related species; self-pollination is the transfer of pollen between flowers of the same cultivar. Most fruit crops require pollination and subsequent fertilization of the egg for the flower to set and develop into a fruit. Flowers that are not pollinated usually wither and abscise. The exceptions are cultivars that set parthenocarpically or without seeds, for example, the Bartlett pear under some conditions, Hachiya persimmon, and Mission fig.

Cultivars of most apricot, loquat, peach, and walnut, which can set seeds by self-pollination and/or cross-pollination, are considered to be self-fruitful and/or cross-fruitful. Almost all commercially important cultivars of sweet cherry and almond are self-unfruitful and self-incompatible, although their pollen grains are viable and functional. Within these species, cultivars whose pollen will not set fruits following reciprocal cross-pollination are termed cross-incompatible.

During the evolution of higher plants, certain genetic mechanisms arose whereby inbreeding by selfing or by cross-pollination between closely related cultivars gave way to outcrossing. Inbreeding leads to homozygosity and tends to weaken the species, whereas outcrossing affords greater opportunities for the species to remain heterozygous and promotes hybrid vigor.

The process whereby inbreeding is prevented is called gametophytic incompatibility. A mechanism known as sporophytic incompatibility prevents natural hybridization between unrelated individuals. It also provides stability to species (Heslop-Harrison, 1975).

GAMETOPHYTIC INCOMPATIBILITY

The existence of the incompatibility gene, S (formerly called the sterility factor), is responsible for gametophytic incompatibility. The S factor occurs as a multiple allelomorphic series, that is, several S genes or alleles occupy the same locus on a given chromosome. Researchers at the John Innes Institute in England reported that at least six S alleles for incompatibility occur on the same sweet cherry chromosome (Crane and Lawrence, 1934). On this basis, 12 incompatibility groups and a group 0, the universal donors that can pollinize all members of the other groups, were identified (Knight, 1969). Subsequently, another incompatibility group was discovered.

The name of the S allele was changed from sterility factor to "incompatibility gene" because in self- and cross-incompatible species carrying the S allele, such as the almond and sweet cherry, both pollen and embryo sacs are functional. Their fruits produce viable seeds; therefore, by definition, cultivars carrying the S alleles are not sterile. The terms *self-* and *cross-sterile* are used to describe the cultivars in which one or the other gamete is defective, so that they are incapable of producing seeds.

The biochemical basis of gametophytic incompatibility within sweet cherry and almond is still unknown. The following examples explain how the S factors work. Bing cherry is assigned the genes S_3 and S_4; its pollinizers, Black Tartarian and Van, carry the alleles S_1 and S_2, and S_1 and S_3, respectively (Fig. 5.14). During meiosis in the pollen mother cell, the alleles at one locus normally segregate at random so that half the pollen will receive one allele and the remaining half the other.

A pollen tube carrying the same allele as the somatic stylar cells grows more slowly than one with an unlike allele. Thus, if self-pollination occurs on a Bing flower (Fig. 5.14A), the pollen carrying S_3 and S_4 alleles grow a short distance into the style and then degenerate. If a Bing flower is cross-pollinated by Van pollen grains carrying S_1 and S_3 alleles, both kinds will germinate. But growth of the pollen tube carrying S_3 is inhibited, whereas that possessing S_1 is not (Fig. 5.14B). The pollen tubes of Black Tartarian, having either S_1 or S_2 alleles, proceed un-

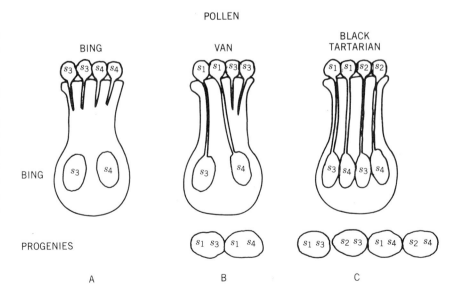

Figure 5.14
Gametophytic incompatibility in sweet cherry is governed by the S allele, the self- and cross-incompatibility factor. The (A) Bing pistil, S_3,S_4, is self-incompatible with its own pollen, S_3,S_4; partially cross-compatible with (B) Van pollen, S_1,S_3; and completely cross-compatible with (C) Black Tartarian pollen, S_1,S_2. (From Crane and Lawrence, 1934)

hindered down the Bing style until one reaches the embryo sac (Fig. 5.14C). Statistically, pollen carrying S_1 or S_2 is equally capable of causing fertilization.

Growth of an incompatible pollen tube becomes arrested about halfway down the style. The morphology of the thin-walled pollen tube tip changes from being densely packed with organelles to being swollen and vacuolated. In some species, callose is deposited in cells of the stylar tissue in front of the incompatible pollen tube, barring its progress.

SPOROPHYTIC INCOMPATIBILITY

In contrast to gametophytic incompatibility, sporophytic incompatibility involves the recognition and subsequent rejection by the stigmatic cells of a foreign protein from a taxonomically unrelated pollen. According to Heslop-Harrison (1975), proteins of the tapetum are absorbed within the cavities of the exine, the hard outer cover of the pollen grain, during the late stage of microsporogenesis. These sporogenous proteins are diploid in origin because they are derived from the tapetum, a sporophytic tissue, whereas those within the pollen are haploid. When a pollen grain is transferred to a stigma, the tapetal proteins spill out on the stigmatic surface, which is usually covered with fine hairlike structures, the papillae. If the pollen is from a taxonomically unrelated species, the proteins of the stigma "recognize" these foreign tapetal proteins and arrest or delay pollen germination. When a foreign pollen does germinate, its tube often grows weakly and is soon filled with callose, a complex carbohydrate. The synthesis of callose can continue from the tip portion of the pollen tube back into the grain itself. Thus, further growth of the tube is blocked and fertilization of the egg cannot occur.

Among the various means of overcoming cross-incompatibility are (1) using "recognition" or "mentor" pollen, (2) denaturing the tapetal proteins, (3) removing the style, and (4) exposing the flowers of self-incompatible cultivars to an elevated concentration of carbon dioxide prior to pollination. Why these techniques are successful with some species and not others is unknown.

With the recognition pollen technique, a mixture of killed compatible pollen and viable incompatible pollen is placed on the stigma of a cross-incompatible flower. Under certain conditions, stigmatic cells are unable to distinguish between the killed and incompatible pollen grains. Apparently, an exudate from the killed pollen overrides the inhibitory reaction, allowing the incompatible pollen to germinate and grow. Treating incompatible pollen grains with ethanol and other organic solvents denatures the tapetal proteins. The success of this technique has been attributed to the inability of the stigmatic cells to recognize the pollen as being "foreign" or taxonomically unrelated.

With certain cultivars of lily, the practice of removing the stylar column and placing a cross-incompatible pollen on the cut surface of the ovary has yielded viable seeds, indicating that the incompatibility factor resides in the stylar tissue.

Hybridization among species unrelated because of incompatibility has been accomplished by protoplast fusion, a biotechnological method. Although cell fusion between species in cell and callus cultures has been achieved, successful regeneration of plants from fused products is rare. Discussion of these techniques is beyond the scope of this textbook.

POLLEN TUBE GROWTH AND FERTILIZATION

If conditions are favorable, a compatible pollen becomes hydrated on the stigmatic surface within an hour of pollination and germinates. The pollen tube emerges through the germ pore and grows into the papillose epidermis. The stigma, being a secretory tissue, exudes a fluid that is rich in stigmasterol, a steroid, and phenolic compounds. The osmotic concentration of the stigmatic fluid approximates that of the pollen tube; hence the pollen tube does not absorb excess water and explode.

The pollen tube releases the exogenous hydrolytic enzymes, cutilase and pectinase, onto the stigmatic surface. They dissolve the cuticle and the pectin in the middle lamella, thus allowing the pollen tube to grow through the stigma and between the cells of the style. In some genera, as in *Actinidia* and *Lilium*, in which the style is tubular, the pollen tube grows down the side of the canal.

Upon reaching the ovary, the pollen tube continues down through the ovary wall and enters the locule. During transit, the older part of the pollen tube is filled and sealed by deposits of callose. Thus, the elongating pollen tube, isolated from the pollen grain on the stigma, must absorb nutrients from the stylar and ovarian tissues. The tube nucleus degenerates, and the generative nucleus divides into two sperm nuclei in binucleate pollen types. Eventually, the pollen tube enters the ovule via the micropyle, grows through a few layers of nucellar cells, and finally discharges the sperm nuclei into the embryo sac, the *female gametophyte* (Fig. 5.15). A pollen tube normally enters the ovule via the micropyle and rarely through the integuments. One sperm nucleus fertilizes the egg to form the zygote, and the other unites with the two polar nuclei to form the triploid endosperm. The antipodal and synergid cells then degenerate. The process by which the zygote, the new *sporophytic generation*, and endosperm are formed is called double fertilization. The term *triple fusion* is used to describe the formation of the $3n$ or triploid endosperm.

THE EFFECTIVE POLLINATION PERIOD

The effective pollination period (EPP), a term coined by Williams (1965), is the difference between the number of days the pollen requires to germinate and effect fertilization of the egg and the number of days the egg remains viable and receptive after anthesis (Fig. 5.16). If growth of

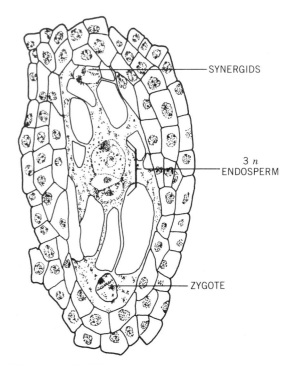

Figure 5.15
Embryo sac (the female gametophyte) undergoing double fertilization. The gametophyte is surrounded by the nucellus.

the pollen tube is rapid, the EPP is longer than if growth is slow, provided the longevity of the egg remains constant. Another interpretation is that pollination can occur several days after anthesis and still result in fertilization, provided the sperm nucleus in a fast-growing tube reaches the embryo sac while the egg is still viable. The sperm nucleus in a slow-growing tube will not arrive in time to effect fertilization. The longevity of the egg and the rate at which the tube grows depend on nutrient supplies and the respiratory rates of the egg, pollen tube, and stylar tissues.

Germination of compatible pollen results in increased growth-promoting substances in the style, but their role in pollen tube growth is uncertain. Egg longevity is prolonged in apple flowers by supplying a nitrogenous fertilizer to trees. The amount and time of nitrogen applica-

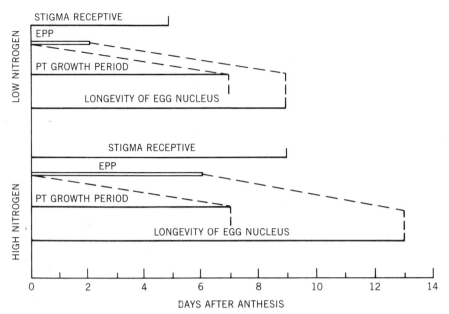

Figure 5.16
Effective pollination period (E.P.P.), of apple flowers on trees receiving small and large amounts of nitrogen fertilizer. Although the time required for pollen tubes (PT) to grow and enter the embryo sac is constant, EPP is lengthened because the large amount of nitrogen extends egg longevity. (From Williams, 1965)

tion are important cultural considerations because these two factors dictate how much and when the element becomes available to the flower buds.

Pollen tube growth is hastened by moderate temperatures (Fig. 5.17), thus shortening the time required for fertilization to occur. Supraoptimal temperatures affect EPP adversely, however, and result in poorer fruit set because respiratory rates of pollen tube, stylar tissues, and egg

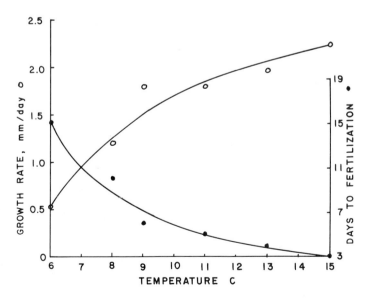

Figure 5.17
Relation between the temperature and growth rate of Comice pollen tubes in Bartlett pear pistils and the number of days required for fertilization of the egg. (From Lombard et al., 1972)

Anthesis and Zygote Formation

apparatus increase, inhibiting tube growth and shortening egg longevity. Cold temperature may extend the viable period of the egg, but it decreases the growth rate of pollen tubes. Consequently, the opportunities for fertilization to occur may be reduced.

FACTORS INFLUENCING POLLINATION AND FERTILIZATION

Many fruit species require bees or other insect vectors as pollinators for pollination. Pollen grains of these species are usually large and tend to clump together. Wind-pollinated species, for instance, the catkin-bearing nut crops and pistachio, have small, light pollen. Plants that supply pollen are called pollinizers or pollenizers. In some species, as in the male kiwifruit, flowers have very fine pollen grains that aggregate. Hence, pollen grains are carried by wind, but pollination is better if bees and syrphid flies are present in the plantation.

Floral Morphology and Insect Visitation. Insects, especially honeybees, seeking nectar and pollen are attracted to flowers primarily by their coloration. Thus, they do not approach flowers of apetalous cultivars or those emasculated by fruit breeders in preparation for hybridization. Nectar glands are usually located on the inner sides of floral tubes of stone fruit flowers and near the base of styles in apple. To gather nectar from a stone fruit flower, bees crawl between the whorls of stamen and the pistil, transferring pollen in the process.

Some peach cultivars have broad showy petals and others short, narrow, nonshowy ones. The size and shape of petals of stone fruit flowers do not influence how bees "work" the flower. The stigmas of nonshowy flowers often protrude beyond the petals before anthesis, but they are rarely cross-pollinated because bees are not attracted to unexpanded flowers. Cleistogamy or self-pollination within an unopened flower rarely occurs in stone and pome fruit species.

Bees take advantage of differences in the floral morphology of apple cultivars while gathering pollen and nectar. Some cultivars have flat, spreading petals and upright stamens, as in Delicious apple; others have cupped petals, as in Whitney, or stamens that droop, as in Wealthy and Transparent. Two morphological characteristics that influence how bees extract nectar from apple flowers are (1) the relative width of the gaps between bases of filaments and (2) the flexibility of stamens. In cultivars with upright stamens and wide gaps, sideworking bees land on the petals and gather nectar without contacting stigmas (Robinson, 1979). In cultivars with cupped petals such as Whitney, bees are forced to crawl over and down the anthers and stigmas to obtain nectar and pollen. Thus, many more of these are pollinated as compared to Delicious flowers (Roberts, 1945).

Pollen Sterility. A genetic aberration known as pollen sterility exists in *Prunus* species, notably in the peach. The pollen grain does not fill during microsporogenesis. The anthers of these pollen-sterile cultivars appear flat and pale; they lack the anthocyanins possessed by normal anthers which contain viable pollen. Cultivars such as J. H. Hale and Alamar are pollen-sterile and require cross-pollination to set a crop.

Nonviable Pollen. Triploid apple cultivars do not produce any viable pollen during microsporogenesis. Some diploid cultivars, for example, Golden Delicious, also produce nonviable pollen grains but enough viable ones for cross-pollination of other cultivars.

Abnormal Embryo Sac Development. Not only do apple cultivars such as Delicious produce flowers with varying amounts of nonviable pollen grains, but they also form abnormal female gametophytes because of an anomalous chromosome division during meiosis. Therefore, even if a normal pollen tube with its sperm nuclei reaches the embryo sac, fertilization does not occur. When too few embryos develop in fruits, they are unable to compete with adjacent fruits with more seeds so that they eventually abscise. The formation of abnormal gametes is not a factor in the fruit set of Delicious apples because sufficient numbers of normal functional gametes are produced in most years.

Dichogamy. Dichogamy is a condition in which the period of stigma receptivity does not coincide with that of pollen shedding. Most monoecious walnut cultivars shed pollen either before or after the majority of pistillate flowers become receptive (see Fig. 13.1). Most commercial walnut cultivars are protogynous; their stigmas are receptive before the catkins expand and dispense pollen. Others are protandrous; they shed pollen before the stigmas are fully receptive. Dichogamy is rarely so complete as to prevent the setting of a fair crop through self-pollination. However, the presence of a pollinizer that overlaps the bloom period of the main cultivar always improves fruit set. The same is true for filbert or hazelnut. Bloom periods of commercial cultivars of female and male pistachio trees overlap enough to assure good fruit set, but a wrong combination of cross-pollinizers could severely reduce yields.

Dichogamy is not manifested by all catkin-bearing species. Although chestnuts bear bisexual catkins and pollen shedding and stigma receptivity are synchronized, cross-pollination is essential because most members of the species are only partially self-unfruitful.

ENVIRONMENTAL CONDITIONS

Weather during the bloom period is one of the crucial factors determining how many flowers will set and develop into fruits. Under ideal weather conditions, as many as 15,000 flowers on an individual peach tree could set. To reduce the crop to 1500 fruits of marketable size is expensive. But indeed, inclement weather prior to and during bloom can occasionally reduce fruit set to a point at which harvesting is not profitable.

Lack of Winter Chilling. When deciduous fruit trees are not exposed to sufficient chilling temperatures during the winter, not only do some of their flower buds abscise prior to bloom, but those that persist open at different times. Thus, bloom may extend over several weeks, reducing the potential for good pollination and fruit set. After an exceptionally warm winter, Lambert cherry trees may bear green and ripe fruits at the same time.

Wind. Although winds are necessary for wind-pollinated species, strong, dry winds desiccate the stigmas so that pollen grains will not hydrate and germinate. Furthermore, honeybees do not leave their hives when the wind velocity exceeds 25 to 40 kilometers (15 to 25 miles) per hour.

Extreme Temperatures. When the temperature drops below $10°C$, bees remain in their hives. Pollen grains may germinate on the stigma but grow slowly at this temperature. Bartlett pear pollen tubes grow one-fourth as rapidly at $6°C$ as those at $15°C$. At $6°C$ the pollen tube growth is so slow that fruit set could be minimal. When pollination is delayed or the pollen tube grows too slowly, the egg within the embryo sac may lose its viability before fertilization occurs. At moderately high temperatures ($22°$ to $25°C$), Lu and Roberts (1952) observed that Delicious apple flowers abscised.

COMPETITION FROM WEEDS

Flowers of some fruit tree species, for example, pear, produce little or no nectar. If a cover crop is allowed to flower, honeybees will often be attracted to its flowers rather than to those of the pear trees. The cover crop should be mowed or turned under before the trees bloom in the spring. Another reason for eliminating the cover crop during this period is to let sunlight warm the soil. Reradiation of this stored heat at night from the earth to the sky lessens the chance of frost damage (see Means of Reducing Frost Damage, Ch. 7).

BEE POPULATION

Insect-pollinated crops, especially those that require cross-pollination, should have beehives in the orchard. The number of hives and their location depend on the species. With almond, apple, and sweet cherry, five to eight hives per hectare (two to three hives per acre) are recommended. Because nut size is not important to the almond trade, the hives are kept in the orchard as long as blooms persist to maximize set. In sweet cherry orchards, beehives may be kept in the orchard if inclement weather should set in during the pollination period. If, however, the days are clear and the daytime temperatures are moderate, beehives ought to be withdrawn relatively early to avoid a heavy crop and hence small fruits.

Female kiwifruit flowers are insect- and wind-pollinated. Honeybees are not attracted to kiwifruit flowers, especially the female flowers, if other sources of pollen and nectar exist. Because harvest size is directly proportional to seed count (see Fig. 6.8), 15 to 20 hives per hectare should be set out to assure that sufficient pollination occurs.

CULTURAL PRACTICES THAT FAVOR FRUIT SET

Four cultural practices commonly used to increase fruit set are (1) using rest-breaking agents after a mild winter, (2) moderating tree vigor by withholding nitrogen and water, (3) girdling or scoring the trunk or scaffold limbs, and (4) adjusting the ratio of pollinizers to the main cultivar.

Rest-Breaking Agent. After a mild winter, abscission of apricot and peach buds can be very severe and the bloom period is often extended. Dormant oil sprays are applied four to six weeks before budbreak to alleviate the abscission problem and to shorten the bloom period. The treatment should be applied after a rain while the bark is completely wet. If the trees are dry, they should be sprayed with water before the oil is applied, for oil penetrates a dry bark and kills the cambial cells.

Moderate Tree Vigor. If the tree is too vigorous and the flowers are not setting, nitrogen and water can be withheld to reduce vigor. Girdling the trunk or scaffold limbs reduces flower and fruit drop in many species. All three treatments are commonly practiced on vigorous persimmon trees, especially the Hachiya variety grafted on the invigorating *Diospyros lotus* rootstock.

Adjusting Pollinizers and Pollinators. When pollination is inadequate, additional pollinizers are grafted onto existing trees, or bouquets from compatible pollinizers may be placed in containers of water and set in tree crotches or between trees. In some high-density apple plantings, crab apples are planted solely to supply pollen. Additional hives of bees might also be distributed in advantageous sites, as under a pergola of a kiwifruit planting. This forces the bees to visit the underside of the canopy where the majority of the pendulous flowers are located.

Pollen inserts—shallow containers filled with pollen—have been placed at the exits of specially designed beehives. The objective of the inserts is to force bees to pick up the pollen as they leave and pollinate the desired crops.

Mechanical and Hand Pollination. Helicopters and air blasts from sprayers have been used in orchards to disperse pollen. The success or failure of these methods is difficult to assess statistically because under field conditions there are many variables, such as planting distances, tree vigor, proximity of other pollinizers, and tree to tree variations that cannot be controlled.

Hand pollination of apple, kiwifruit, and Asian pear flowers in conjunction with blossom thinning has been used successfully in Japan to improve fruit set and size. In walnut orchards where nut set is consistently low, catkins have been distributed throughout the planting. For this purpose, immature walnut catkins are collected and placed in a humidifier to mature. Then they are placed in a weighted cheesecloth bag to which a long string is attached. The string is used to sling the cloth bag high into a tree where the string catches on branches. The catkins mature fully and the anthers dehisce, liberating numerous tiny pollen grains. Such practices are tedious and expensive and may not be cost-effective.

POLLINATION-RELATED PHENOMENA

Apomixis or Parthenogenesis and Polyembryony

Apomixis (*apo*, without; *mixis*, mixing in Greek), or parthenogenesis (*parthenos*, virgin in Greek; *genesis*, to be born in Latin), is a phenomenon in which individual plants arise without fusion of gametes (apogamy). There are two forms of apogamy: (1) haploid individuals arising from the egg nucleus without benefit of fusion with a sperm nucleus, and (2) embryoid structures derived from diploid cells of the nucellus protruding into the endosperm of the embryo sac, a phenomenon known as nucellar embryony.

Nucellar embryony commonly occurs in ovules of *Actinidia*, *Citrus*, and *Rubus* spp. in which these diploid embryoids develop into embryos, similar in appearance to the zygotic embryo. (There is only one zygote per ovule.) The formation of these nucellar embryos is attributed to the unique nutritional environment. In a single *Citrus* seed, several nucellar embryos may form; hence, the condition is called polyembryony. These nucellar embryos are vegetative propagules or clones of the parent plant because they arise by mitosis instead of meiosis. That they express the same juvenile characteristics as does the hybrid progeny during the seedling stage complicates the citrus breeding program because the nucellar seedlings are often indistinguishable from the true hybrid.

Xenia and Metaxenia

The transmission by pollen of a phenotypic characteristic of a parent seed directly to the seed of the offspring is known as xenia. A classical example of this phenomenon is the transmission of various colors by corn pollen. The pollen that carries the gene for a certain reddish pigment, upon double fertilization within the embryo sac of a pure white or yellow corn, transmits the color code to the endosperm. This causes the aleurone layer (the outermost cells of the endosperm) to exhibit the pigmentation. Cross-pollination from plants that are heterozygous for pigment results in multicolored ears known as Indian corn.

Occasionally, almond kernels harvested from a cultivar that normally has sweet kernels will be bitter. This undesirable taste is due to the presence of amygdalin (mandelonitrilegentiobioside) which is transmitted to the kernel by pollen from a nearby tree that bears bitter almonds. Pollination of small-seeded chestnuts and date palm flowers by pollen from large-seeded cultivars promotes the growth of developing hybrid seeds within the pollinated flowers. Hybrid seeds usually need to be grown into mature trees before characteristics of the pollen parent are visible.

Metaxenia is the influence of the pollen on fruit characteristics, for example, size, color, and flavor. A metaxenia-like phenomenon exists in persimmon. In the pollination-variant persimmons, the formation of seeds causes a change in flavor and color of the mesocarp tissue adjacent to the seed. In pollination-constant phenotypes, seed formation has no effect on the flavor and color of the pulp.

These changes in pollination-variant fruits are the result of polymerization of tannins and their subsequent oxidation to a brown pigment in the large specialized cells adjacent to the developing seeds (Fig. 5.18). Polymerization causes the tannins to lose their astringency, and the brown portion becomes sweet. Mesocarp tissues surrounding an unseeded locule remain yellowish-orange and astringent. The enzyme responsible for the oxidation of tannins to a brown pigment is polyphenoloxidase. The polymerization and loss of astringency are attributed to the secretion of ethanol and acetaldehyde by the developing seeds (Sugiura and Tomana, 1983). The amount of eth-

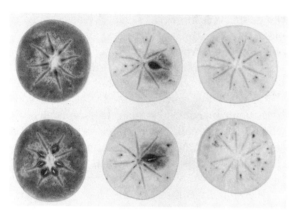

Figure 5.18
Metaxenia-like response of a pollination-variant persimmon cultivar. Only the region about the seeded locule becomes brown and sweet; the yellow portion remains astringent.

anol and acetaldehyde produced by the seed seems to be controlled by the fruit tissue.

This two-step reaction exhibited by pollination-variant persimmons is not considered a true exhibition of metaxenia because any pollen that results in seed formation is effective in causing the condensation of tannins. A puzzling observation about this phenomenon is that Hachiya fruit, which is normally pollination-constant, occasionally has browning of the flesh about the seed.

PARTHENOCARPY

Parthenocarpy (*parthenos*, virgin, and *carpos*, fruit in Greek) is the development of seedless fruits. In vegetative parthenocarpy, fruits set and develop without pollination, as in Hachiya persimmon, Bartlett pear, and Mission fig. In stimulative parthenocarpy, fruits require pollination but not fertilization of the egg by the sperm nucleus. In Thompson Seedless grape, embryos do form but abort shortly after fertilization. This type of "seedlessness," called stenospermocarpy, should not be included under parthenocarpy.

Bartlett pear, a self-sterile cultivar, sets without pollination when grown in the Sacramento River district of California. Elsewhere in California and the rest of the world it requires cross-pollination. That flowers of Bartlett pear, Concord Seedless, and Washington Navel orange are able to set parthenocarpically is attributed to their relatively high endogenous level of hormones. Immature Bartlett pear grown in the Sacramento River district contains more native gibberellins than does Winter Nelis pear which requires seed formation to set (Gil, 1971).

Hormonally Induced Parthenocarpy

Gustafson (1937) was able to set parthenocarpic tomato fruits by applying pollen extracts to the sides of the ovary. Auxin applications induced the same result, which suggests that pollens produce auxins that are necessary for setting fruits. Subsequently, Bartlett pears, peaches, apricots, cherries, and many other seedless fruits have been induced to set with gibberellic acid. Figs have been set with parachlorophenoxyacetic acid (a synthetic auxin), gibberellic acid, and cytokinins (Crane, 1964).

Seedless apple and pear have been induced by low and often freezing temperatures during bloom and the subsequent fruit-setting period. Currently, there is no good explanation for this occurrence.

Early Fruit Ripening as a Cause of Embryo Abortion

Fruits of early-ripening cherry and peach cultivars have aborted seeds. In the sweet cherry cultivar Early Burlat, the embryo aborts early in its development, coincidentally with pit hardening and mesocarp enlargement. Lignin deposition in the endocarp, accumulation of solutes, and cell wall expansion by the mesocarp cells require hormones, considerable amounts of current photosynthates, and other reserve foods. Apparently, the endosperm, which develops while the embryo is still microscopic, supplies the mesocarp with necessary hormones for growth. The endosperm is exhausted of nutrients and energy by the time the young embryo begins to develop. The inability of the endosperm to continue mobilizing food results in embryo abortion.

Seed Development and Physiology

Seeds have been the mainstay of the human diet since the dawn of history. Rice and wheat, which can be harvested, dried, and stored, have been a primary source of starch, whereas seeds of other species, such as cotton, corn, bean, olive, safflower, and sunflower, have yielded oil and protein. Cotton seed has also been an important source of fiber. Among pomological crops, the seeds of almond, chestnut, filbert or hazelnut, pecan, pistachio, and walnut have enriched our standard of living and contributed to the economy of fruit-growing districts.

If nuts are properly refrigerated after harvest, kernels retain their vitamins and mineral nutrients and are, therefore, nutritionally superior to processed nuts. Almond and chestnut kernels that are roasted or pulverized into paste or flour lose some nutritional value. Rancidity is a serious postharvest problem with nuts that contain a large amount of oil. This section deals with the ontogeny, morphology, and physiology of seeds and the relation between seed formation and fruit size.

OVULE AND EMBRYO ONTOGENY

Within a developing stone or pome fruit pistil, an ovule primordium usually appears on the inner side of each ventral margin (see Fig. 5.12) As the ovary swells after pollination and fertilization of the egg, the ovule also enlarges, keeping pace with the increase in locule size. The ovule consists primarily of the two- or three-layered integuments or seed coats, with a network of vascular bundles and the enclosed nucellus made up of thin-walled watery cells. The endosperm and embryo remain small and rather quiescent during the early stage of ovule development. As the ovule attains full size, the endosperm undergoes free nuclear divisions and begins to absorb the contents of the nucellus. The microscopic zygote forms suspensor cells, usually linearly, which push the zygote deep into the endosperm.

In stone fruits the endosperm undergoes cytokinesis, that is, the $3n$ nucleus with its cytoplasm is enveloped by cell walls, coincidentally with the onset of lignin deposition in the endocarp (pit-hardening stage). Consequently, the endosperm goes from a free nuclear to a cellular phase and acquires a rubbery texture. The enlarging endosperm consumes the nucellus from within, while the endosperm itself is consumed by the developing embryo which becomes macroscopic (Fig. 5.19).

A dicotyledonous embryo normally develops two cotyledons attached to an embryonic axis consisting of a plumule and radicle. Occasionally, stone fruit embryos possess cotyledons that are split longitudinally. It is not uncommon to find two embryos develop within a single peach ovule, a form of polyembryony.

In apple and peach ovules, the endosperm and nucellus are nearly consumed by the time the embryo is full-sized. The remnants of these two nutrient-rich tissues become thin membranous layers lining the integuments. In other species, such as the sweet cherry, part of the endosperm may remain in the mature seed. The persimmon endosperm consists of thick-walled cells in which a tiny embryo is embedded (Fig. 5.20).

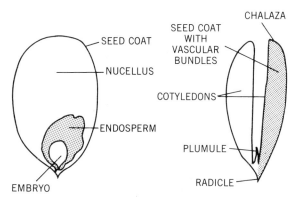

Figure 5.19
Longitudinal section of a developing peach ovule with an immature embryo (left) and a mature seed with part of the seed coat removed (right).

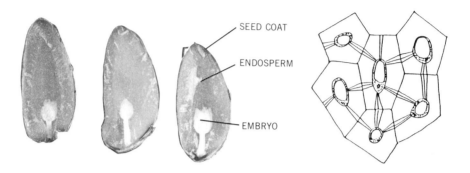

Figure 5.20
Longitudinal section of a persimmon seed with an embryo embedded in the horny endosperm (left); an enlarged section of the endosperm consisting of thick-walled cells interconnected with numerous plasmodesmata (right).

DEVELOPMENTAL PHYSIOLOGY

Embryo Nutrition. Photosynthates and nitrogenous compounds are translocated to the developing ovule through the funicular bundle that divides at the chalaza of the ovule, forming a reticulation of vascular bundles within the thin seed coat. By the process known as phloem unloading, substances being translocated in sieve tubes are actively drawn into the bundle sheath cells and then transferred to adjacent parenchyma cells. Phloem unloading and loading occur in all parts of plants. Starch grains are formed in parenchyma cells surrounding these vascular bundles, indicating that a portion of the photosynthates is temporarily transformed into reserve food in the seed coats. Other constituents of the seed coat of apple and peach are metabolized, presumably in the bundle sheath during the unloading process. Sorbitol, a sugar alcohol, and sucrose concentrations are relatively high in peach seed coats but low in the nucellus–endosperm tissues. Furthermore, the activity of sorbitol oxidase, the enzyme that transforms sorbitol to glucose, is high in the seed coat and low in the interior tissues. Since the sucrose level in the nucellus is low, the enzyme invertase, presumably present in the seed coat, hydrolyzes sucrose to glucose and fructose. The soluble carbohydrates, including inositol, and nitrogenous compounds diffuse into nucellus–endosperm tissues, and ultimately to the embryo. The cotyledons and the embryonic axis absorb these metabolites and convert them to fats, proteins, and carbohydrates which are stored or utilized for structural purposes. The embryonic axis uses the reserve food for growth during germination.

Food Storage by the Embryo. A mature seed consists of the seed coat or testa, remnants of the nucellus and endosperm, and the embryo. Cotyledons of pistachio, stone fruits, and walnut accumulate 40, 55, and 65 percent oil, respectively. The remainder of the embryo consists of proteins and cell wall fractions. Seeds of chestnut and macademia store primarily starch in the cotyledons. The persimmon seed stores a glucomannan, a polysaccharide, in the thick cell walls of the endosperm which makes up 90 percent of the seed weight. These cells are interconnected by prominent plasmodesmata (Fig. 5.20). During germination the embryo embedded in the endosperm secretes enzymes that hydrolyze the cell wall to simple sugars.

Influence of the Seed on the Fruit. The formation of seeds within the pericarp has a profound effect on fruit size and development. In apple, peach, kiwifruit, and grape, the size and shape of the fruit are a function of the number of seeds that are formed. Fruits that develop a full complement of seeds are larger and more symmetrical than those with fewer seeds. Seeded Bartlett pears are tastier than parthenocarpic ones (see Ch. 6). In pollination-variant persimmons, the color and flavor of the flesh change with seed development.

Seed Germination and Storage

FACTORS INFLUENCING GERMINATION

Immaturity of Seeds. Hybrids of early-ripening stone fruits are difficult to obtain because their embryos often abort or do not mature sufficiently before the fruit is harvested. Seeds obtained from peach cultivars that ripen between 120 and 150 days after full bloom rarely germinate. These immature embryos can grow and develop into normal seedlings, if they are excised early and are placed in a sterile, artificial, nutrient medium, a technique known as embryo rescue. That the embryos can be rescued indicates that they are potentially viable, but because they are unable to compete with the enlarging mesocarp tissue for food and mineral nutrients, they abort.

As the embryo grows and attains maximum size, the plumule goes into a resting stage from which it is not released until the seed is exposed to a period of chilling temperature. The length of the chilling period depends on the species, a heritable characteristic similar to that of buds.

Inhibition by Seed Coat and Endocarp. Seed morphology is one of several factors that affect the germination of seeds. Freshly harvested mature seeds of stone fruits will germinate immediately if they are fully imbibed with water and their seed coats are removed. Lipe (1966) found that removing the seed coat only from the chalazal end of the seed does not result in radicle growth, whereas removing a small portion of the integument surrounding the radicle end of the seed or puncturing the seed coat in this area is sufficient to induce the radicle to elongate and emerge. He detected abscisic acid in the chalazal and radicle ends of the seed coat. Thus, the inhibition of radicle growth is not attributable to this hormone. That the seed coat and the membranous remnants of the nucellus and endosperm may be impeding free exchange of gases necessary for germination to occur has been postulated.

Seeds with an impermeable, thick testa often require scarification (scratching) or dipping in concentrated sulfuric acid for a short period to hasten germination. These treatments which rupture, etch, or dissolve part of the testa allow rapid penetration of water and exchange of air.

An endocarp with densely packed, highly lignified sclereids or stone cells impedes water movement into the locule. At least three days of soaking are required for water to penetrate the Nemaguard peach endocarp. If the embryos are not fully imbibed, their rest requirement cannot be satisfied during cold storage because moisture is needed for physiological changes to occur in cells.

Went (1948, 1949) demonstrated with desert annuals that a certain minimal amount of rainfall is required for their seeds to germinate successfully and complete their life cycles. It seems that during the evolution of these desert species the testa accumulated germination inhibitors as an adaptation for surviving under arid conditions. By the time these compounds were leached from the seed coats, soil moisture was sufficient for germination and for the resulting plants to grow, flower, and disperse their seeds.

Soaking seeds with highly lignified testa or endocarp in large volumes of water or running water for about four days will leach growth and germination inhibitors. These inhibitors include abscisic acid and cinnamic acid derivatives; the latter are lignin precursors. Apparently, the stony endocarp serves as a physical barrier against animals and as a storage tissue for germination inhibitors. The seed coat seemingly acts as a chemical barrier against soil microorganisms because healthy intact seeds are rarely attacked by soil organisms.

Lack of Chilling. Although freshly harvested seeds will germinate if their seed coats are removed, their epicotyls will form rosettes similar to those of seedlings derived from seeds that did not receive sufficient chilling (Fig. 5.21). The more chilling the seed perceives, the longer the stem will be before a rosette is formed. Basal leaves on the epicotyls are normal in appearance, whereas the crinkled leaves in the rosette often have chlorotic streaks. Daily administration of gibberellic acid to these rosetted tips will cause the terminal buds to reinitiate normal growth, but another rosette is formed when the treatment ceases. If, however, the rosetted apex is removed and a lateral bud is forced into growth, the lateral shoot is normal. The terminal bud will also grow out of the rosetted condition if the seedling is forced into dormancy and is chilled.

Roles of Growth Regulators. Even adequately chilled seeds will give rise to rosetted seedlings if, prior to stratification, the seeds are steeped in a solution of daminozide, a growth retardant that interrupts gibberellic acid biosynthesis. Chilling somehow promotes synthesis of gibberellins necessary for normal seedling development, but blockage of the synthetic pathway

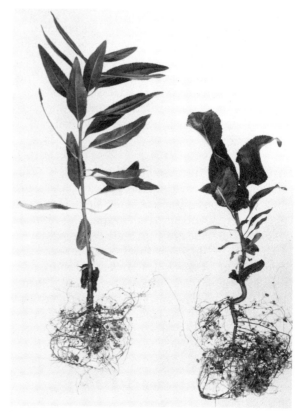

Figure 5.21
Almond seedling with a rosetted shoot terminal (right) as a result of insufficient chilling during stratification, compared to a normal one in which rest was completely satisfied.

negates the cold effect and results in rosette formation.

There is evidence for and against this idea. The level of gibberellin-like substances increases while the abscisic acid level decreases in walnut kernels during the stratification period (Martin et al., 1969). In hazelnut seed, however, gibberellic acid is not detectable during stratification. The hormone becomes detectable only after the seeds are brought to room temperature, and seedlings begin to grow (Ross and Bradbeer, 1968).

With some species such as grape, the rest requirement can be overcome by dipping freshly harvested seeds in water maintained at 35° C for

30 minutes. Soaking seeds in solutions of potassium nitrate or thiourea has been shown to enhance the germination rates of some seeds, but the mechanisms by which these treatments promote germination are not understood.

Fermentation and Spoilage. Another factor contributing to poor germination of seeds is the delay between the time the fruit is harvested and the seed is separated from the pulp. Extended delays without refrigeration often lead to fermentation and anaerobiosis, which decrease seed viability. If any mesocarp tissue is left adhering to the endocarp, there may be enough organic substrates to allow mold growth. The fungus eventually invades the locules and attacks the seed.

Allelopathy. Juglone, a toxic substance found in leaves and nuts of walnuts, suppresses the growth of tomato and other garden plants. Some desert plants produce volatile terpenes that prevent seeds of other species from germinating, thereby eliminating competition. Diffusates from peach pits inhibit the germination and growth of cucumber seeds; the inhibition is reversible with gibberellic acid. These suppressions of germination and growth of one species by another are known as allelopathy.

STORAGE OF SEEDS

Filbert, pecan, and walnut intended for seeding in the nursery are separated from their involucres, cleaned, and stored, preferably in a cool, dry place. Stone fruit seeds can be stored relatively dry, but their capacity to germinate decreases markedly after two or three years. Embryos can be stored for several years at 0° to 2° C if the moisture content is reduced to below 8 percent. The most common and serious problem with oily nuts is rancidity.

Chestnuts should be cleaned and steeped in clean water overnight before being cold-stored in a humid atmosphere. The large amounts of moisture and sugar that they contain make them highly susceptible to infection by molds. Persimmon and citrus seeds must be kept moist after removal from the fruit because they do not germinate if they become desiccated.

References Cited

BAXTER, P. 1970. The Flower Process—a New Theory. Ed. C. J. Carr. Proc. 7th Int. Conf. Growth Subst. Canberra, Australia. Springer-Verlag. New York.

BORTHWICK, H. A., AND M. W. PARKER. 1953. Light in relation to flowering and vegetative development. Rept. 13th Int. Hort. Congr. Vol. II. 801–810. London.

BRUNDELL, D. J. 1975. The effect of chilling on the termination of rest and flower bud development of the Chinese gooseberry. *Scientia Hort.* 4:175–182.

CARR, K. J., D. M. REID, AND K. G. M. SKENE. 1964. The supply of gibberellins from the root to the shoot. *Planta* 63:382–392.

CHAILAKHYAN, M. KH. 1968. Internal factors of plant flowering. *Ann. Rev. Plant Physiol.* 19:1–36.

CHAN, B. G., AND J. C. CAIN. 1967. The effect of seed formation on subsequent flowering in apple. *Proc Amer. Soc. Hort. Sci.* 91:63–68.

CRANE, J. C. 1964. Growth substances in fruit setting and development. *Ann. Rev. Plant Physiol.* 15:303–326.

CRANE, M. B., AND W. J. C. LAWRENCE. 1934. *The Genetics of Garden Plants.* Macmillan & Co. Ltd. London.

DARROW, G. M. 1936. Interrelation of temperature and photoperiodism in the production of buds and runners in the strawberry. *Proc. Amer. Soc. Hort. Sci.* 34:360–363.

DAVIS, L. D. 1957. Flowering and alternate bearing. *Proc. Amer. Soc. Hort. Sci.* 70:545–556.

FRITZSCHE, R., B. KRAPF, AND L. HUBER. 1964. Dungungsversuch mit Apfel- und Kirschbaumen in Gefassen. IV. Einfluss auf die Blutenknospenbildung. I. Apfelbaume. *Schw. Zeits. fur Obstund Weinbau.* 73:579–586.

GARNER, W. W., AND H. A. ALLARD. 1920. Effect of length of day on plant growth. *J. Agric. Res.* 18:553–606.

GIL, GONZALO, F. 1971. Studies on the development of parthenocarpic and seed pears, *Pyrus communis* L., in relation to synthetic and natural growth regulators. Doctoral thesis, University of California at Davis.

GOURLEY, J. H., AND F. W. HOWLETT. 1927. Ringing applied to the commercial orchard. Ohio Agr. Exp. Sta. Bull. 410.

GOURLEY, J. H., AND R. W. HOWLETT. 1947. *Modern Fruit Production.* Macmillan Co. New York.

GRANT, J. A. 1983. Influence of in-vine light environment on components of yield of kiwifruit (*Actinidia chinensis*, Planch.). Master of Science thesis, University of California at Davis.

GRASMANIS, V. O., AND G. W. LEEPER. 1967. Ammonium nutrition and flowering of apple trees. *Austr. J. Biol. Sci.* 20:761–767.

GROCHOWSKA, M. J. 1968. The influence of growth regulators inserted into apple fruitlets on flower bud initiation. *Bul. Acad. Pol. Sci. Ser.* 16:581–584.

GROVER, R. M. 1980. The effect of exogenous GA on floret differentiation in *Vitis vinifera* L. 'Zinfandel.' Master of Science thesis, University of California at Davis.

GUSTAFSON, F. G. 1937. Parthenocarpy induced by pollen extracts. *Amer. J. Bot.* 24:102–107.

GUTTRIDGE, C. G. 1959. Evidence for a flower inhibitor and vegetative promoter in the strawberry. *Ann. Bot.* n.s. 23:351–360.

GUTTRIDGE, C. G. 1969. Fragaria. In *The Induction of Flowering*. Ch. 10. Ed. L. T. Evans. Cornell University Press. Ithaca, N.Y.

HACKETT, W. P., AND H. T. HARTMANN. 1964. Inflorescence formation in olive as influenced by low temperature, photoperiod, and leaf size. *Bot. Gaz.* 125:65–72.

HARTMANN, H. T. 1947. Some effects of temperature and photoperiod on flower formation and runner production in the strawberry. *Plant Physiol.* 22:407–420.

HARVEY, E. M., AND A. E. MURNEEK. 1921. The relation of carbohydrates and nitrogen to the behavior of apple spurs. Oreg. Agr. Exp. Sta. Bull. 176.

HESLOP-HARRISON, J. 1975. Incompatibility and the pollen–stigma interaction. *Ann. Rev. Plant Physiol.* 26:403–425.

HEWETT, E. W., AND P. F. WAREING. 1973. Cytokinins in *Populus* × *robusta* (Schemeid): Light effects on endogenous levels. *Planta* 114:119–129.

HOAD, G. V. 1979. Growth regulators, endogenous hormones and flower initiation in apple. Rept. Long Ashton Res. Sta. Bristol, England.

HOAD, G. V., AND S. P. MONSELISE. 1976. Effects of succinic acid 2,2-dimethylhydrazide (SADH) on the gibberellin and abscisic acid levels in stem tips of M26 apple rootstocks. *Scientia Hort.* 4:41–47.

HOOKER, H. D., JR. 1920. Seasonal changes in the chemical composition of apple spurs. Mo. Agr. Exp. Sta. Bull. 40:1–51.

HUET, J. 1964. Compte-rendu des travaux poursuivis, 1954–64. Sta. Rech. Arb. Fruitiere. Angers, France.

HUET, J. 1972. Etude des effets des feuilles et des fruits sur l'induction florale des brachyblastes du Poirier. *Physiol. Veg.* 10:529–545.

KNIGHT, R. L. 1969. Incompatibility Groups: Sweet cherry (*P. avium*). Appendix III. Abstr. Bibl. Fruit Breed and Genetics to 1965: Prunus. Commonwealth Agric. Bur. England.

KNOTT, J. E. 1934. Effect of localized photoperiod on spinach. *Proc. Amer. Soc. Hort. Sci.* 31:152–154.

KOBEL, F. 1954. Lehrbuch des Obstbaus auf Physiologischer Grundlage. Springer-Verlag. Berlin.

KRAUS, E. J., AND H. R. KRAYBILL. 1918. Vegetation and reproduction with special reference to the tomato. Ore. Agric. Coll. Exp. Sta. Bull. 149:1–90.

LANG, A. 1952. Physiology of flowering. *Ann. Rev. Plant Physiol.* 3:265–306.

LIN, J., B. SHEBANY, AND D. E. RAMOS. 1977. Pistillate flower development and fruit growth in some English walnut cultivars. *J. Amer. Soc. Hort. Sci.* 102:702–705.

LIPE, W. N. 1966. Physiological studies of peach seed

dormancy with reference to growth substances. Doctoral thesis, University of California at Davis.

LOMBARD, P. B., R. R. WILLIAMS, K. G. SCOTT, AND C. J. JEFFRIES. 1972. Temperature effects on pollen tube growth in styles of Williams' pear with a note on pollination deficiencies of Comice pear. Proc. Sym. on "Pear Growing." *Proc. Int. Sco. Hort. Sci.*, Angers, France.

LU, C. S., AND R. H. ROBERTS. 1952. Effect of temperature upon seeding of Delicious apples. *Proc. Amer. Soc. Hort. Sci.* 59:177–183.

LUCKWILL, L. C. 1970. Control of growth and fruitfulness in apple trees. In *Physiology of Tree Crops*. Ed. L. C. Luckwill and C. V. Cutting. Academic Press. London, New York.

MAGNESS, J. R. 1927. Pruning investigations. Second Rept. Studies in fruit-bud formation. Oreg. Agr. Exp. Sta. Bull. 146.

MANIVEL, LAKSHMANAN. 1973. Influence of growth regulators on photosynthesis. Doctoral thesis, University of California at Davis.

MARTIN, G. C., M. IONA, R. MASON, AND H. I. FORDE. 1969. Changes in endogenous growth substances in the embryo of *Juglans regia* during stratification. *J. Amer. Soc. Hort. Sci.* 94:13–17.

MAXIE, E. C., AND J. C. CRANE. 1967. Effect of ethylene on growth and maturation of the fig, *Ficus carica* L. fruit. *Proc. Amer. Soc. Hort. Sci.* 92:255–267.

MOLISCH, HANS. 1921. Pflanzen-physiologie als Theorie der Gartnerei, 4th Ed. Gustav Fischer. Jena.

MULLINS, M. G. 1966. Test-plants for investigations of the physiology of fruiting in *Vitis vinifera* L. *Nature* 209:419–420.

MURNEEK, A. E., AND R. O. WHYTE (EDS.) 1948. Vernalization and photoperiodism, a symposium. *Chronica Bot.* Waltham, Mass.

POLITO, V. S., AND J. A. GRANT. 1984. Initiation and development of pistillate flowers in *Actinidia chinensis*. *Scientia Hort.* 22:365–371.

POLITO, V. S., AND N-Y. LI. 1985. Pistillate flower differentiation in English walnut (*Juglans regia* L.): a developmental basis for heterodichogamy. *Scientia Hort.* 26:333–338.

PRATT, C. 1971. Reproductive anatomy in cultivated grapes—a review. *Amer. J. Enol-Vit.* 22:92–108.

RAMIREZ, H. 1979. Effects of growth substances on some physiological processes in apple in relation to flower initiation. Doctoral thesis, University of Bristol, Bristol, England.

RAMIREZ, H., AND G. V. HOAD. 1978. Effects of succinic acid 2,2-dimethylhydrazide (SADH) and hormones on flower initiation in apple. *British Plant Growth Regulator Group Monograph* 2:37–47.

ROBERTS, R. H. 1945. Blossom structure and setting of Delicious and other apple varieties. *Proc. Amer. Soc. Hort. Sci.* 46:87–90.

ROBINSON, W. S. 1979. Effect of apple cultivar on foraging behavior and pollen transfer by honey bees. *J. Amer. Soc. Hort. Sci.* 104:596–598.

ROSS, J. D., AND J. W. BRADBEER. 1968. Concentrations of gibberellin in chilled hazel seeds. *Nature* 220:85–86.

RYUGO, K. 1976. Gibberellin-like substances in the endosperm-nucellus tissues of the developing almond, *Prunus amygdalus* Batsch. cv. Jordanolo. *J. Amer. Soc. Hort. Sci.* 101:565–568.

RYUGO, K. 1985. Promotion and inhibition of flower initiation and fruit set by plant manipulation and hormones, a review. Proc. 5th Intl. Sym. on "Growth Regulators in Fruit Production." Rimini, Italy.

RYUGO, K., AND D. E. RAMOS. 1979. The effects of defoliation and pruning on flower bud initiation and differentiation of 'Chico' walnut (*Juglans regia* L.). *HortScience* 14:52–54.

RYUGO, K., AND S. SANSAVINI. 1972. Effect of succinic acid 2,2-dimethyl hydrazide on flowering and gibberellic acid contents of sweet cherry (*Prunus avium* L.). *J. Hort. Sci.* 47:173–178.

SACHS, J. VON 1888. Uber die Wirkung der ultravioletten Strahlen auf die Blutenbildung. *Arb. des. Bot. Inst. in Wurzburg*. 3:372–388.

SACHS, R. M., AND W. P. HACKETT. 1983. Source–sink relationships and flowering. In *Strategies of Plant Reproduction*. Ch. 16. Ed. Werner J. Meudt. BARC Sym. #6. Allanheld, Osmum, Ottawa.

SRINIVASAN, C., AND M. G. MULLINS. 1979.

Flowering in *Vitis:* Conversion of tendrils into inflorescences and bunches of grapes. *Planta.* 145:187–192.

SUGIURA, A., AND T. TOMANA. 1983. Relationships of ethanol production by seeds of different types of Japanese persimmons and their tannin content. *HortScience* 18:319–321.

SUGIURA, A., N. UTSUNOMIYA, AND T. TOMANA. 1976. Induction of inflorescence by CCC application on primary shoots of grapevines. *Vitis* 15:88–95.

TUFTS, W. P., AND E. B. MORROW. 1925. Fruit bud differentiation in deciduous fruits. *Hilgardia* 1:3–14.

VISSER, T. 1964. Juvenile phase and growth of apple and pear seedlings. *Euphytica* 13:119–129.

WATANABE, K., AND B. TAKAHASHI. 1984. Flower bud differentiation and development of kiwi (*Actinidia chinensis* Planch.). *J. Japan. Soc. Hort. Sci.* 53:259–264.

WELLINGTON, R. 1924. An experiment in breeding apples. II. New York. Agric. Exp. Sta. (Geneva) Tech. Bull. 106.

WENT, F. W. 1948. Ecology of desert plants. I. Observations on the germination in the Joshua Tree National Monument, California. *Ecology* 29:242–253.

WENT, F. W. 1949. Ecology of desert plants. II. Effect of rain and temperature on germination and growth. *Ecology* 30:1–13.

WESTWOOD, M. N., AND H. O. BJORNSTAD. 1978. Winter rainfall reduces rest period of apple and pear. *J. Amer. Soc. Hort. Sci.* 103:142–144.

WILLIAMS, R. R. 1965. The effect of summer nitrogen applications on the quality of apple blossom. *J. Hort. Sci.* 40:31–41.

WOOD, B. W., AND J. A. PAYNE. 1983. Flowering potential of pecan. *HortScience* 18:326–328.

WORLEY, R. E. 1979. Fall defoliation date and seasonal carbohydrate concentration of pecan wood tissue. *J. Amer. Soc. Hort. Sci.* 104:195–199.

ZIMMERMAN, R. H. 1971. Flowering crabapple seedlings. Methods of shortening the juvenile phase. *J. Amer. Soc. Hort. Sci.* 96:404–411.

ZIMMERMAN, R. H. 1972. Juvenility and flowering in woody plants. A Review. *HortScience* 7:447–455.

CHAPTER 6

Fruit Growth and Development

Fruit Set

Fruit production is limited in any one season by the number of (1) flower buds that differentiate; (2) buds that bloom and proceed to anthesis; and (3) flowers that subsequently set and develop into mature fruits. The reasons for abscission of buds and flowers were discussed previously, but they are reiterated briefly as an introduction to the ontogeny of fruits.

In pome and stone fruit there are three waves of abscission of buds and flowers. The first drop takes place as buds begin to swell in late winter, if the winter has been too mild to satisfy the chilling requirement of buds. The second wave of abscission is that of unpollinated flowers just after full bloom. About ten days later a third wave consisting of pea-sized fruitlets occurs. It is attributed to lack of fertilization of eggs and polar nuclei. The stimulus of pollination is sufficient to cause pistils to grow slightly, but because the zygote and endosperm do not develop, these pistils soon wither and abscise.

The remaining flowers that set possess pistils with their ovaries and accessory parts, if any, that develop into fruits. The ovary and the developing ovule within are complex organs, each consisting of several tissues. The complexity and innumerable variations among fruit types, such as pomes, as in apple, pear, and pomegranate;

drupes, as in almond, cherry, and peach; and berries, as in grape, persimmon, and kiwifruit, make it difficult to describe the many morphological changes that fruits undergo during their ontogeny from tiny pistils in small flowers to large, mature fruits.

Similarly classified fruits for the most part exhibit a common developmental pattern or growth periodicity with slight variations. The pattern is a reflection of the integrated morphological changes that occur within the fruits. Minor variations among the species and cultivars are attributed to the duration of cell division, the time of tissue differentiation, the crop load, the time of fruit ripening, and whether the fruit had set with seeds or parthenocarpically.

Cell Division and Tissue Differentiation

Cell division in the pistils of stone fruits begins as their primordia appear on the receptacle tissue in late summer and continues for a period of two to four weeks after anthesis. In apple and pear, cell division continues in the cortical and ovarian tissue for as long as eight weeks after full bloom. Epidermal cells continue to divide for a short period thereafter. The duration of the cell division period can be extended by reducing the number of blossoms or fruitlets on the tree. When cell division is promoted, a large increase in the number of cells per fruit results. With increasing fruit size, mesocarp volume in immature Paloro peaches was proportionately greater than that of the endocarp (Ross, 1952; Fig. 6.1). In small fruits such as currants and bushberries, cell division ceases about the time of petal fall.

Cells with dividing chromosomes are detectable in the pulpy receptacle of mature strawberry fruits. Cell division in the green portion of the kiwifruit ceases about 60 days after full bloom. It continues in the columella, a white parenchymatous tissue, until the berries are ready for harvest in the fall.

The duration of cell multiplication is deter-

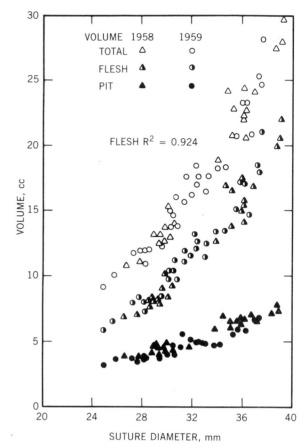

Figure 6.1
Relation between volume and suture diameter of Paloro peach and its parts at the onset of Stage II. (From Ross, 1952)

mined by noting the frequency of dividing chromosomes in fruits or analyzing them for DNA, a constituent of chomosomes (Yamaki, 1983; Fig. 6.2).

As the cell division phase ends, individual cells enlarge and so does the entire fruit. At the onset of the cell enlargement phase, cells are still small, tightly compressed, and rich in cytoplasmic substances; and they possess small vacuoles. As cells enlarge, the primary cell wall and the cytoplasmic layer become relatively thinner. Vacuoles begin to occupy a greater proportion of the cell volume. Nevertheless, the total cell wall and cytoplasmic

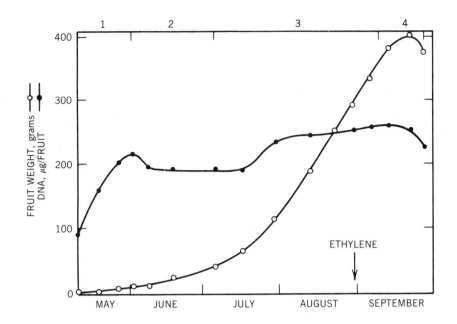

Figure 6.2
Changes in DNA content (●) and fruit weight (○) during development and ripening of Japanese pear, cv. Hosui. The numbers at the top indicate the stage of cell division, preenlargement stage of cells, enlargement stage, and ripening stage, respectively, The arrow indicates the onset of ethylene evolution. (From Yamaki, 1983)

content per fruit increases until near harvest because additional cellulosic, pectic, and proteinaceous substances are synthesized as cells increase in size. The different tissues in the fruit become visibly distinguishable from one another as individual cells within each tissue differentiate, acquiring their mature morphological characteristic.

GROWTH PATTERNS

Patterns of fruit growth and tissue differentiation and the physiological changes that accompany them vary greatly from one species to another. When their cumulative increases in diameter, length, and fresh or dry weight, or volume, are plotted, their growth patterns consist of a single, double, or triple sigmoid curve. To facilitate the explanation of events that gave rise to this pattern, Blake (1926) divided the double sigmoid growth curve of the peach fruit into three stages (Fig. 6.3).

Stage I is a period of rapid fruit growth following anthesis during which the volume of the endocarp accounts for most of the increase (Fig. 6.3). Cell division usually ceases about midpoint in Stage I. During Stage II the fruit grows slowly, but in Stage III it resumes rapid growth until harvest. The slope and duration of Stage II depend on the extent of overlap between Stages I

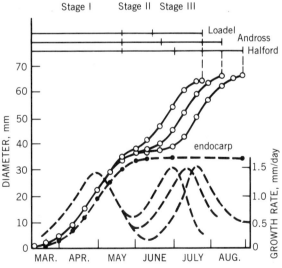

Figure 6.3
Growth stages, cumulative growth curves, and growth rates of three clingstone peach cultivars that ripen at different times.

and III, a function of the time when the fruit ripens. Fruit growth during the final period is by enlargement of the mesocarp cells.

Cumulative Growth Curves and Growth Rates

Fruits vary in shape; their growth in equatorial diameter is not always proportional to that of the polar diameter. Therefore, care must be taken when interpreting growth based on linear measurements. Fruit volumes are not always exact measures of actual growth because the interior of some fruits, for example, squash, becomes hollow with age; pit cavities of freestone peach, nectarine, and apricot, likewise, fill with air beginning about ten days before harvest when the mesocarp separates from the endocarp. Similarly, the thin-walled pericarp of pistachio enlarges rapidly after full bloom for about 30 days, but the locule remains hollow until July when the ovule begins to develop (Crane and Al-Shalan, 1973). Dry weight increases thereafter but with little increase in volume. Loss of moisture and changes in specific density are also not reflected by volume measurements.

The difficulty in determining weight or volume of intact fruit has led horticulturists to measure equatorial or polar diameters for evaluating treatment and environmental effects or for estimating the physiological stages of fruit development. These kinds of studies revealed that the strawberry, European pear, and walnut exhibit a single sigmoid growth curve (Fig. 6.4).

Drupes of stone fruits (except the almond), olive, and pistachio and drupelets of bushberry, mulberry, and fig have double sigmoid curves, as do berries such as the persimmon and both seeded and seedless grapes (Fig. 6.4). The kiwifruit has a triple sigmoid curve with the final swell occurring just prior to harvest.

Although the growth curve of the apple fruit is said to manifest a single sigmoid curve, diameter measurements of numerous individually tagged fruits recorded at close intervals by Japanese horticulturists during the growing season have disclosed a slight but definite reduction in the growth rate, comparable to Stage II of stone fruits, during early summer. This period coincides with the onset of the cell enlargement phase when embryos are developing. Similar growth retardation patterns are noted by Abbott (1984) on individual fruits (Fig. 6.5), but they are less noticeable when their average diameters are plotted. The diminution of the magnitude and duration of Stage II are attributed to differences in the physiological age and growth potential of fruits within a sampling group. Physiological ages vary

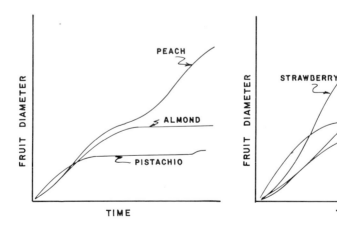

Figure 6.4
Cumulative growth curves of different fruit species based on fruit diameter.

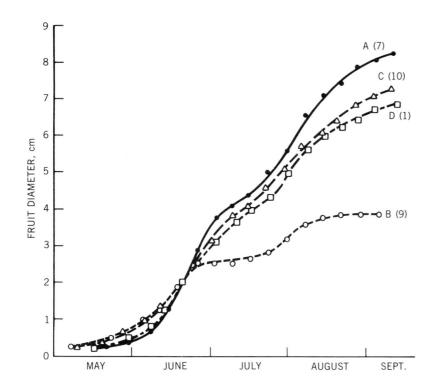

Figure 6.5
Growth curves of tagged apple fruits. The numbers in parentheses are seed counts. (A) Largest fruit, one of two on a spur; (B) smallest fruit, single fruit on a spur; (C) largest seed count, single fruit on a spur; and (D) smallest seed count, single fruit on a terminal cluster. (From Abbott, 1984)

because apple flowers do not bloom simultaneously. The growth potential of a fruit is governed by its position on the tree, its proximity to other fruits competing for photosynthates and mineral nutrients, and its seed count. The influence of position on the tree overrides that of seed count (Fig. 6.5).

The growth curves of European pear also reveal a single sigmoid curve with minor irregularities. As with the apple growth curve, data points that do not fit into a smooth "S" curve are attributed to environmental factors and to measurement errors. Asian pear cultivars manifest a double sigmoid curve, the period of growth retardation coinciding with the time when numerous sclereids differentiate about the core zone. In *Pyrus calleryana*, the ovarian tissue becomes highly lignified, resembling a tiny endocarp. Evolutionists have debated whether or not this may be a step toward the formation of an endocarplike structure in this genus to provide protection during seed dispersion.

Growth Rate

The growth rates (Fig. 6.3) of fruits are derived from these cumulative growth curves and are expressed as net change in diameter, weight, or volume per unit time interval, as shown by the following equation,

$$R_g = \frac{W_2 - W_1}{t_2 - t_1} = \frac{dw}{dt}$$

where R_g is the growth rate; W_2 and W_1 are size attributes at times t_2 and t_1, when the measurements were taken. The growth rates of peach fruits derived from diameter measurements reveal that there are two peaks. The first coincides with the time when the increase in endocarp volume constitutes more than 80 percent of the size and the second with mesocarp enlargement.

If the weight and volume measurements are available, fruit density can be calculated by dividing the weight of the fruit by its volume and expressing the quotient in grams per cubic centimeter or equivalent units.

Relative Growth Rates

Relative growth rate (RGR) is derived from growth curves by dividing the growth increment per unit time interval by the initial size and expressing the rate of growth as a percentage increase over the initial weight as shown by the following equation:

$$\text{RGR} = \frac{(W_2 - W_1)/W_1}{(t_2 - t_1)} \times 100$$

These values reveal how rapidly a given unit of fruit is adding onto itself. For example, a peach fruit will double in volume within four to five days, shortly after full bloom; in July, during Stage III, as many as 20 days are required for it to double in size (Fig. 6.6). The actual amount of reserve material or photosynthates assimilated by the fruit during July is, however, several times greater than the amount added at the earlier date.

Growth Curves as Tools

Growth curves have proved to be valuable tools for evaluating the effectiveness of certain horticultural treatments, for example, chemical thinning, application of fertilizers, and girdling of limbs. Under irrigated conditions, the growth patterns of clingstone peach cultivars are predictable year after year. A biological occurrence marking the physiological stage of peach fruit development is the onset of pit hardening. The date of pit hardening or tip change is determined by cutting thin, transverse slices from distal ends of green peaches starting about the time the endocarp attains full size. On that day about 80 percent of the endocarp tips are flinty and resist cutting by a sharp knife. The reference date is arbitrarily set as ten days after tip change (Fig. 6.6).

Harvest size can be estimated with a high degree of accuracy by measuring the suture diameters of fruits on a reference date in any one season (Davis and Davis, 1948). The correlation between fruit size at reference date and harvest is used to gauge how much fruit thinning is necessary to maximize the marketable crop.

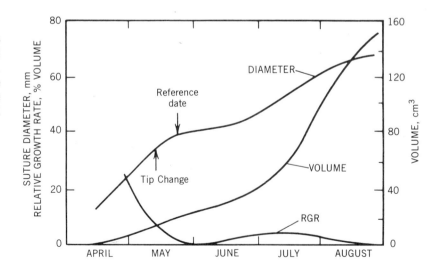

Figure 6.6
Growth patterns of a peach fruit based on diameter, volume, and relative growth rate (RGR). Reference date and tip change are noted on the cumulative growth curve. (From Davis and Davis, 1948)

The time of tip hardening has no correlation to the time when the fruit ripens. Tip hardening of Dixon and Elberta peaches occurs about a week after that of Halford, yet Halford ripens three weeks later than Dixon. Although tip change of a single variety may vary from one growing district to another in any given season, the canning industry uses a single reference date to avoid confusion. Once reference date is set, harvest dates can also be predicted because the interval between the two dates is nearly constant for a given cultivar. Canneries and their suppliers of sugars, sweeteners, labels, cans, packaging materials, and fuel, as well as labor organizations and transportation firms, use this information to synchronize their operations.

ENDOGENOUS FACTORS AFFECTING FRUIT GROWTH

Number of Cells per Fruit. Ultimate fruit size depends on the number of cells per fruit and their potential to grow. Thus, endogenous and exogenous factors that extend the duration of the cell division period and promote cell enlargement play an important role in growth.

Leaf : Fruit Ratio. A mature peach tree may bear as many as 25,000 flowers, and mature apple and Japanese plum trees could have more than 100,000 flowers at full bloom. Fortunately, many of them usually do not set, and of those that set, a percentage of them will abscise at June drop. The abscission is attributed to competition between developing fruits and the vegetative growth that is taking place simultaneously. Hence, if the number of fruits is reduced before June drop begins, a smaller percentage of fruits will abscise during June drop. In areas where spring comes early, June drop occurs in mid-May, just prior to the time when pit hardening begins in stone fruits; young apple and pear fruits abscise about the same time.

After June drop, a mature, clingstone peach tree may set as many as 6000 fruits and could have about 60,000 leaves by the end of the growing season. If such a peach tree is not thinned, the leaf : fruit ratio will be 10 : 1. A standard peach or apple tree on seedling rootstock requires a leaf:fruit ratio of 50 : 1 to 70 : 1 to attain marketable-sized fruits (Table 3). If the crop is not thinned early in the season, the fruits will be small and unmarketable; the excessive weight of the crop may break branches. Such overbearing in one season could predispose trees of some species to bear in alternate years.

Reserve Food Supply. Initial energy and food for vegetative growth and flowering come from reserve foods in storage cells of branches, trunk, and roots. These reserves are allocated among the rapidly dividing lateral cambium and its derivatives, breaking vegetative buds, and floral parts during anthesis. The reserve food is utilized by these tissues and organs until the newly formed leaves become independent and begin exporting photosynthates to meet the demands of the various sinks. If the demand is high before sufficient leaves are formed, shoot growth is arrested early in the spring so that fewer and smaller leaves are produced for that season. If fruits set heavily when the reserve supply of food is limited, cell division in the pericarp also ceases early, further limiting the growth potential of fruits.

Time of Ripening of the Crop. Competition between the developing crop and vegetative growth for reserve food and current photosynthates is keenest in cultivars that ripen in late spring to early summer. Harvest of these cultivars coincides with the grand period of shoot elongation when approximately 60 percent of the final leaf area is present. Thus, the leaf : fruit ratio is still small. The internal competition for photosynthates is so great that these early-ripening peach and plum cultivars do not attain marketable fruit size unless the crop is reduced to attain a leaf : fruit ratio of about 60 : 1. With the same leaf : fruit ratio, cultivars that ripen in August, when vegetative growth has ceased, are

Table 3. *Influence of Leaf: Fruit Ratio on Fruit Size*

Cultivar Species	Leaf:Fruit Ratio	Mean Leaf Area, cm²	Total Leaf Area, cm²/fruit	Fruit Volume, cm³
Golden Delicious apple	10	17.1	171	131.4
	20	18.6	372	167.4
	30	19.5	585	225.5
	40	20.3	812	227.2
	50	19.3	965	227.2
Johnathan apple	10	20.6	206	141.6
	15	21.3	320	167.4
	25	21.3	534	199.0
	40	22.2	888	216.1
Elberta peach	5	44.0	219	46.3
	10	44.0	438	68.7
	20	44.0	877	89.8
	30	44.0	1316	90.7
	40	44.0	1754	110.1
	50	44.0	2199	119.4
	75	44.0	3300	133.8

Apple data from Magness et al., 1931.
Peach data from Weinberger, 1932.

able to carry about 50 percent more fruit of equal size and quality.

Seed Formation and Distribution. Size is proportional to the number of mature, viable seeds in developing fruits that normally produce many seeds, for example, apple (Fig. 6.7), blackberry, fig, kiwifruit (Fig. 6.8), mulberry, pear, pomegranate, and strawberry (Fig. 6.9). Commonly, pome fruits are misshapen if seeds form in some locules and not in others (Fig. 6.10). However, the exceptions are cultivars that set parthenocarpic fruits because they are inherently rich in growth-promoting substances. Similarly, peaches with one ovule per fruit are slightly smaller and less symmetrical than those with two ovules.

The ratio between fruit growth and number of seeds is attributed to the amount of hormones produced by each seed. Because developing achenes are the source of auxins, growth of a strawberry fruit is not impaired if naphtoxyacetic acid in lanolin is smeared over an area where the achenes were removed (Nitsch, 1950; Fig. 6.11). Similarly, exogenous applications of GA to parthenocarpic apples and Asian pears enhance cell division and enlargement, indicating that the effusion of GA from seeds promotes fruit growth (Bukovac and Nakagawa, 1968; Nakagawa et al., 1968).

Crane (1969) proposed that seeds serve as a focal point for the mobilization of hormones and photosynthates to the fruit. His proposition is based on the ability of synthetic auxin, cytokinins, and gibberellins to induce fruit set and development in the absence of fertilization. Apparently, double fertilization resulting in the formation of an endosperm is a prerequisite for fruit set and development. But embryo maturation is not essential for some fruit to develop to maturity. Early-ripening stone fruits develop to maturity, although the embryos abort owing to their com-

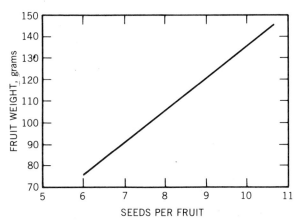

Figure 6.7
Relation between apple fruit weight and seed count. (From Mollisch, 1921)

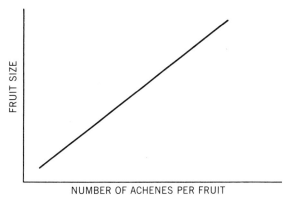

Figure 6.9
Relation between strawberry fruit size and number of pollinated achenes per fruit. (From Nitsch, 1950)

petition for nutrients with the developing mesocarp. A fruit will also ripen even though its endocarp splits, resulting in a ruptured funicular bundle and an aborted embryo. Experimentally, apple fruits have been found to continue growing even though their seeds are excised, provided the locules are filled with auxin-soaked cotton wads.

What evidence exists that the hormone-rich triploid endosperm tissue is a source of exportable hormones? A peak production of auxinlike substances in Halehaven peach ovules coincides with endosperm and embryo enlargement. Two other peaks of auxin activity corresponding to Stages I and III of fruit growth were observed (Powell and Pratt, 1966; Fig. 6.12). The level of growth-promoting substances increases in bearing apple spurs several days after a maximum peak of like substances in seeds. A comparable increase in nonbearing spurs was not detected (Luckwill, 1970). Early Burlat, a sweet cherry cultivar in which all embryos abort, has peaks of gibberellin-like and cytokinin-like substances three weeks after anthesis. These are similar to but of less magnitude than those of Rainier and Bing which produce viable seeds (Williams, 1979). These peaks coincide with the time when the endosperm and nucellus are developing and

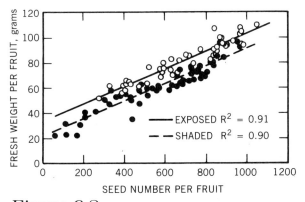

Figure 6.8
Relation between kiwifruit weight and the number of seeds per fruit and that between size and location of fruits in the canopy. (From Grant and Ryugo, 1984)

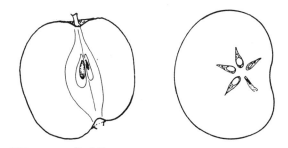

Figure 6.10
Cross and longitudinal sections of apples indicating that the part of a fruit that lacks seeds does not develop as well as the seeded part. (From Molisch, 1921)

Cell Division and Tissue Differentiation

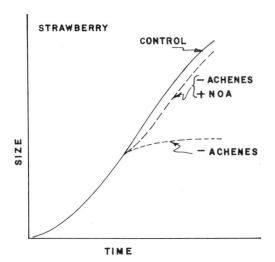

Figure 6.11
Effect of achene removal on strawberry fruit growth (−achene) and the restoration of growth by application of naphthoxyacetic acid (−achene + NOA) compared to the growth of the intact control fruit (+achene). (From Nitsch, 1950)

are presumably exporting growth-regulating substances to the mesocarp. Analyses of almond endosperm revealed that the tissue is rich in gibberellins. There is evidence that these are synthesized *in situ* rather than imported with photosynthates.

When radioactive mevalonic acid was administered to endosperms of wild cucumber and almond, gibberellic acid precursors were found in their extracts, suggesting that the tissue possesses enzymes necessary for gibberellin biosynthesis (Graebe et al., 1965; Ryugo, 1976).

ENVIRONMENTAL FACTORS

Temperature. Although the time of maturation is a heritable factor, the period from full bloom to harvest for a given cultivar is shorter in warmer fruit-growing districts than in cooler areas. Enclosing apricot branches and heating them at night with a small light bulb advanced ripening by several days (Lilleland, 1935), but these heated branches failed to initiate any flower buds for the following season.

The concept of heat units as an integrated measure between temperature and time and expressed as degree-days or degree-hours was subsequently established. Heat units between budbreak and harvest are determined by (1) computing the average daily temperature (maximum plus minimum temperatures divided by two) from which the critical temperature is deducted to yield degree-days; and (2) integrating daily the number of hours above a critical temperature from a recording thermometer chart to obtain degree-hours. The critical temperature for stone fruits and pomes is taken as 7° C (45° F); for grapes, it is 10° C (50° F). By using the first method, Winkler (1948) determined that Thompson Seedless grapes needed 2025 to 2081 degree-hours for the berries to attain acceptable eating quality. Similar correlations were developed for apricot (Brown, 1952) and Bartlett pear (Lombard et al., 1971). Thus, the brevity of the growing season and the relatively cool summer temperatures in northern latitudes limit the growing of such late-maturing apple cultivars as Granny Smith, which requires many heat units to mature.

Even though warm temperatures are needed to mature most temperate zone fruit crops, excessively hot temperatures are detrimental to fruit growth and development. Most species display some physiological disorder when fruits are exposed to high temperatures during their ripening process. Side-cracking of prunes, blackening of peaches, water core of apples, and pit-burning of apricots are examples of heat stress (see Physiological Disorders later in this chapter).

When heat-sensitive grape cultivars are exposed to 40° C for four days, the gibberellin content of berries decreases. The berries lose their ability to accumulate soluble solids, and their growth rate slows down. Nevertheless, anthocyanins are synthesized so that part of the ripening process continues. These berries do not attain their best quality for wine making or for

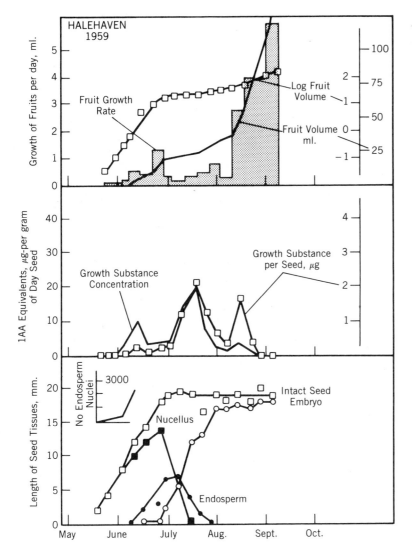

Figure 6.12
Seasonal patterns for total growth substances extracted from developing peach ovules, and the corresponding developmental curves for the ovules and entire fruits. (From Powell and Pratt, 1966).

fresh consumption. If, however, the clusters are sprayed with a mixture of gibberellic acid and benzyladenine before the onset of or during heat stress, the berries retain their ability to accumulate sugars and resume growth (Matsui et al., 1985). Heat stress seemingly upsets the hormonal balance and reduces the solute mobilizing capacity and growth potential of cells.

The apparent inability of cells to accumulate sugars may result from a rapid increase in respiration and a cessation of importation of photosynthates. Concurrent to the inhibition of berry development, heat stress for four days induces chlorosis, wilting of leaves, and cessation of transpiration (Matsui et al., 1985). Stomatal closure in wilted grape leaves is attributed to a fortyfold increase in abscisic acid content within a matter of minutes (Loveys and Kriedemann, 1973).

Conversely, a cool spell during the growing

period shortens storage and shelf life of Bartlett pears. These fruits, exposed to cool temperatures for 30 or more days prior to harvest, develop pink calyxes and ripen quickly after harvest (Wang et al., 1971).

Water Stress. Maximum shoot elongation and fruit growth rates are maintained when the range of soil moisture tension is kept between −40 and −60 centibars. Should the soil become any drier, fruit growth rate is diminished. When trees are grown in soil continuously moist, fruits are larger, but the texture is softer and their flavor not as good as that of fruits gathered from trees that are slightly stressed between irrigations (Assaf et al., 1975; Kumashiro and Tateishi, 1966).

Winds. Winds and high temperature are detrimental to fruit growth, for they cause an increase in the transpiration rate of leaves and respiration rates of leaves and fruits. Under these circumstances, even in a recently irrigated orchard, fruit cells are unable to retain water and maintain turgor. Not only is growth arrested, but fruits may occasionally shrink in size. If the stress is temporary, the original growth rate is slowly restored, but any loss in size is rarely made up by harvest time.

Light. There are very limited data indicating that green, immature fruits carry on photosynthesis and contribute to their own dry weight. It is well known that intense direct solar radiation causes sunburning and inhibits fruit development on the exposed side, but little is known about how sunlight influences fruit growth. Fruits growing in the shade are usually smaller than those growing in well-lit zones of the canopy. The smaller size is attributed to the shading of nearby leaves, thereby limiting photosynthesis. The size of sweet cherries growing on shaded spurs, however, is not markedly smaller than that of sweet cherries growing on limbs exposed to sunlight. Individual cherries that were covered with aluminum sleeves at the beginning of the pit-hardening period were much larger, but they contained less soluble and insoluble solids than adjacent exposed fruits at harvest time. Thus, intense light inhibits fruit growth or, conversely, darkness or etiolation favors growth, presumably through an auxin-sparing mechanism. Fully ripe, darkened fruits were slightly firmer but so bland and insipid as to be inedible, suggesting that shading may enhance growth but not flavor. Some direct exposure is needed for sweet cherries to develop their best eating quality.

Possible Interaction Between Temperature and Lengths of Day and Season. Horticulturists have no satisfactory explanations for some growth observations. For example, fruits grown in the cooler, northern latitudes differ in shape from those grown in warmer southern latitudes of the Northern Hemisphere. Similar differences are noted in fruits grown at higher and lower altitudes.

Yellow Newton Pippin apples grown in the Watsonville district of California are truncate; those grown in Hood River, Oregon, are cordate. The Red Delicious apples grown in the Yakima Valley of Washington and western Canada are elongated and have prominent lobes; those grown in the more southern latitudes and along Lake Michigan are shorter and lack the pinched appearance. Some apple cultivars grown in Scandinavia do not fill out as well as those grown in the more southern apple-growing districts of Europe. Bartlett pears grown in the Sacramento River delta district of California have a larger diameter : length ratio than those grown at higher elevations.

Spraying pome flowers at full bloom with a mixture of gibberellins and cytokinin induces fruit elongation. This kind of response evokes the following questions. Do fruits in districts where the summer is short but the days are very long synthesize more of these hormones than those grown in other areas? Does light quality that changes with elevation and latitude have an influence?

Blossom and Fruit Thinning

MEANS OF RELIEVING THE CROP LOAD

Pliny the Second wrote in 50 A.D. that if the number of berries in a grape cluster or cherries on a tree were reduced, the remaining fruits would be larger than if the crop were left unthinned. Thus, the practice of fruit thinning has been known for at least 2000 years, and to this day various crops are still thinned by hand. Only recently were chemicals that cause flowers and fruits to abscise introduced.

Although fruits on unthinned trees are edible, they are not comparable in quality to large fruits from well-thinned trees. The ratio of the edible portion of a peach or apple to the endocarp or core, respectively, is relatively greater for larger fruits. Furthermore, fruits of some species, as, for example, Bartlett pear, do not become palatable unless they attain a certain minimum size. To assure that consumers are provided with an edible product, federal and state agricultural codes for some crops specify minimum fruit sizes depending on the sales outlet. The minimum diameter for Bartlett pear and canning clingstone peach according to the California codes is 2⅜ inches (60.3 millimeters); peaches for pickling must be between 2 and 2¼ inches. The code also specifies that table grapes must attain a minimum soluble solids content to be marketed (see Grape Thinning, later in this chapter).

To meet these minimum standards, growers must adjust crop loads by dormant pruning and then thinning the flowers or young immature fruits. The following crops are nearly always thinned: apple, apricot, table grape, peach, Asian and European pear, plum, and persimmon.

Although thinning increases fruit size in most crops, some are not thinned. At present, nut crops are not thinned because kernel size is not an important factor in marketing of almond, walnut, pecan, and pistachio. The industry accepts small kernels provided they are sound. Small nuts and fragments are sliced or chipped for confectionery usage. Overcropped sweet and sour cherry trees are not thinned because hand-thinning is slow and expensive; furthermore, the size response is negligible. If conditions for pollination are good, beehives may be withdrawn early from cherry orchards to avoid oversetting of small fruits.

Although small prunes are rejected by the market, prune trees in California are rarely thinned because enough fruits attain minimum acceptable size in most years. The cost of handling and dehydrating small unmarketable fruits can be reduced by eliminating undersized fruits in the orchard by ataching a sizing screen on the mechanical harvester. Currently, some small dried prunes are diverted for processing into juice or for confectionery uses, but the market for these by-products is relatively small.

Kiwifruits have been thinned, but the reduction in crop load has little effect on harvest size because fruit size depends on seed count and on exposure of leaves to light. However, removing misshapen, small, and damaged fruits early in the season may have some beneficial effect on the vine because photosynthates are diverted to current shoot growth and reserve food supply for the following season. Elimination of culls in the field hastens the harvesting and subsequent packaging operations.

Two cultural means by which the crop load can be adjusted or relieved are (1) dormant pruning, removing during the winter months shoots that grew during the previous season, and (2) thinning flowers or immature fruits early in the spring.

Dormant Pruning

Peaches and nectarines bear their crops on one-year-old shoots, so that removing 75 percent of the weaker and ill-positioned shoots will still assure a sufficient number of fruitful shoots for the following season. In spur-bearing species, as in plum, apple, pear, and apricot, the pruner may thin out all extension shoots (not spurs) without materially reducing the cropping potential of the trees.

Pruning during the dormant season removes numerous potential growing points from the aboveground portion of the tree while leaving the roots untouched. Doing so alters the shoot:root ratio. The reserve food in the remaining parts of the tree will now be divided among fewer potential sinks when growth resumes in the spring. A small shoot:root ratio gives the top a big initial impetus so that shoots and flowers on these dormant pruned trees grow more vigorously than those on unpruned ones; leaves, flowers, and fruits, therefore, tend to be larger. Although reasonable harvest size of fruits might be attained by dormant pruning alone, it is not commercially feasible because yields would be reduced to a point where the orchard becomes unprofitable and noncompetitive. (For principles and details regarding pruning of fruit trees, see Ch. 9).

Fruit Thinning

The purpose of thinning, whether it is done manually, chemically, or mechanically, is to reduce the crop load. To that end, no chemical or robot has replaced an experienced person because judgment is required to space fruits evenly along a branch or to leave only one fruit per spur and then space these bearing spurs more or less evenly along a large limb.

Hand-Thinning. Thinning by hand is an expensive and time-consuming cultural operation, especially if the fruit set is heavy. It must be completed within a short period during which a large crew must be assembled and supervised. An experienced thinner may require 15 to 20 minutes to thin a peach tree bearing about 10,000 fruits so that the remaining 1200 peaches will attain marketable size.

When peach trees are so large that ladders are required, an extra person with a long pole usually follows the crew to remove fruits out of the workers' reach. Rubber-tipped poles, 4 to 6 feet long, have been used to thin canning clingstone peaches when labor was scarce. This technique is not recommended for spur-bearing species or for crops destined to the fresh market because spurs are likely to be broken and remaining fruits are often bruised.

Time of Thinning. The earlier fruits are thinned and the leaf:fruit ratio is altered, the larger fruits will be at harvest (Fig. 6.13). In blossom thinning, a portion of flowers is removed prior to or at full bloom, whereas in fruit thinning, young fruits are spaced as evenly as possible along limbs. Both practices increase the leaf:fruit ratio, and thereby increase the potential of the remaining fruit to grow. Better fruit size is obtained by blossom thinning because competition among developing fruits and elongating shoots and roots is relieved early. However, the practice is risky because inclement weather may set in during the postbloom period and further reduce fruit set.

In Japan, apple and Asian pear clusters are blossom-thinned and the remaining flowers simultaneously hand-pollinated to assure a good set. Furthermore, the clusters are covered with pesticide-impregnated paper sleeves to reduce the need for spraying.

Degree of Thinning. Standard apple and peach cultivars require leaf:fruit ratios between 40:1 and 75:1 to obtain saleable harvest size (Table 3). Stone fruit cultivars that ripen within 75 to 90 days after full bloom must have their leaf:fruit ratios adjusted about seven weeks after full bloom to attain marketable size. New spur-type apple cultivars grafted on dwarfing rootstocks are able to size their crops with leaf:fruit ratio of about 25:1 because (1) leaves are more efficient photosynthetically and/or (2) fewer assimilates are distributed for vegetative growth and more for fruit growth. Because of this sizing capacity, fruits on these scion–stock combinations can be spaced closer.

Species bearing relatively small fruits, such as apricot and plum, are spaced 10 to 15 centimeters

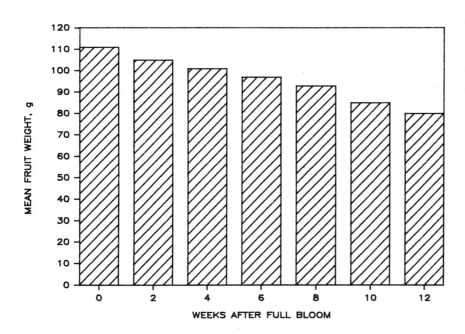

Figure 6.13
Relation between the time of thinning and peach fruit size at harvest. (Adapted from Havis, 1947)

apart. Those bearing larger fruits, such as the apple, pear, and peach, should be spaced 20 to 30 centimeters apart to attain suitable size.

Attaining excessively large fruit by thinning is not always desirable because with an increase in average size there is a decrease in the number of fruits per tree. Not only does total yield decline, but income to the grower usually decreases. Buyers are not likely to pay a premium for excessively large fruits that are difficult to market. Hence, a tree should be thinned only to the extent that a saleable yield is maximized.

Mechanical Thinning. Limb shakers, normally used for mechanical harvesting, have been employed to thin peaches and plums. The force of the shaker is transmitted to the limb at the point of contact. This sets the branch in motion while fruits tend to remain stationary because of their inertia. The force causes the heavier fruits and those closest to the point of impact to separate from the spurs. Thus, the remaining crop is poorly distributed.

If the operator of the shaker is inexperienced or the machinery is poorly designed, the limbs can be set into a whipping motion, causing much damage to the fruits, especially in cultivars with long peduncles. Excessive shaking also separates the bark from the wood, creating large wounds which are entry points for various organisms.

Chemical Thinners

Several registered chemicals effectively thin flowers and young fruitlets (Fig. 6.14), but they also produce side effects, some beneficial and others undesirable. Their modes of action may differ, and each may have more than one mode.

Dinitro-ortho-cresol (DNOC; Elgetol) is a caustic material that causes necrosis of flower parts; hence, it undoubtedly stimulates ethylene evolution. The material is also a pollenicide at high concentrations and an inhibitor of pollen tube growth at low concentrations.

DNOC is sprayed when trees are approaching full bloom. The material kills the parts of the flower that it contacts, but to be effective, the spray droplets must fall on pistils. Hence, a conservative grower would put on an application when 50 to 60 percent of the flowers are open,

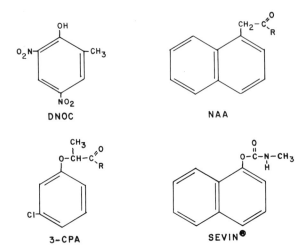

Figure 6.14
Chemical structures of four blossom thinners. DNOC is 4,6-dinitro-ortho-cresol; NAA is naphthaleneacetic acid and 3-CPA is 3-chlorophenoxypropionic acid, R = OH for acid, NH_2 for amide; and Sevin is 1-naphthyl N-methyl carbamate.

with the anticipation of hand-thinning fruits that set. The daring grower would wait until 90 percent of the bloom is out to put on the spray to reduce the cost of hand-thinning. With some cultivars of Japanese plums, removing 90 percent of the blossoms would still leave about 10,000 unopen flowers to survive.

Naphthaleneacetic acid (NAA), naphthalene acetamide (NAAm), 3-chlorophenoxypropionic acid and its amide (3-CPA), and 1-naphthyl N-methyl carbamate (carbaryl, marketed as Sevin) are synthetic auxins. Carbaryl was initially introduced as an insecticide and is still used as such.

One mode of action of these auxinlike substances is to interfere with phloem transport of hormones and photosynthates in and out of fruits. Normally, an auxin gradient between a leaf or fruit and the spur defers the formation of an abscission layer. Exogenous applications of synthetic auxins stop and/or reverse the auxin flow and induce fruit drop (Addicott and Lynch, 1951; Crane, 1969). Radioactive carbaryl, applied to developing apples, accumulates with photosynthates in the pedicels, suggesting that this material interrupts the transport of food and hormones (Williams and Batjer, 1964). This implies that auxinlike thinners must be absorbed at or translocated to the site of action.

The second mode of action of auxinlike substances is through the evolution of ethylene (Walsh et al., 1979). If the evolution of ethylene by NAA or NAAm is prevented by treatment with aminoethoxyvinylglycine (AVG), an inhibitor of ethylene generation, the thinning effect is negated (Williams, 1981). The gas is evolved spontaneously by auxins but is not translocated. Ethylene lowers membrane integrity, thereby allowing substrates to react with enzymes that are normally kept separate or compartmentalized in the cytoplasm. Thus, ethylene derived from ethephon, (2-chloroethyl)phosphonic acid, an ethylene-generating compound, induces gummosis of shoots. Secretion of gum into the xylem increases resistance to water movement (Olien and Bukovac, 1982) and results in increased xylem water potential, which may be responsible for the fruit abscission. An overdose of ethephon causes premature leaf senescence and abscission.

NAA, NAAm, and carbaryl are applied when apples are between 9.5 and 12 millimeters in diameter. Fruitlets reach this stage of development about 15 to 21 days after petal fall. With spur-type Red Delicious strains that are difficult to thin, NAA or ethephon combined with carbaryl is effective. One beneficial effect of chemical thinning is the increase in return bloom.

Two undesirable side effects of NAA and NAAm when applications are poorly timed are a reduction in the size of the fruits and the formation of "pigmies," which are fruits smaller than 40 millimeters in diameter. NAA may occasionally increase the harvest size of apples without noticeably thinning the crop (Rogers and Thompson, 1983) or thin the crop without increasing fruit size (Luckwill, 1953), suggesting that the material has a direct effect on fruit growth. Carbaryl does not cause pigmy formation, but it does induce russeting of Golden Delicious. Russeting may be reduced or avoided by combining GA with NAAm.

Leaves on peach trees sprayed with 3-CPA

often wilt and/or exhibit epinasty. At high dosages, margins of leaves become scorched. These stress symptoms may be caused by auxin-induced ethylene production which inevitably accompanies injuries. Wilting causes a manifold increase of abscisic acid in water-stressed grape leaves (Loveys and Kriedemann, 1973). Since ethylene and abscisic acid are associated with induction of an abscission layer, the mode of action of 3-CPA is probably multipronged.

Factors Affecting the Efficacy of Chemical Thinners

Although apples are routinely thinned in many districts, satisfactory, uniformly consistent results have not always been achieved because (1) sensitivity to a specific thinning agent varies from clone to clone, and (2) climatic conditions enhance or diminish the efficacy of chemicals. Chemical thinners have not been widely used for other fruit tree species, with the possible exception of pears in some areas and the Japanese plum.

Varietal Susceptibility. The reasons why some cultivars are easier to thin chemically than others have not been investigated systematically. The ease or difficulty of thinning may be due to varietal differences in (1) cuticle thickness and stomatal density that regulate the rate of absorption by leaves and fruits, (2) water retention by leaves (see Peach, Ch. 12), (3) flower strength or vigor and stage of fruit development, (4) rapidity of transport of the chemical to the abscission zone, and/or (5) the rate at which the thinner is metabolized.

Cuticle composition and thickness of apple fruits vary according to (1) cultivars, (2) location of fruits on the tree, and (3) the temperature and humidity of the orchard site. A thinning agent must be absorbed through the cuticle and act *in situ* or be transported to the site of action. The molecular configuration of the thinner must be such that it can penetrate several chemical barriers. The epidermis is coated with a waxy cuticle that repels water. The cell wall below the cuticular layer is cellulosic and, therefore, hydrophillic or wettable with water. The outermost cytoplasmic layer, the plasma membrane, is proteinaceous and lipophillic; polar compounds, including water, do not readily penetrate this layer. Thus, the thinner must diffuse through these various tissue layers and finally enter the cytoplasm, the metabolic realm of cells, in order to function.

Another factor that may contribute to inconsistent thinning is the chilling requirement of different cultivars. Those requiring many hours of chilling usually bloom over a longer period than those requiring few hours of chilling. Timing of the spray application is not critical if the bloom period is extended as compared to a bloom that is completed in a few days.

Another varietal characteristic that influences chemical thinning is flower size. Peach cultivars vary in petal size, ranging from large showy petals as in Fay Elberta to short nonshowy types as in Halford. Thus, the blossom or target size on which spray droplets must contact varies from cultivar to cultivar. The short petals of Halford also allow the stigma and part of the style to protrude from the flower bud before anthesis. This floral trait makes the pistils more vulnerable to caustic substances than those covered by large petals of showy flowers.

Developmental Stage of Fruitlets. Chemicals applied at a given dosage are usually more effective on younger than on older fruits. During the postbloom period when the endosperm and nucellar tissue are developing within the ovule, endogenous hormone levels are constantly fluctuating. The endosperm is especially rich in hormones while it is undergoing cytokinesis. This is the transition period when the endosperm changes from the free nuclear to the cellular state. If the chemical thinner is applied at this stage of fruit development, the total level of the exogenous synthetic material and native compounds exceeds what tissues can tolerate and eventually causes the fruitlets to abscise. As the hormonal level in the endosperm and nucellar tissues decreases, larger dosages of exogenous auxins are required to accomplish the same degree of thinning.

Weather Conditions: Fog, Rain, and Wind. Elements of weather, fog, rain, and wind, affect the rate at which water evaporates and, therefore, alter the efficacy of the thinning agent. Depending on the relative humidity at the time the chemical thinners are applied, as little as 10 to 25 percent of the active ingredient penetrates the leaf. Therefore, if fog should roll in after the trees have been sprayed, some of the chemical residue on the leaf surface will dissolve and be reactivated. It will penetrate the leaf, which could cause excess thinning of the crop. A light shower would have the same effect, but a heavy rain would wash off the surface residue and diminish the effectiveness of the compound.

Wind has a drying effect so that even with the slightest wind, more of the chemical is left on the leaf surface as an inert residue. Although the residue loses its effectiveness with time, the chemical can be reactivated by fog or rain following the spray treatment. Spraying on windy days is not recommended because coverage is likely to be poor. If the ground is too wet following a rainstorm, spray applications are often made by airplanes or helicopters. Aerial application is illegal in some localities if the wind speed exceeds a certain velocity. Materials drifting onto nearby crops for which the chemical is not registered can create problems.

Chemical Instability. Other mechanisms may interfere with the stability of chemicals that promote or prevent abscission. Endogenous IAA breaks down readily when tissues are exposed to light (Kawase, 1965); other indole compounds are rapidly conjugated with amino acids or sugars. Breakdown of ethephon is pH-dependent. The variability in the pH of water used for diluting the chemical may be one factor responsible for the erratic results reported with the use of this compound. Kinetin, which is effective in preventing walnut flower abscission, decomposes to at least four by-products within 24 hours on any surface.

Surfactants or Wetting Agents. Surfactants are added to thinners to increase their absorption by fruits and leaves. The two must be compatible and be water-soluble or form a fine emulsion. Surfactants reduce the surface tension of a solution so that spray droplets, instead of remaining as beads on the leaf surface, spread out and form a thin layer. Increasing the surface area of the solution in contact with the leaf surface allows more thinning agent to penetrate the epidermal cells, but the solution also evaporates more rapidly.

GRAPE THINNING

Table grapes, like other fruits for the fresh market, must be large and colorful to attract consumers. The crop or clusters must be thinned to attain these characteristics. The two common methods of reducing the crop load are berry thinning and cluster thinning. Both methods are based on the bearing habits of the cultivars.

Figure 6.15
Comparison between gibberellic acid-treated Thompson Seedless grapes (right) and untreated control berries (left).

Those that produce large clusters are berry-thinned by reducing the number of berries on a cluster. Varieties that form numerous clusters are cluster-thinned. Berry size is not important with wine grapes, but the crop is thinned if the vine is overloaded. Berries on overcropped vines are slow to accumulate soluble solids and may not yield an optimum-quality wine.

Cultivars that consistently set large numbers of berries and produce "tight" clusters or bunches are berry-thinned by removing small branches of the rachis with needle-nosed thinning shears. On long clusters, removing the distal one-half to one-third of the rachis with one snip of the shears may be sufficient to loosen up the cluster. Distal berries usually ripen last so that berry thinning results in clusters with a uniform maturity. If large clusters are not thinned, berries in the interior of clusters are squeezed and eventually crack. Mold grows on the spilled cellular contents, ruining the cluster. Even if the interior berries do not crack, they become angular and do not color well. Tight clusters tend to be bulky and difficult to pack for storage and shipment. Cultivars that produce many clusters per shoot are cluster-thinned by removing poorly filled rachises after estimating the size of the set.

With the introduction of gibberellic acid, thinning and sizing practices have changed markedly, especially with seedless types such as Sultanina and Centennial. With these cultivars, an application of 25 to 40 ppm gibberellic acid several days prior to bloom causes the rachis to elongate. The hormone is administered again at full bloom as a blossom thinner to reduce the number of florets. Another application about five days after full bloom induces cells to enlarge and increases berry size (Fig. 6.15). Berries have an even greater size when the trunk is girdled.

With repeated applications of gibberellic acid, canes have poorly formed inflorescences the following season. Thus, the hormone is not recommended for seeded cultivars, especially wine grapes.

Source–Sink Relationships

Leaves and, to a lesser extent, green stems and immature fruits supply the rest of the plant with organic carbon compounds. However, other organs such as the roots and seeds synthesize substances that are exported to other parts of trees. Thus, organs and tissues that synthesize and export substances are called sources, whereas those that import substances are called sinks. Some tissues such as the ray cells in the xylem serve as sinks when they are accumulating carbohydrates and storage proteins, but they become sources when these same foods are exported later to other organs. A young immature leaf can simultaneously be a source and a sink because it fixes carbon and utilizes the assimilates for its own growth.

Horticulturists consider the source–sink relationship between leaves and fruits to be important because the crop is a strong competitor for reserve foods and photosynthates, and this competition influences the entire tree physiology. When leaves of bearing apple spurs were enclosed and administered radioactive carbon dioxide, most of the labeled carbohydrates were translocated to fruits on the same spurs (Hansen, 1967a,b, 1970). Radioactive photosynthates from nonbearing spur leaves moved to fruits on adjacent spurs, provided the distance was not far. Radioactive carbohydrates synthesized by leaves on vigorously growing terminal shoots remained principally within the treated shoot. In grapevines, photosynthates from fully expanded mature leaves beyond the twelfth to fifteenth node are exported acropetally to nourish the elongating shoot terminal. Basal leaves meet the demands of the nearby grape clusters, and the remainder is translocated to the roots. These translocation patterns demonstrate that the relative ability of organs and tissues to mobilize photosynthates (sink strength) depends on their size and stage of development. These factors and the distance between the source and sink determine the rate at which substances move from the source to the sink. The distance and rate make up the source:sink gradient.

Chandler (1934) found that the dry weights of bearing apple trees and grapevines were lighter than those of nonbearing plants after several years, but their cumulative total dry matter, including leaves and fruits, was larger. He concluded from these data that the leaves of bearing trees fixed carbon more efficiently than those of nonbearing trees. Avery (1975) found that leaves on bearing apple spurs assimilated carbon dioxide 50 percent faster than those on nonbearing ones, confirming the findings of Chandler. These findings indicate that if the sink becomes strong and the gradient between the source and the sink steepens, the source is stimulated to become more productive.

A three-way competition among storage cells in one-year-old stems subtending ripening stone fruits, the fruit itself, and the elongating shoot is another manifestation of a source–sink relationship. Stone fruits rapidly accumulate comparatively large amounts of soluble solids during the 10- to 14-day period prior to harvest. The starch in the storage cells of xylem and phloem tissues in branches subtending the ripening fruit is then hydrolyzed to sugars. These sugars are mobilized to the fruit along with current photosynthates. The amount of starch and the extent to which hydrolysis occurs down the branch depend on the size of the crop and the time of ripening.

The clingstone peach Coronado ripens in mid-July, concurrently with the grand period of shoot elongation. Starch does not accumulate much in the bearing stems prior to harvest but increases rapidly in amount afterward (Fig. 6.16). Elberta ripens in early August when shoot growth is beginning to cease. Some starch is accumulated in the bark tissue in May. The level decreases slightly as the endocarp begins to lignify; it increases slightly during Stage III but decreases again as the crop ripens. After harvest, the wood storage cells accumulate starch for nearly two months until the leaves fall. Corona is harvested in early September after shoot growth has completely stopped. Starch in the wood is depleted steadily from budbreak to early June when pit hardening takes place. In contrast, bark cells accumulate starch during early fruit development; starch content increases rapidly for a short period after the endocarp hardens. The ripening fruits mobilize only that stored in the bark cells, but starch is rapidly accumulated in the bark and wood as soon as the fruits are harvested. Phloem storage cells are more responsive to the source–sink gradient because they are closer to the sieve tubes which transport carbohydrates.

In French prune a heavy crop diverts considerable amounts of carbohydrates from being translocated to roots (Hansen et al., 1982). Root growth is limited and there is reduced nutrient uptake, especially of potassium, so that the trees exhibit a distinct pattern of chlorosis, a symptom of potassium deficiency. In incipient cases, leaves recover, but in severe cases, leaves become scorched and eventually abscise and the shoot tips die (see Plates 6,7).

A similar phenomenon occurs in some pecan cultivars. When there is a heavy set of fruits, potassium deficiency symptoms appear in the terminal leaves which subsequently abscise. The appearance of symptoms coincides with the period when the shuck increases in weight and accumulates large quantities of potassium (Sparks, 1977).

Another manifestation of internal competition in pecan is associated with its alternate bearing tendency. In the "on year," when the terminal bud laden with pistillate flowers elongates into a shoot, the distal portion of lateral shoots at lower nodes bearing pistillate flowers elongates slightly and then abscises, leaving behind a leafless stub with several catkins attached. This abscission of the lateral shoot tip is attributed to internal competition. If the terminal bud is removed during the dormant season, thereby removing a potentially strong sink consisting of a shoot with its large inflorescence, the shoot tips of these lower lateral buds do not abscise but develop normal

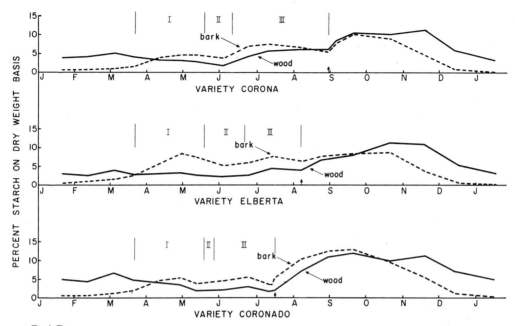

Figure 6.16
Seasonal changes in starch content of three peach cultivars that ripen at different times of the season. Roman numerals represent growth stages. (From Ryugo and Davis, 1959)

inflorescences of pistillate flowers (Wood and Payne, 1983). In Persian walnut, immature catkins on bearing spurs abscise in mid-August as the nuts on the terminal begin to fill (see Fig. 5.13).

Maturation of a heavy nut crop in Kerman pistachio results in a concurrent abscission of inflorescence buds and yellowing of leaves. The greater the number of fruits produced by a branch in one year, the shorter the terminal growth the next year. This decreased growth in the "off year" is attributed to the depletion of food and nutrients by the maturing crop the previous fall. When leaves on bearing and nonbearing shoots were exposed to radioactive carbon dioxide, inflorescence buds on bearing shoots accumulated less radioactivity than developing embryos on the same shoots or buds on nonbearing shoots (Taketa et al., 1980). Hence, the abscission of inflorescence buds is attributed to internal competition between embryos and buds, the embryos being by far the stronger competitor, even though they are farther from the source than buds. Chlorosis or senescence induced in leaves by the maturing nuts is another evidence of internal competition for nutrients as well as for photosynthates.

Both fruits and vegetative tissues of apple and pear accumulate starch concurrently. A heavy crop limits fruit, shoot, and root growth. The reduction in diameter growth limits the numbers of xylem and phloem ray parenchymna cells that are laid down by the lateral cambium, thereby decreasing the storage capacities of the bark and wood. Conversely, if the crop is light, vegetative growth is enhanced so that the tree has a large leaf:fruit ratio. Consequently, mineral nutrients in the transpiration stream are diverted toward leaves instead of fruits. In large apples this diversion results in calcium deficiency, leading to the disorder known as bitter pit. (See Bitter Pit under Growth-Related Disorders at the end of this chapter; also see Calcium, Ch. 8, and Plate 8.)

Blossom and Fruit Thinning

Morphological, Physical, and Chemical Changes

DRUPES

Drupes are fruits that possess a thin exocarp, a fleshy mesocarp, and a stony endocarp. All stone fruits are drupes including the almond which is so classified, even though the mesocarp does not enlarge during Stage III and becomes leathery rather than juicy toward harvest. The mesocarp and endocarp of immature almonds are fleshy, prior to pit hardening. In many Arabic countries the fruit is consumed at this stage of development.

Tissue differentiation begins early in stone fruits, even while cell division in the pericarp or ovary wall is progressing. Cells in these various tissues also mature at different stages of fruit development. The endocarp appears translucent during Stage I, whereas the mesocarp is thin and has a light green cast. The exocarp may be pubescent, as in peach and apricot, or glabrous, as in nectarine and plum. The smooth surface of plum gradually develops a white, waxy cuticle or bloom. As the endocarp approaches full size, the secondary cell walls of the future stone cells fill with cellulosic substances by apposition and the tissue becomes opaque. Fruits destined for "June drop" do not progress beyond this stage of development but turn from green to brownish yellow and soon abscise. The remaining fruits enter Stage II.

Endocarp

Lignin deposition is initiated within individual sclereids or stone cells, starting from the middle lamella and progressing centripetally toward the cell lumen. The distal cells at the tip of the peach endocarp lignify first, followed by cells on the ventral side and then by those on the dorsal side (Fig. 6.17). Within five to seven days, a peach endocarp becomes completely hard and difficult

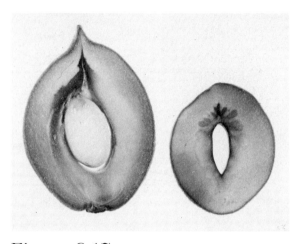

Figure 6.17
Longitudinal (left) and cross (right) sections of peach fruits at the onset of lignin deposition in the endocarp. Sections were treated with acidified alcoholic vanillin solution.

to cut with a sharp knife, but lignin and other cell wall components continue to be deposited for another 30 to 40 days (Fig. 6.18). During this lignification process, the specific density of stone cells increases from 1.15 to 1.44 grams per cubic centimeter. Lignin synthesis and deposition require much energy because stone cells consist of 35 percent lignin on a dry weight basis, and lignin has a heat content of 6.3 kilocalories per gram compared to 3.7 kilocalories per gram for cellulose. Therefore, when thinning is delayed until the pit is even partially lignified, photosynthates that could have gone into fruit sizing are dropped to the ground and wasted.

Sclereids are normally isodiametric, but those lining the inner surface of the endocarp are elongated. The endocarp of soft-shelled almond does not lignify except for cells in and about the vascular bundles and the inner lining. Almonds with hard shells develop much like the endocarp of peach. The pit-hardening pattern of plum and cherry is similar to that of peach. The surface texture varies from that of cherry, which is smooth, to that of plum, which is rough; the endocarp surface where the vascular bundles emerge in the peach is deeply pitted. In apricot,

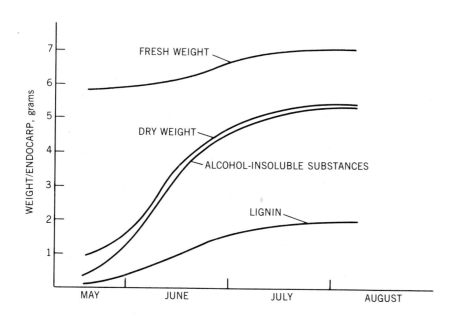

Figure 6.18
Seasonal increases in fresh weight, dry matter, and lignin by the Elberta peach endocarp.

cells of the inner and outer layers lignify before those between these layers (Fig. 6.19). This pattern of lignification may account for subsequent expansion of the apricot endocarp without splitting or cracking as in peach. Lignified endocarps from ripe fruits contain lignin precursors and metabolites, such as syringaldehyde, coniferaldehyde, ferulic acid, coumarin, and vanillin; some of these are germination and growth inhibitors. These and other aromatic constituents of the endocarp enrich the flavor and impart a delicate perfume to canned products. Thus, apricots

Figure 6.19
Cross (left) and longitudinal (right) sections of Tilton apricot fruits revealing pattern of lignin deposition. Outer and inner layers of cells lignify before the interior cells. Notice the round passageway for the dorsal bundle.

Morphological, Physical, and Chemical Changes

canned whole have a better flavor than halves preserved without pits.

Mesocarp

As the mesocarp cells enlarge during Stage III, their cell walls and cytoplasmic layers become thinner and the vacuoles increase in volume. While the primary cell walls are stretching and becoming thin, additional cellulosic materials that prevent the walls from rupturing are deposited. Intussusception is the process by which cellulose microfibrils are synthesized and linked into existing ones as they "slide" past one another. Cytoplasmic materials are also synthesized so that both cell wall and protein components of a fruit increase throughout its development.

The composition of the cell wall changes as the mesocarp enlarges and the fruit approaches full maturity and/or ripens. Protopectin in the middle lamella gradually dissolves, yielding soluble pectins and pectic acids. The dissolution of protopectins allows the cells to separate along the middle lamella. Intercellular spaces are thus formed. These spaces are filled with air enriched with carbon dioxide emanating from respiring cells. The formation of intercellular spaces tends to decrease fruit density, but the accumulation of soluble solids in the vacuoles compensates for the decrease. Pectinaceous and mucilaginous compounds in the cell sap add to its viscosity.

Cell enlargement causes an alteration in the shape of mesocarp cells, depending on their proximity to the endocarp. The number of mesocarp cells on the endocarp surface becomes fixed after cell division ceases midway through Stage I. When the mesocarp cells begin to enlarge in Stage III, those on the endocarp surface can only enlarge radially because tangential or lateral expansion is restricted by adjacent cells. Hence, these cells acquire a cylindrical configuration. Cells more distant from the endocarp have more space in which to expand, and they gradually become less cylindrical and more isodiametric.

The texture of stone fruits is classified as being melting or nonmelting. Examples of the melting type are Black Tartarian cherry, Royal apricot, and Elberta peach. Bing cherry and Halford peach have a nonmelting texture.

Separation of the mesocarp from the endocarp in freestone peach, apricot, and some plum, such as President, occurs seven to ten days before harvest. This separation leaves a cavity which indicates that the mesocarp tissue had a large growth potential even before separation occurred. In some soft-shelled almond cultivars, the exocarp and mesocarp tissues become dry and split along the ventral suture. Because the endocarp adheres to the mesocarp, it is pulled apart, exposing the embryo. Although splitting facilitates shelling of the "nut," it also allows birds, rodents, and insects to feed on the embryo.

These internal stresses created by differential growth rates of the endocarp, mesocarp, and exocarp lead to physiological disorders known as split pit in peach and cherry and skin cracking of prune.

The mesocarp accumulates mineral elements at different rates depending on whether they are imported via the xylem only, as are calcium and magnesium, or via both the xylem and phloem, as is potassium (Fig. 6.20). The differential importation of these cations leads to differences in the calcium : potassium and magnesium : potassium ratios in fruits.

Exocarp

Epidermal cells must accommodate a large increase in surface area without an increase in cell number after cell division ceases. Thus, they become stretched laterally into flat disk shapes. Epidermal and subepidermal cells of young, immature fruits contain chlorophyll, synthesize starch, but contribute small amounts of photosynthates to the fruit.

In ripe freestone peaches the skin peels easily. This characteristic is linked with the melting texture of the mesocarp and the ease of separating the mesocarp from the endocarp. The epidermis of clingstone peach does not separate freely from the mesocarp, which has a nonmelting flesh.

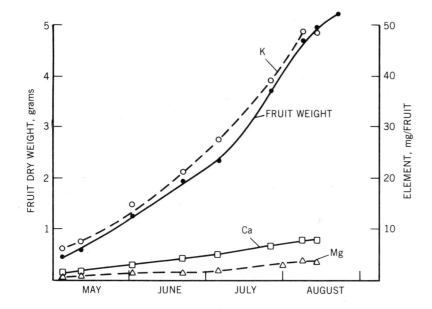

Figure 6.20
Seasonal accumulation of calcium, magnesium, and potassium in relation to growth of French prune fruit. (From Hansen et al., 1982)

Ovule

Ovule growth keeps pace with that of the endocarp, but the embryo within does not become visible to the naked eye until pit hardening commences. As the cotyledons enlarge, they consume the endosperm, which is rich in nutrients and hormones, and the developing endosperm consumes the nucellus. While the embryo is absorbing nutrients from the endosperm, the endosperm presumably exports hormones through the same vascular system to the mesocarp and to the stem subtending the fruit. Embryos of early-ripening cultivars normally abort. Cultivars that ripen later in the season have greater percentages of seeds that germinate.

DRUPELETS

Drupelets of raspberry and blackberry fruits undergo developmental changes that are similar to those of drupes of stone fruits. At one time brambles with red drupelets and a characteristic delicate aroma were known as raspberries, whereas those with black aggregate fruits were called blackberries. This distinction by color and flavor no longer holds because yellow-fruited cultivars as well as many interspecific hybrids have been developed (see Bushberries, Ch. 14).

Drupelets of the fig and mulberry accumulate sugars as the multiple fruits approach full size. In the fig, cells of the thick peduncular tissue enlarge simultaneously while becoming sweet and highly pigmented. After attaining maximum size, the fruit loses moisture and shrivels under hot, arid growing conditions and eventually abscises. If the fruit is kept clean, it retains its flavor because its constituents do not ferment.

Depending on the species, mulberry fruits with their fleshy sepals may remain white as chlorophyll disappears or accumulate red or purplish-black pigments as they ripen.

POMES

Among pomes, the apple and pear have received the widest attention because of their delectable eating qualities, relatively long storage life, and adaptability to different climates and growing conditions. These fruits develop from flowers with inferior ovaries derived from five fused carpels.

Morphological, Physical, and Chemical Changes

Cell division in apple and European pear continues for six to eight weeks after anthesis. The increase in DNA content per Asian pear fruit during this period is attributed to the multiplication of cells with their nuclear material (see Fig. 6.2). Cell enlargement is responsible for subsequent growth. In some cultivars, seed formation plays an important role in these growth processes. An apple can have 15 or more seeds, but if any locule lacks seed because of poor pollination, development of this side of the fruit, in some cultivars, is arrested (see Fig. 6.10). But the symmetry of some cultivars is independent of seed formation because (1) their fruits are inherently high in hormones and tend to set parthenocarpically; (2) the fruits develop seeds that abort late, after effusing hormones to complete the setting and shaping processes; or (3) the few seeds that form may effuse relatively large quantities of hormones to the nonseeded portion, enabling the fruit to develop symmetrically. Thus, seeded and parthenocarpic Bartlett pears are equally heavy at harvest, but their diameter:length ratios are different.

As apple and pear fruits enlarge, the water-insoluble dry matter content of the fruit decreases, partly because the starch in the cortical cells is hydrolyzed to glucose. Stone cell aggregates in a pear break down into smaller units; the number of these units per cubic centimeter of edible accessory tissue or stone cell density decreases as the fruit approaches maturity (Fig. 6.21). The sclerified nature of the individual stone cells does not change, however, because lignin does not dissolve.

While the fruits continue to grow, the cortical parenchyma cells become isodiametric as the cells separate from one another along the middle lamella. Pomes that are left on the tree too long become overmature; when bitten, their individual cells tend to separate and retain their vacuolar contents rather than rupturing and spilling on the tongue. Thus, the tongue perceives a mealy texture.

Fruit density decreases as the volume of intercellular air spaces increases. The air in these spaces is relatively rich in carbon dioxide and poor in oxygen because of cell respiration and the slow rate of gaseous exchange with the surrounding atmosphere.

Protein content in apple fruit decreases to a minimum, two to three weeks before harvest, but increases again as fruits approach full maturity. According to Hulme (1936), this increase in protein level just prior to harvest may signal the onset of the maturation and senescence phases of the fruit. During this period, a shift occurs in the kinds of enzymes being produced by the cells. In

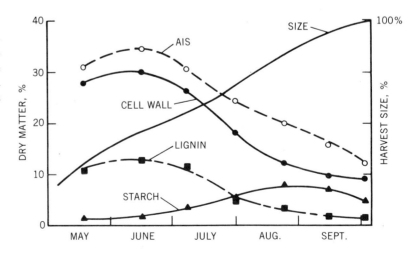

Figure 6.21
Seasonal changes in the composition of an Asian pear. The alcohol-insoluble substances (AIS) is the sum of starch and the cell wall fraction, including lignin. (From Ryugo, 1969)

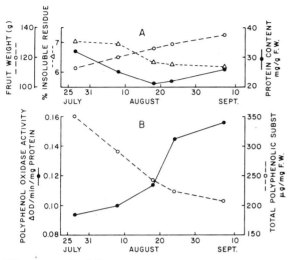

Figure 6.22
(A) Seasonal changes in size and composition of maturing Golden Delicious apples. (B) Inverse relation between polyphenoloxidase activity and substrate concentration. (From Zocca and Ryugo, 1975)

Golden Delicious apples, polyphenol oxidase activity increases concurrently with a decrease in the phenolic substrate level (Fig. 6.22). Starch content decreases as apples mature. Those intended for extended storage should be harvested when the starch content is still relatively high. However, apples should possess no starch when ripe.

The accumulation of mineral elements by apple fruit is proportional to the rate of increase in dry matter (Fig. 6.23). Potassium and nitrogen make up 1 and 0.25 percent, respectively, of the dry matter, whereas calcium, phosphorus, and magnesium constitute an even smaller fraction of the total mineral nutrients. These values are similar to those for prunes and kiwifruit.

BERRIES

Although the cumulative growth curves of seeded Muscat of Alexandria and Emperor and seedless

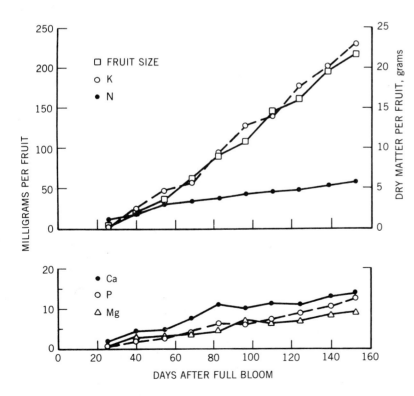

Figure 6.23
Average seasonal accumulation of mineral nutrients in the flesh of Delicious and Winesap apple fruits. (Adapted from Rogers and Batjer, 1954)

Morphological, Physical, and Chemical Changes

Sultanina and Corinth grape berries based on volume manifest a double sigmoid pattern (Coombe, 1960; Fig. 6.24), the curve of parthenocarpic Delaware berries treated with gibberellic acid has a single sigmoid pattern (Ito et al., 1969). The rapid growth during Stage I is attributed to relatively high auxin and gibberellin levels. However, these hormones are not detectable during Stage III when cell enlargement is occurring. GA-treated seedless Delaware berries do not manifest Stage II, but their fluctuation of GA content is similar to that of seeded berries.

It is tempting to ascribe the growth retardation of seeded fruits during Stage II to the large amount of reserve foods being assimilated by the relatively large seeds (in proportion to the entire fruit), but it would not explain the growth retardation of seedless berries. Berries are relatively resistant to distortion by external pressure during Stage II, indicating that the cell walls are somewhat rigid and inelastic until berries approach veraison or the ripening stage.

Seeded persimmons also have a double sigmoid curve. The large, green, fleshy sepals of persimmon fruits play an important role in fruit development. Poor growth of persimmon fruit has been observed when the sepals are damaged by limb bruising. When varying numbers of sepal lobes are removed experimentally, growth is inhibited proportionally (Fig. 6.25). When all lobes were removed, fruits abscised (Nakamura, 1967). This

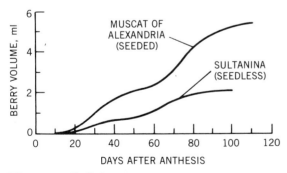

Figure 6.24
Cumulative growth curves of berries from seeded and seedless grape cultivars. (From B. G. Coombe, 1960)

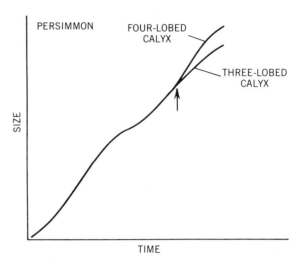

Figure 6.25
Growth pattern of a persimmon fruit with a four-lobed calyx and that of a fruit from which a calyx lobe was removed. The arrow indicates the date of sepal excision. (From Nakamura, 1967)

abscission may be related to the hormonal contents of the calyxes. The cytokinin level in the calyx of the cultivar Hachiya, which is notorious for dropping its fruit prematurely, is lower than that of the cultivar Fuyu, which abscises a lower percentage of fruits. Application of kinetin to sepals of Hachiya fruits reduces fruit abscission.

The dry weight accumulation by the kiwifruit has a single sigmoid curve, whereas growth based on diameter or fresh weight manifests a triple sigmoid curve (Fig. 6.26). Growth in diameter is slightly retarded when seeds begin to accumulate dry matter. Subsequent growth resumes quickly until another plateau is reached about a month before harvest. A small but noticeable increase in diameter occurs just prior to harvest. This final increase in diameter coincides with the onset of starch hydrolysis and a rapid increase in glucose content (Fig. 6.27). Quinic, citric, and malic acids are the main organic acids in immature fruits; ascorbic and quinic are the predominant acids in mature fruits (Fig. 6.27).

With the exception of water-soluble nitrogenous compounds which increase until early August and then decrease, kiwifruit accumulates

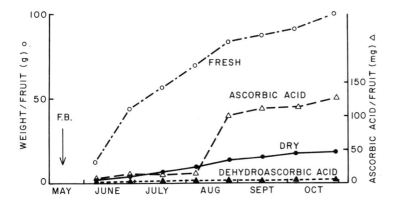

Figure 6.26
Relation between fresh and dry weight increases of Hayward kiwifruit and the accumulations of dehydroascorbic and ascorbic acids. (From Okuse and Ryugo, 1981)

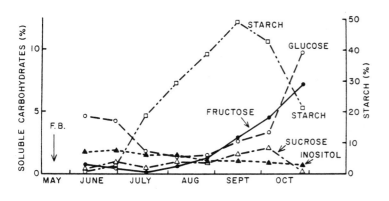

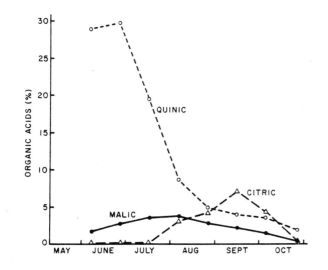

Figure 6.27
Seasonal changes in carbohydrates (upper) and organic acids (lower) during development of Hayward kiwifruit. (From Okuse and Ryugo, 1981)

Morphological, Physical, and Chemical Changes   135

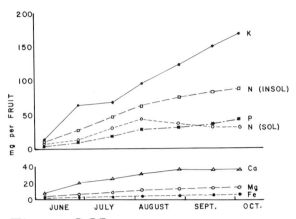

Figure 6.28
Seasonal accumulation of mineral nutrients by Hayward kiwifruits. (Courtesy of I. Okuse)

other mineral elements gradually from full bloom until harvest (Fig. 6.28). Among the elements, potassium, which is imported with soluble carbohydrates as the fruit develops, accumulates to the highest concentration, as it does in prunes and apples. Certain specialized kiwifruit cells synthesize large amounts of fine raphides, the needle-shaped calcium oxalate crystals.

STRAWBERRY, AN AGGREGATE FRUIT

The edible portion of the strawberry is the receptacle tissue on which the true fruits, the achenes, are appressed. The final fruit size depends on the number of seeded achenes that develop (see Fig. 6.9). When achenes are not uniformly pollinated, the distorted fruit is called a "catface." Depending on the cultivar, a ripe strawberry may lack sucrose, but it is rich in glucose and ascorbic acid. A dessert fruit, prime for eating, should have a well-balanced sugar : acid ratio and a rich aroma.

NUTS

A true nut, such as the walnut, pecan, and chestnut, is characterized by the presence of an involucre. In walnut and pecan the involucre consists of bracts that are fused to the pericarp wall. The involucre, pericarp, and embryo develop sequentially in these species (Fig. 6.29). In chestnuts and filberts, the bracts and pericarp are not fused; in filberts (and acorns), the involucre does not completely envelop the pericarp.

After anthesis, the involucre and the pericarp of Persian walnut enlarge rapidly; the ovule, with its watery nucellus-endosperm tissues, keeps pace. About the time the nut reaches full size in mid-July, the shell, which is the inner part of the endocarp, begins to lignify and become woody. The accumulation of dry matter by the involucre and shell becomes retarded, but that of the whole continues because the embryo begins to develop and enlarge. Initially, the endosperm and nucellus are watery and rich in nutrients diffusing from the vascular bundles in the seed coat to the embryo. The placental tissue at the base of the endocarp and the septa that separate parts of the ovule are soft and spongy. As the zygote develops into an embryo, it begins to accumulate dry matter which consists of about 65 percent oil (Fig.

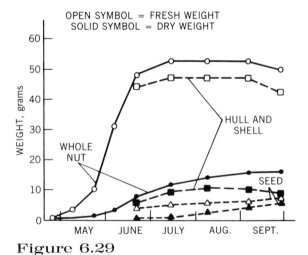

Figure 6.29
Seasonal increases in fresh and dry weights of the Persian walnut. The hull and shell were weighed separately from the seed beginning June 20. (Courtesy of B. Marangoni)

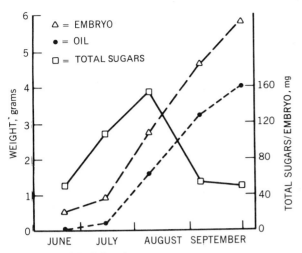

Figure 6.30
Growth curve of Persian walnut embryo and its changes in sugar and oil contents. (From Ryugo et al., 1980)

6.30). The placental tissue and the septa atrophy during maturation and become thin, leathery protrusions which separate the lobes of the embryo; the remnants of the dry septa and placenta are called the packing tissue. As the nut matures on the tree, the hull and shell decrease in dry weight and moisture content. When the packing tissue turns brown, and the green, pithy hull begins to split and separate from the shell, the crop is ready to be harvested. In most cultivars the integument or pellicle becomes browner and less astringent the longer the harvest is delayed. Since the market prefers and pays a better price for light-colored kernel, cultivars such as Chandler are favored, for their seed coats retain the light tan color even when harvest is delayed for several days.

Although the green spines of the chestnut bur assimilate carbon dioxide, the ultimate size of the nut depends on the number of pistils that are pollinated and develop embryos. The four important *Castanea* species all express the phenomena of xenia and metaxenia. That is, the size of the seed and fruit, including the bur, is determined by the nut size of the pollen parent. The cotyledons of mature seeds contain 53 to 65 percent moisture, 27 to 40 percent starch, 2 to 8 percent total sugars, and 2 to 5 percent protein on a fresh weight basis.

Ripening and Maturation Processes

Horticulturists make a distinction between the terms *ripe* and *mature*. A ripe fruit is one that is ready to be eaten, whereas a mature fruit is one that has reached such a stage of development on the plant that it will ripen following harvest. Thus, fruits such as peaches, cherries, figs, apples, and loquats will ripen on the trees. However, ripe fruits cannot be stored very long even under refrigeration. Both apples and pears will continue to grow if they are left on the tree, but their keeping and eating qualities diminish after they pass their optimal maturity. Therefore, pears are harvested when still green and firm, and they are stored at least a week at 0° C before being transferred to 20° (68° F) and 85 percent relative humidity to ripen.

Generally, as fruits begin to ripen, levels of soluble solids in the cell vacuoles increase (Fig. 6.31) while acidity decreases. Concurrently, ripening fruits change in texture from firm and crisp to tender and juicy. The degree to which the mesocarp of stone fruits becomes soft is a heritable characteristic. The mesocarp of canning

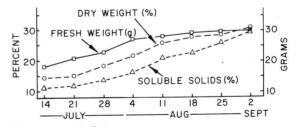

Figure 6.31
Concurrent increases in fresh weight, dry weight, and soluble solids content of French prunes.

clingstone peaches is firm or nonmelting when prime for canning, whereas that of dessert varieties is soft or melting when ripe.

CARBOHYDRATE METABOLISM

The common sugars in stone fruits and pomes are glucose, fructose, and sucrose. Ripe stone fruits are relatively rich in sorbitol, a sugar alcohol, whereas mature apples and pears contain mere traces of this compound. Sorbitol and sucrose are the two main translocatable carbohydrates in the Rosaceae family.

Developing stone fruits synthesize starch grains in the choloroplasts of epidermal and subepidermal cells, but as fruits ripen and chlorophyll is degraded, starch is metabolized. Mesocarp cells accumulate relatively large quantities of sugars within a short period prior to harvest. Since this demand by the fruits is usually not met by current photosynthesis, starch in branches subtending fruits is concurrently hydrolyzed to glucose, which, in turn, is reduced to sorbitol and/or is synthesized to sucrose. These are then loaded into sieve tubes and translocated with current photosynthates to the ripening fruit. Although the exact mechanism by which a ripening peach activates the hydrolysis of starch grains deposited in xylem and phloem parenchyma cells several centimeters away is not known, it is postulated that this mobilization of sugars results from a concentration gradient established between the cells of the mesocarp and those in the stem.

In immature French prune fruits, sorbitol and sucrose are readily converted to other metabolites upon entering the fruit tissue (Fig. 6.32). When mesocarp extracts of ripening plums were examined after allowing the fruits to draw radioactive sorbitol through their peduncles, the chromatogram of extracts revealed several radioactive substances, the major by-product being sucrose (Hansen and Ryugo, 1979). When radioactive sucrose was introduced into comparable fruits, labeled glucose was recovered but very little radioactive sorbitol was detected.

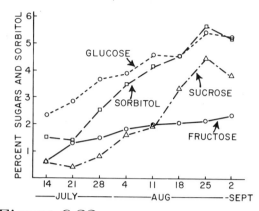

Figure 6.32
Seasonal changes in soluble carbohydrates in the mesocarp cells of French prunes. (From Ryugo et al., 1977)

Starch is not hydrolyzed to a great extent in spurs subtending pome fruits, such as apple, pear, and quince, possibly because the immature fruits themselves accumulate large quantities of starch in the ovarian and accessory tissues. Much of the starch in the maturing fruit is hydrolyzed before the onset of the climacteric rise in respiration takes place.

Starch makes up nearly 50 percent of the dry matter of mature kiwifruit. It is stored in the green carpellary tissue and the white columella. As the fruits mature, starch is hydrolyzed to glucose, which is metabolized to other compounds or accumulated in the vacuole. This hydrolysis may be a signal that the maturation process has begun because the soluble solids content now increases gradually. When it reaches 6.5 to 7 percent, the fruits are ready for harvest.

At the same stage of maturity, seeded Bartlett pears accumulate more soluble solids and organic acids in their vacuoles than do parthenocarpic ones borne on the same tree. Hence, seediness improves Bartlett pear flavor (Griggs et al., 1960).

Immature stone fruits contain relatively large quantities of gums and mucilages. These complex polymers of 5- and 6-carbon sugars are also found in leaves and stems. Whether they are synthesized *in situ* or translocated from the leaves to the fruits and roots has not been demonstrated. Gum pockets are often found in fruits of plums and canning peaches, usually on the distal end if the crop is light. Injuries to stems and fruits by birds, insects, or hailstones cause exudation of gums and evolution of ethylene from the wounded tissue. Dry, amorphous gums have not been reported to support mold growth; therefore, gums are thought to serve as a protective barrier against infection by microorganisms.

ORGANIC ACID METABOLISM

Malic and citric acids are the predominant acids in fruit tree species; tartaric acid is the principal one in grapes. Molar ratios of malic to citric acid in ripe peach fruits range from 5:1 to 80:1 and are independent of the total titratable acidity of the juice. A small fraction of organic acids occurs as salts of potassium and calcium (Ryugo, 1964). Some calcium is imported into fruits presumably as a citrate complex with other photosynthates, but most of the calcium is passively carried into fruits via the xylem.

The seasonal trends of organic acids in peach and nectarine depend on the relative acidity of the cultivar at harvest. Fruits that are highly acid at harvest accumulate citric and malic acids after the endocarp begins to lignify, attaining maximum levels just prior to harvest. Those that are bland at harvest attain a low maximum level early in the season, and this level declines gradually until harvest. Seasonal trends of commercial cultivars with a well-balanced acidity at harvest lie between these extremes (Fig. 6.33).

When reciprocal grafting was made between flower buds from high-acid, yellow-fleshed cultivars and low-acid, white-fleshed peach and nectarine cultivars, fruits that developed from these buds were of the same size, color, and acidity as those growing on their own shoots and nourished by leaves of identical genetic makeup. This indicates that even if the quantity and types of organic acids translocated from the leaves may be

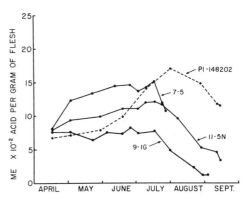

Figure 6.33
Seasonal fluctuations in organic acid levels in three selected peaches—Plant Introduction (PI) 48202, 11-5N, and 9-1G—and a nectarine seedling, 7-5. They were selected on the basis of variation in acid content at harvest. (From Ryugo and Davis, 1958)

different from those of fruits, cells of the pericarp tissues are able to metabolize these substrates to an equilibrium programmed by genes for their particular stage of development.

When radioactive fructose was fed to strawberry fruits via their peduncles, it was metabolized to various organic acids, especially citric and malic (Markakis and Embs, 1964). In a similar experiment with green grape berries, labeled glucose was split into 2- and 4-carbon fragments, the 4-carbon fragment giving rise to tartaric acid. Ripe berries reversed the reaction, transforming labeled malic and tartaric acids to radioactive glucose (Ribereau-Gayon, 1968). Because malic and citric acids participate in the Krebs tricarboxylic acid cycle, other components of the cycle and even some amino acids become labeled if radioactive malic and citric acids are injected or fed to rapidly respiring fruits.

Immature kiwifruits are rich in quinic acid. As it disappears in July, ascorbic acid increases rapidly. This does not necessarily mean that quinic acid is the precursor of ascorbic acid. Citric acid increases with ascorbic acid, but citric acid peaks in mid-September and then decreases steadily until November when it is barely detectable (see Fig. 6.27). Ascorbic acid reaches a plateau of about 100 milligrams per 100 grams of fresh weight until fruits are harvested.

PIGMENT FORMATION AND DEGRADATION

Fruit pigments (Fig. 6.34) undergo considerable changes during fruit development. The fat-soluble pigments, carotenes and xanthophylls, are synthesized and accumulated in the chloroplastids or chromoplastids. The common water-soluble pigments, anthocyanins, are accumulated in the vacuoles.

Chlorophyll

This pigment is present in young immature fruits, especially in the epidermal cells and to a lesser level in cortical cells of pomes. In some apples and other types of fruit that remain green when mature, chlorophyll persists, although its concentration decreases with maturity. In those fruits that turn yellow or red as they mature, chlorophyll is degraded and disappears so that other pigments become visible.

The chlorophyll content in the carpellary tissue of the commercially grown kiwifruit is high while the fruit is firm; the intensity of the green pigment fades as the fruits soften after a relatively long period of storage.

The rate of chlorophyll degradation is hastened by treatment with ethylene or ethylene-generating compounds and some growth retardants such as chlormequat (cycocel) that interfere with gibberellin synthesis. Green tissues treated with cytokinins tend to retain chlorophyll and thus delay the onset of senescence, whereas mature Washington Navel oranges treated with gibberellins will resynthesize chlorophyll and turn green again.

Anthocyanins

Anthocyanins, which are flavonoid compounds, range from hues of pink through deep red to purple in the plant kingdom. There are many kinds of anthocyanins and anthocyanidins, the latter being aglycones of the anthocyanins. The common names, such as peonidin, pelargonidin, and delphinidin, were assigned to these compounds because they were first extracted from petals of peony, pelargonium, and delphinium plants, respectively. But petals of many species contain more than one kind of anthocyanins.

Stems of the red-leafed plum, *Prunus cerasifera* var. Psardii, and cells surrounding vascular bundles in Red Delicious apple and some peach cultivars possess anthocyanins. The postulation that anthocyanins are synthesized *in situ* and not translocated was tested by reciprocally side-grafting spurs of Bing cherry and Yellow Spanish; nearly all leaves on the spurs were removed as they emerged. The partial defoliation was to ensure that the fruits on the scions received most of their assimilates from stock leaves. At harvest, fruits on the grafted spurs were identical in appearance and flavor to those borne on the respective scion mother plants. These results support another horticultural axiom: genetic characteristics are not transmitted from the rootstock to scion by grafting.

Anthocyanins are accumulated in the exocarp and mesocarp of Red Heart, Elephant Heart, and Laroda plums but are confined to epidermal layers of Queen Anne and Nubiana plums. Plum cultivars range from deep blue, purple, blood red to pink, depending on the kinds and concentrations of anthocyanins.

Red-fleshed sweet cherries borne on spurs located on the periphery of a canopy contain greater amounts of anthocyanins than those on shaded spurs in the interior of the canopy (Patten and Proebsting, 1986). The white-fleshed cultivars, such as Bigarreau Napoleon, syn. Royal Ann, Emperor Francis, and Rainier, form a blush only on the side exposed to light and contain the identical pigments chrysanthemin and keracyanin, in approximately the same proportions as the red phenotypes.

When immature Bing cherry fruits, growing on well-lit spurs, were wrapped in aluminum foil sleeves to exclude light, they grew larger but

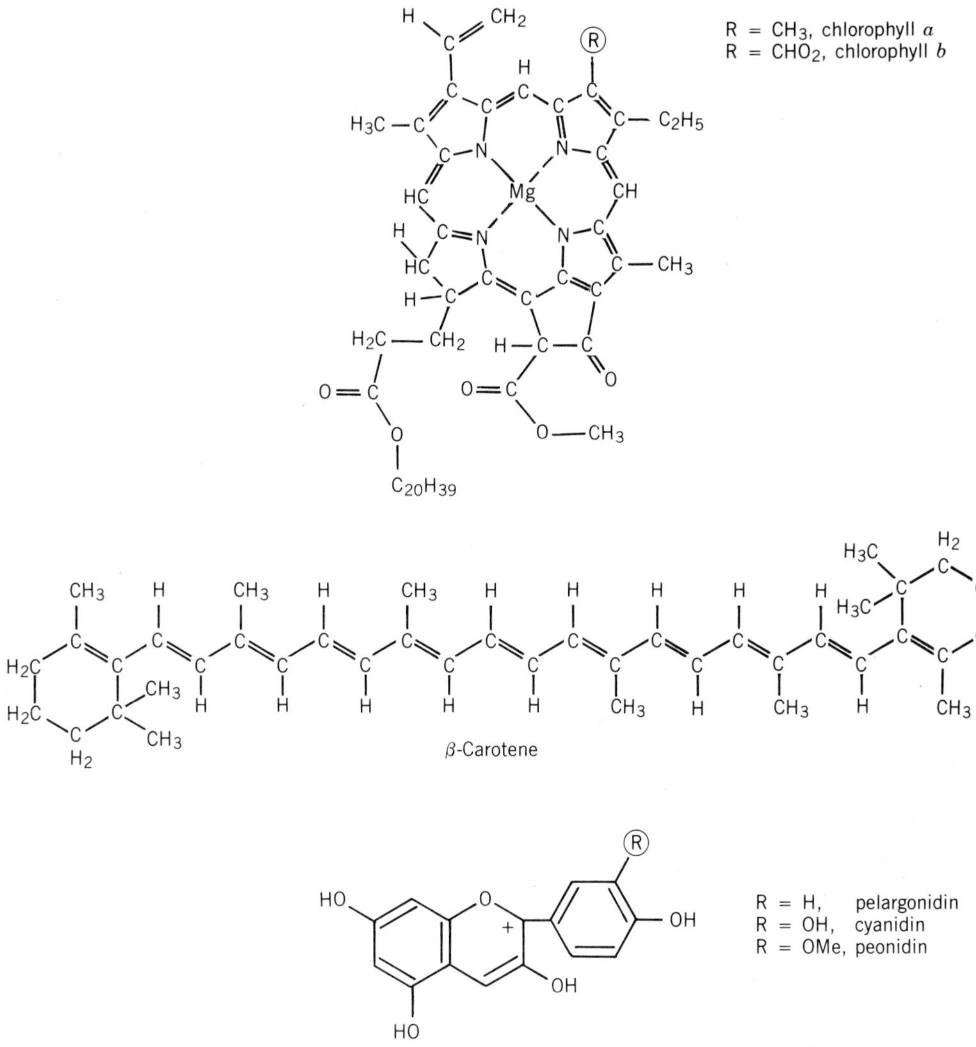

Figure 6.34
Structural formulas of chlorophyll (upper), carotene (middle), and the aglucone moiety of three anthocyanins (bottom).

synthesized 15 percent less anthocyanins than uncovered fruits on the same spurs. Much of the reduction in pigment concentration could be accounted for by the increase in water content. Aluminum-wrapped fruits of Royal Ann and Emperor Francis failed to form any anthocyanins. Thus, red-fleshed cherries do not require light to form anthocyanins, provided adjacent leaves perceive light and supply fruits with substrates that promote pigment synthesis. Most white-fleshed cherries must be exposed to light to produce a blush.

A similar mechanism for synthesizing anthocyanins seemingly exists in strains of Red Delicious and highly pigmented peach and nectarine cultivars. The skin of these cultivars will turn deep red even when fruits are situated in the darkest part of the foliar canopy, whereas epidermal cells of light-skinned apple and peach cultivars will develop a red blush only when exposed to direct sunlight. Often a silhouette is observed where a leaf has overlain maturing fruits.

Anthocyanin synthesis is enhanced in apples when day temperatures are relatively warm and night temperatures drop to 7° to 10° C. Cool night temperatures reduce respiratory loss of carbohydrates, especially in the epidermis, whereas light favors pigment formation.

Growers in northern Japan induce anthocyanins to form in Mutsu and Golden Delicious apples by individually bagging the fruit or covering the tree with a tent about a month prior to harvest. The covers are removed about two weeks later to expose fruits to direct sunlight and to reflected light, by laying sheets of aluminized cloth beneath the trees. Mutsu fruits turn brilliant red, whereas Golden Delicious apples become rosy pink. The relation between etiolation caused by bagging and the formation of anthocyanins has not been determined.

The application of such a treatment in areas where solar radiation and temperatures are relatively high could lead to sunburning and possibly a higher incidence of water core.

In semiarid regions, sprinkling trees of red apple cultivars just before sunset, beginning a few weeks prior to harvest, greatly enhances anthocyanin production. The evaporative cooling effect of the water lowers the fruit and ambient temperatures and thus the fruit respiration rate, thereby conserving energy. Sprinkling also removes dust particles from leaves and improves their photosynthetic capacity. In areas where humidity is high, the practice of sprinkling the trees toward dusk could compound the disease problem because it creates an ideal environment for fungi, such as mildew, to flourish.

The anthocyanin content of some strawberry cultivars is reputed to increase during transit to market, which indicates that light is not necessary provided sufficient substrates are previously stored in the receptacle. Such a change may improve the appearance of the strawberries but not their flavor.

Since monosaccharides and disaccharides are a moiety of anthocyanin molecules, treatments that enhance photosynthesis, for example, summer pruning, or that cause carbohydrates to accumulate, for example, girdling, increase pigment concentration. Various growth retardants that are applied to fruit trees enhance pigment formation but do not necessarily advance maturity. Dixon peaches treated with Alar develop anthocyanins and carotene and soften as much as four to seven days earlier than untreated, control fruits. Thus, ripening is advanced in peaches by growth retardants. Alar-treated Bing cherries turn red early, but the soluble solids content increases at the same rate as that in untreated cherries. Some Alar-treated red apple cultivars also develop better color and firmer flesh, but the soluble solids content remains the same as that of untreated apples. Whether the application of Alar advances the maturity of red apples and cherries then depends on which harvest criterion one uses: surface color, flesh firmness, or amount of soluble solids.

Some white peaches, commonly known as "Indian" peaches in midwestern states, have streaks of anthocyanin in the mesocarp. The contrast of anthocyanins against a white or yellow background of the mesocarp and pit cavity of ripe freestone peaches greatly enhances the attractiveness of fresh peach halves or slices. Anthocyanins are undesirable in canning peaches because the pigments turn brown when processed. Moreover, the presence of red streaks in the mesocarp of yellow peaches often indicates that the fruit is overripe.

Carotenes and Xanthophylls

Carotenoids range in color from light yellow to deep orange to tomato red. As chlorophylls become degraded in maturing yellow-fleshed fruits, the chloroplasts become chromoplasts. The cells then begin to synthesize carotenes and their dihydroxy analogues, the xanthophylls. A given cultivar may contain more than one kind of carotene. Carotenoid pigments in peach are accumulated rapidly during the final weeks before harvest. Most peach cultivars are overripe when they develop a deep, apricot-orange color, but some, like Kakamus and Golden Queen, are at their optimum ripeness at this color intensity.

Heavy applications of nitrogenous fertilizers to yellow peach and apricot cultivars delay the degradation of chlorophyll and the onset of carotene synthesis. If harvest is delayed to allow further color development, a heavy preharvest drop of straw-colored but physiologically mature fruits may occur.

Anthocyanins are not astringent or bitter per se, but they are associated with leucoanthocyanins and tannins that impart these flavors. Thus, the skins of some red apple cultivars, for example, Red Delicious, are somewhat bitter, and red wines are relatively more astringent than white wines. When making white wines from red-skinned cultivars, enologists exercise care not to crush the skin, for the anthocyanin would introduce an undesirable hue, and the tannins could impair the taste of the finished product. When fruits containing anthocyanins are bruised, the pigments are oxidized by anthocyanase to a brownish pigment.

Tannins and Polyphenolic Substances

Tannins and polyphenolic substances are not pigments per se, but when plant tissues are injured, these chemicals are oxidized to brownish-black pigments. Tannins, by definition, are any chemicals that tan leather by reacting with proteins, thereby making the finished product pliable. Some common polyphenolic substances found in fruits are chlorogenic and caffeic acids and quercetin. In intact healthy tissues, these compounds are compartmentalized in different parts of cells. When a fruit is bruised, cells rupture and these substances come into contact with the enzymes polyphenoloxidase and peroxidases, which cause the browning reaction. This browning reaction is retarded by antioxidants such as ascorbic acid (vitamin C), which is preferentially oxidized by polyphenoloxidase. The kiwifruit does not discolor because the mesocarp tissue is rich in ascorbic acid and poor in phenolic substrates.

The polyphenoloxidase activity in peach and apple increases as fruits ripen, but the phenolic substrate levels decrease concomitantly (see Fig. 6.22). Cultivars with high phenolic content or those with a white skin show bruises readily because the discoloration of oxidized tannins creates a sharp contrast to the ground color.

Tannins are astringent to taste because they react with the proteins on the surface of the tongue. The astringent or puckery factor in persimmon is a leucoanthocyanidin located in large, specialized tannin cells. As the fruit ripens, the tannins polymerize into large globules and lose their astringency.

RESPIRATION AND ETHYLENE EVOLUTION

That living cells respire, giving off carbon dioxide while consuming oxygen, has been known for nearly two centuries, but Kidd and West (1930), two English pomologists, were the first to report that the respiration rates of apples decrease and then increase as the fruits approach harvest maturity. They termed this preharvest acceleration in respiration rate the climacteric rise and the maximum the climacteric peak. After attaining the climacteric peak, the rate gradually decreases.

The initial decline in the respiration rate of immature fruit is accompanied by a steady disappearance of starch. The subsequent climacteric rise corresponds to the period when apples are prime for harvest, storage, and consumption. Kidd and West postulated that the postclimacteric decline in respiratory activity signals the onset of senescence as eating and keeping qualities of apples diminish.

Hulme (1954) noted that protein synthesis is resumed during the low ebb of respiration. The increase in protein parallels the climacteric rise, which he attributed to cellular reorganization. Enzymes necessary for maturation and ripening are now being elaborated rather than those previously needed for growth.

Other fruits such as pear, plum, apricot, and peach exhibit a climacteric rise and peak (Fig. 6.35), the duration and height of which differ from species to species and even within the same species. Cherry, strawberry, and citrus, which do not have such respiratory fluctuations as they approach eating ripeness, are called nonclimacteric fruits.

Since sugars are being respired in a process utilizing oxygen and evolving carbon dioxide, Kidd and West sought a means of slowing this reaction. They stored apples in an atmosphere with increased carbon dioxide and decreased oxygen concentrations; the oxygen level was lowered by introducing additional nitrogen. They indeed found that the ripening process was delayed, which led to the concept of controlled-atmosphere storage. Since then, by maintaining constant temperatures while reducing oxygen and increasing carbon dioxide concentrations in gastight storage facilities, postharvest researchers have established, albeit empirically, the optimal temperatures and carbon dioxide and oxygen concentrations for different apple cultivars (Table 4). These discoveries have made it possible to store some apple cultivars from one season to the next with little or no loss in eating quality. Whether or not it is economical to store fruits, for example, Emperor grapes, for such long periods is questionable, for large quantities of Italia and Thompson Seedless grapes are exported from the Southern to the Northern Hemisphere.

A disadvantage of a controlled-atmosphere storage room is that once it is opened to store additional fruits or to withdraw a quantity of fruit, the room must be purged and recharged with the optimal gas concentrations.

In 1901 the Russian scientist Neljubow described how certain gases caused epinasty and yellowing of plants. Twenty-three years later Denny (1924) demonstrated that it is ethylene gas that causes degradation of chlorophyll. He was able to explain why burning of kerosene heaters in a California citrus warehouse caused yellowing of lemons; ethylene is a combustion product of kerosene. Ethylene is also emitted with exhaust gases of internal combustion engines. Therefore, only forklift tractors with electric motors are allowed in and around modern cold-storage facilities.

Upon learning that ethylene gas would induce early ripening of fruits, horticulturists engaged in postharvest research began referring to it as a ripening and senescence hormone. It was not until the gas chromatography apparatus was invented and refined that the concentration of this gas could be measured in parts per billion in fruit tissue and its physiological activity evaluated.

Many fruits evolve both ethylene and carbon dioxide simultaneously as they approach their climacteric peak. When unripe but mature fruits are stored together with ripe fruits, ethylene evolved from ripe fruits initiates an autocatalytic reaction within the unripe fruits, stimulating them to ripen quickly. Elevating the carbon dioxide concentration from 0.03 percent in ambient air to 3 percent or an even higher concentration for a short period inhibits ethylene evolution, but extended exposure of fruits to high carbon dioxide concentrations shortens their shelf life and sometimes diminishes flavor.

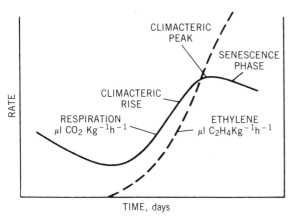

Figure 6.35
Respiration and ethylene evolution curves of a climacteric fruit.

Table 4. *Optimal Ranges of Temperature and Carbon Dioxide and Oxygen Concentrations for Controlled-Atmosphere Storage of Apple Cultivars*

Cultivar	Temperature, °C	Carbon Dioxide, percent	Oxygen, percent
Delicious	1.1–0	1–2	2–3
Golden Delicious	1.1–0	1–2	2–3
Johathan	0	3–5	3
McIntosh	3.3	2–5	3
Rome Beauty	1.1–0	2–3	3
Stayman Winesap	1.1–0	2–3	3
Yellow Newtown Pippin	3.3–4.4	7–8	2–3

Adapted from USDA Handbook 66.

The ripening effect of one lot of fruit on another through the evolution of ethylene can be prevented by refrigerating different lots of fruits in separate compartments and then constantly scrubbing their atmospheres to eliminate ethylene and other undesirable volatiles. Scrubbing is done in commercial storage plants by recycling the enclosed atmosphere through filters impregnated with potassium permanganate and/or containing activated carbon (charcoal). Permanganate oxidizes carbon compounds to carbon dioxide, whereas activated carbon adsorbs the volatile substances, thus ridding the atmosphere of pollutants. Since highly volatile substances given off by apples and pears are responsible for scald and the transmission and absorption of off-flavors, agricultural chemists and horticulturists are constantly seeking more efficient means of eliminating volatiles.

A laboratory exercise that demonstrates the effect of ridding ethylene from the atmosphere is sketched in Figure 6.36. Scrubbers S1 and S2 contain vermiculite saturated with potassium permanganate. Scrubber S3 contains only vermiculite. Container A is filled with ripe pears, bananas, or avocados which give off ethylene. Containers B and C are filled with unripe pears, bananas, or astringent persimmons. As the air passes through S1, any ethylene in the atmosphere is eliminated but is reintroduced by ripe fruits in container A. As the contaminated air flows through S2, it is freed of ethylene by permanganate, but the portion that passes through S3 to container C remains contaminated. After a

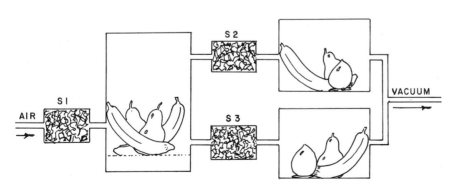

Figure 6.36
Arrangement of containers and gas scrubbers to demonstrate the effect of ripening fruits on other less ripe fruits. S1 and S2 contain vermiculite impregnated with potassium permanganate; S3 contains plain vermiculite.

few days in container C, pears and bananas will become yellow; persimmons will lose their astringency.

FLAVOR AND AROMATIC COMPOUNDS

We sense some flavors through the taste buds on our tongue and others through our olfactory nerves in the nasal passages. Those we can taste are sweetness, imparted by sugars and some amino acids, and sourness owing to organic acids, including ascorbic acid. Astringency or puckeriness is from tannins, whereas bitterness is often associated with alkaloids; piquancy is characteristic of oils in chili peppers. Volatile compounds that affect flavor and aroma are often esters of various organic acids and alcohols, for example, amylacetate, which has the odor of ripe bananas. Some phenolic derivatives, for instance, vanillin, are present in endocarps of stone fruits, but they become volatile only after pretreatment with an organic solvent.

These aromatic compounds occur in very minute quantities, but many are detectable by gas chromatography. Most of them escape detection by our olfactory senses because they are masked by the more volatile and/or concentrated substances. The human nose can nevertheless distinguish the aroma of yellow peach from the delicate perfume of a white freestone peach, and the aroma of a ripe apricot from that of a mellow pear.

These distinct generic aromas of different fruit species develop as fruits mature or ripen, which explains why prematurely harvested fruits lack flavor. Aromatic substances readily escape from fruits that are left in the open or are processed. Cooking elevates the vapor pressure of these compounds and causes them to distill, especially when the container is vacuum-sealed and the contents cool. On the other hand, some aromatic constituents that were not noticeable in the raw product become detectable only after the fruits are processed.

Harvest Technology

Tree crops are harvested manually by workers on ladders or on mobile picking platforms or mechanically by machines. The harvesting operation is expensive because it means investments in human resources or in machinery, or both. Furthermore, it must be accomplished carefully within a relatively short period.

Spoilage of fruits through mishandling during harvest, packaging, and transport is the largest contributor to loss of income for the fruit industry. As firm and green as fruits may be at harvest, they are living organisms that do not tolerate abuse. With the present technology, machine harvesting causes more damage than does handpicking. Thus, most crops destined for the fresh or dessert market are hand-harvested, whereas nut crops and fruits that are dried, canned, or extracted for their juice are often mechanically harvested. As the designs of these machines are improved to minimize damage to fruit, and as farm laborers become scarce and difficult to recruit, mechanical harvesters may be more widely used for dessert fruits.

HARVESTING TECHNIQUES

Hand Harvesting

The method used for harvesting the crop depends on the stage and acceptance of agricultural technology. The cost of the equipment and its operation must be weighed against the availability and cost of field laborers. In the United States and other countries where relatively inexpensive labor is still available, ladders are used in many orchards, whereas in parts of Europe, mobile platforms (Fig. 6.37) are used for thinning and harvesting the crop and pruning. Picking platforms reduce heat and energy stress for workers. They are designed to enable pickers to work as a team and to minimize damage to the harvested fruits. How the crops are harvested is dictated by the manner in which fruits are marketed by the

Figure 6.37
Mobile platforms designed for harvesting (upper) and pruning (lower) trees in hedgerow plantings. (Photographs courtesy of S. Sansavini)

trade. Sweet cherries are harvested with their pedicels attached for three good reasons: (1) the fruit is less likely to be squeezed if the picker holds and twists the pedicel when separating the fruit from the spur, (2) the fruit is not torn at the junction of the pedicel and receptacle and its flesh is kept intact, and (3) the green stalk makes the fruit more attractive.

A common source of spoilage while picking ripe freestone peach is the lifting and tearing of the epidermis as it is separated from the peduncle. Peeling of the epidermis can be avoided by harvesting the fruit at the firm-ripe stage. Apple and pear are lifted gently to prevent spur breakage. Elimination of spurs by pickers and the loss of spur vigor from shading are two principal reasons why older branches become barren. Persimmons are clipped from the stem with hand shears because the fruit lacks an abscission zone between the peduncle and the stem.

Harvested fruits should be handled gently to minimize bruising. Bin depths have been reduced for transporting and storing tender fruits such as kiwifruit.

Because fruits in the tops of tall trees ripen before those on the lower limbs, two to three rounds of harvesting are necessary. This difference in maturity between the top and bottom of trees can be reduced by (1) controlling tree size or training trees to a low flat profile, (2) applying less nitrogenous fertilizer, and (3) heavy thinning. These practices improve fruit quality but reduce yield. Time–motion studies reveal that when peach trees are 5 to 5.5 meters tall, pickers spend about half of their time climbing and descending ladders and placing harvested fruits in bins. Similar data were obtained when the time required to harvest a unit weight of cherries from standard trees was compared with the time required to harvest the same weight from trees grafted on dwarfing stock.

Mechanical Harvesters

Almond, filbert, pecan, pistachio, and walnut are harvested by shaking trees and catching or picking up the fallen nuts mechanically (Fig. 6.38). All prunes for drying and about a quarter of the canning cling peach and apricot tonnages are shaken onto catching frames. Sour cherries destined for canning or for maraschino preserves are also mechanically harvested onto catching frames.

Trunk shakers are used for relatively small trees and limb shakers on large trees. Excessive shaking causes irreparable damage to trunks and limbs, especially if the operator does not tightly secure the jaws of clamps. Trees that are harvested early in the summer when the cambium is still active are more prone to shaker damage than those harvested late in the season. More damage is done if the clamp is placed just above the graft union than if it is positioned lower or higher on the trunk. Discontinuing irrigation well ahead of harvest will reduce cambial activity and alleviate damage to the trunk. Withholding water also reduces the possibility of soil compaction by the heavy harvesting equipment.

Fruit is more readily removed from spurs than from long shoots. It is nearly impossible to mechanically loosen peaches borne on long pendulous hangers because the energy of the shakers cannot be imparted to the fruits. The principle of the inertial machine is to impart a sudden force to a branch laden with fruits. The momentum causes the heaviest fruits and those closest to the point of impact to fall while the lighter and more distant ones do not.

Fruits are more readily removed in the morning hours when cells in the abscission zone are turgid. But fruits are also turgid and more likely to be injured at this time than when harvested late in the afternoon.

Mechanically harvested stone fruits and pomes are bruised or punctured as they drop through the tree. Growers should keep trees low, spreading, and well pruned. Then more fruits of uniform maturity are borne on rigid spurs than on hanger branches. This practice reduces the amount of injury to falling fruits. Equipment has been designed with forklike protrusions that penetrate the canopy at different elevations to minimize the distance the fruits must fall. Because mechanically shaken trees are harvested only once, there are always some overripe and underripe fruits that are unmarketable.

Increasing acreages of vineyards are being mechanically harvested, provided the grapes are destined for raisin, wine, or juice. Mechanical harvesting of brambles, currants, and even strawberry plants is also becoming common if bruising is not a problem and the crop is to be processed.

Figure 6.38
Combination tree shaker and catching frame (upper) and a trunk–limb shaker for nut crops (lower). (Photographs courtesy of Orchard Machinery Corporation, Yuba City, Calif.)

Postharvest Handling of Crops

Once harvested, fruits should be removed from the field and cooled as soon as possible to remove *"field heat,"* the energy which fruits acquire from exposure to the ambient temperature. For each hour a fruit remains in the field, its shelf or storage life is shortened by approximately a day, partly because fruits respire more rapidly at high temperatures and will begin to senesce. One rapid means of removing field heat from harvested fruits is by "*hydrocooling*"—dipping them in a trough of refrigerated water or cascading cold water through them.

Nut crops are mechanically hulled or husked, and the in-shell nuts are dried to a uniform moisture content. In-shell nuts are bleached; embryos from cracked shells are packaged as kernels or processed for the confectionery trade. Domestically produced pistachio nuts are not dyed with a colorant but are sold in their natural tan-colored shell. Currently, walnut hulls are waste products, but almond hulls are sold as animal feed because they contain fiber, carbohydrates, minerals, and protein.

MATURATION AND HARVESTING INDICES

Except for a few species, for example, European pears, kiwifruit, and some oriental persimmons, most fruits can be tree- or vine-ripened and eaten immediately when they are at their prime dessert quality. Unfortunately, commercial growers cannot allow their crops to reach this degree of ripeness because the longer fruits are left on the tree, the more tender they become and the more likely they are to be damaged during harvest. Fruits are subject to additional damage during packaging, storage, and shipment.

Fruits must be harvested when mature enough to withstand traveling to distant markets and still be attractive to the consumer. Transit time by refrigerated railway cars from California to New York is nearly ten days; it has been cut to four to six days by truck transports. A ship laden with fruits requires two to three weeks to go from New Zealand to England or from Santiago, Chile, to New York via the Panama Canal. An estimated 25 percent of fresh fruit spoils between the time the crop is harvested and the time it is purchased by the consumer.

The question then is "How early or late can a crop be harvested and yet provide consumers with fruits of the best possible quality?" Some physical–chemical and/or visual attributes such as color, size, firmness, appearance, and soluble solids content have been utilized as indices of maturity or ripeness. Before harvesting criteria could be established many terms needed to be defined, albeit, subjectively. Hence, horticulturists have defined a mature fruit as one that will become palatable after it is harvested. For example, avocados, bananas, and European pears require a ripening period before they become edible. On the other hand, a ripe fruit is one that can be eaten or processed immediately after harvest. Terms such as firm ripe and marketing ripe have been introduced to describe stone fruits that are firm but still mature enough to ripen. Since changes in color, size, and texture or relative firmness have always been associated with the ripening process, these attributes were the first to be used as maturity or harvest criteria or indices.

Other indices are seed coat color, soluble solids content, which approximates the sugar content of the juice, number of days from full bloom to harvest, fruit removal force, tissue ethylene content, and sonic resonance.

Color

Young immature fruits, whether they are drupes, berries, or pomes, usually have chlorophyll in their epidermal and subepidermal cells. Normally, as fruits approach maturity or ripen, chlorophyll is degraded. In fruits that become yellow, the chloroplasts assume the role of chromoplasts by beginning to synthesize the yellow pigments, carotenes and xanthophylls. The greenish-yellow coloration visible through the epidermal cells is called ground color, and the red coloration that is due to anthocyanins is called surface color.

Ground Color. With some cultivars, such as the Bartlett pear, Elberta peach, Quetta nectarine, Gravenstein apple, Royal apricot, and Wickson plum, in which the ground and surface colors are easily discernible, growers have little trouble in deciding when to begin harvest. For cultivars with visible ground color that gradually intensifies with time, color charts are used to determine whether the crop can be picked and marketed. Such a color chart with four gradations of green is used for Bartlett pear. Fruit graders and orchardists use colored plastic rings to determine whether canning clingstone peaches are ready to be harvested and processed. The color of the plastic disk is matched with that of the mesocarp after removing a small portion of the epidermis. Fruits that do not measure up to the minimum color grade are discarded.

With red-skinned mutants and hybrid progenies of apple, pear, peach, nectarine, and plums, in which the ground color is completely masked by the intense surface color, ground color is a useless harvest criterion.

Surface Color. Surface color is used as a harvest criterion for other crops such as shipping plum, strawberry, and sweet and sour cherries. Since the degree of pigmentation varies from one cultivar to another even within species, setting legal limits is very arbitrary and difficult. For example, Santa Rosa plums are ripe when still rosy pink, whereas Elephant Heart plums must develop an intense blood-red color before they attain good eating quality. Other plums, for example, Queen Anne, are prime for eating when the skin acquires a deep blue hue and the pulp becomes slightly yellow. Thus, regulations governing shipping plums differ for each cultivar. Plums for drying and canning, for instance, Petit d'Agen and Italian, are at their best when the epidermis is fully colored and the mesocarp acquires an amber color.

Strawberry and red-fleshed cherry must be completely red before they are harvested and marketed. White-fleshed cherry should have no green pigment but should be yellowish. The red coloration in some strawberry cultivars may become more intense in the dark during transit, but this color change does not enhance the eating quality.

Seed Coat Color. With apple, pear, and walnut the color of the integument is sometimes used as a harvest criterion. This color is not a good indicator for pomes because the seed coats of early- and late-ripening cultivars begin to darken about the same time. In Persian (English) walnut the color of the integument is utilized as a measure of quality as well as maturity. The integument continues to darken in some cultivars but not in others as harvest is delayed.

Packing Tissue Color. The packing tissue in walnut is the remnant of the placenta which turns brown and woody as the hull splits. These two criteria, the time of hull splitting and the browning of the packing tissue, are used to decide when to begin harvest.

Size and Firmness

Since fruits soften as they enlarge, firmness and size have been used as harvesting criteria for pomes and stone fruits. Minimum size is often used to decide whether the fruit is marketable but not necessarily for determining maturity. Under certain agricultural marketing codes, Bartlett pears are considered mature if they are larger than 2⅜ inches (60.2 millimeters) in diameter. Diameter is also used for some apple cultivars, but it is not a good criterion because harvest size varies with such factors as crop load and soil fertility. Diameter is, therefore, not closely related to maturity.

Firmness is measured by subjecting a fruit to external pressure and observing the point at which the pulp is crushed or the degree to which the fruit becomes distorted. Resistance of the tissue to crushing is determined by the Magness–Taylor or Effegi pressure testers (Fig. 6.39). The plunger tip of ⅝-inch (1.59 centimeters) diameter is used for apples and one of ⅜ inch (0.95 cen-

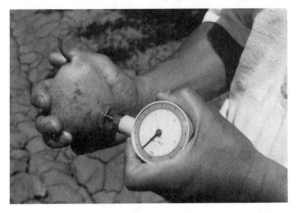

Figure 6.39
The Magness–Taylor (above) and Effegi (below) pressure testers for determining flesh firmness for maturity.

timeter) for pears and stone fruits. A thin circular portion of the skin, about 15 millimeters in diameter, is removed at the maximum equator of the fruit to accommodate a plunger tip. The skin is removed because its resistance varies from that of the subepidermal tissue.

The stationary Allo–Kramer Shear Press uses a mechanically driven piston to measure resistance to shear pressure by diced fruit. The resistance decreases as the fruit matures and softens.

Variations in fruit firmness are considerable within a given tree and among trees in an orchard; thus, an average value may not be a reliable criterion for commencing harvest. Tree vigor and crop load contribute to this variation. Large fruits that have larger cells with thinner walls and more intercellular air spaces are usually softer than small fruits.

Optimum firmness for harvest varies with cultivar and often with the rootstock–scion combination. At equal maturity Bartlett pears are firmer than Beurré Hardy fruits. Bartlett pears grafted on Asian pear rootstocks, *Pyrus serotina*, are harvested when they are firmer than those on *P. communis* rootstocks growing in the same locality.

The degree of distortion as pressure is applied to the fruit may someday be used as a nondestructive criterion of maturity. The method is based on the principle that the more mature a fruit is, the greater will be its degree of distortion at a given pressure. A reliable harvest index might be derived by establishing a correlation between the degree of distortion and flesh firmness.

Another experimental, nondestructive method is the measurement of sonic resonance of fruits at differing stages of firmness. Because fruits of different species resonate at different frequencies, a range of resonances can be established for fruits of optimum maturity. The instrument will sort individual fruits on the basis of their resonance.

Soluble Solids Content

Many different solutes are accumulated in vacuoles of cells as a fruit ripens. After stone fruits are harvested, the accumulation of soluble solids stops because this importation depends on current photosynthates from leaves. Apple and kiwifruit contain much starch that hydrolyzes to sugar as fruits ripen. At the same time, protopectins in the cell walls hydrolyze to soluble pectins. Thus, the soluble solids content of such fruits gradually increases after harvest. Soluble pectins, gums, and mucilages in the vacuole make the juice viscous, but these complex hydrophilic polysaccharides occur in such low concentrations that they contribute little to the total soluble substances.

Soluble solids content is measured by a refractometer or a hydrometer. Hand refractometers designed to be used in the field are calibrated to read percent sucrose content or degrees Brix[1] directly, whereas laboratory models are scaled to read refractive indices of pure solutions. Because fruit juices contain various solutes, refractive indices must be converted to glucose or sucrose equivalents.

A glass hydrometer, a fragile instrument, is used for measuring the density of grape juice; the density is calibrated to read in degrees Brix. A hydrometer requires a relatively large filtered juice sample and a tall cylinder.

Readings of soluble solids content should be made on juice samples as soon as the fruits are harvested. The values tend to increase with time of storage because fruits lose moisture faster than they respire sugar. If a delay is inevitable, error can be reduced by refrigerating the fruits under high relative humidity.

Titratable Acidity and Sugar : Acid Ratio

Titratable acidity and sugar : acid ratios are important criteria for harvesting grapes and citrus fruits. Both total and relative amounts of sugar and acid in wine grapes determine whether the end product will have a desirable balance of alcohol content, sourness, and sweetness. Dessert grapes and sweet oranges require an optimum sugar : acid balance for consumer appeal. Titratable acidity is determined by titrating a known volume of juice with a standard sodium hydroxide solution to a stoichiometric end point, usually pH 8. It is usually expressed as milligrams of citric or tartaric acid per 100 milliliters of juice.

Fruit Removal Force (FRF)

The force required to separate the fruit from its pedicel or the pedicel from the spur is measured with a harness attachment and a pull force gauge (Fig. 6.40). Large variations in the fruit removal

---

[1]The Brix scale is a hydrometer scale for solutions containing sugar, so graduated that its readings at a specific temperature represent percentages by weight of sugar in the solution.

Figure 6.40
Pull force gauge with a cradle for measuring the force required to separate a pedicel from a spur or from a fruit.

force (FRF) exist from fruit to fruit on the same tree; the FRF between pedicels and cherries varies considerably even if the fruits have the same degree of redness. Measurement of the FRF has been used to evaluate the efficacy of chemical looseners for mechanical harvesting (and fruit thinning). If the chemical is effective, an abscission layer is induced to form, thereby reducing the force with which the fruit is held on the spur.

Tissue Ethylene Content

Ethylene content usually increases concurrently with carbon dioxide content in tissues of fruits that show a climacteric rise. The ethylene concentration in the intercellular air spaces of developing fruits is experimentally measured by displacing the internal atmosphere with water. This is accomplished by completely submerging an apple or pear in water and applying a vacuum. The composition of the displaced gas is determined by gas chromatography and plotted against the growth and respiration curves of the fruit sample to ascertain its physiological stage of maturity. Although this technique may be sensitive, it must be applied under laboratory conditions and requires considerable investment in equipment.

Heat Units

Fruits of most species require a certain amount of heat to ripen or mature. The amount of heat required above a given temperature is quantified and calculated as degree-hours and degree-days. The summation of heat units is a useful and accurate means of estimating the harvest date for a particular cultivar (see Temperature under Environmental Factors earlier in this chapter.).

Days after Full Bloom

Within all species, the maturation sequence is nearly constant independent of locality; that is, Gravenstein is ready to pick before Golden Delicious, and Golden Delicious is always harvested before Granny Smith whether the apple cultivars are grown in California or New Zealand. The number of days from full bloom to harvest increases for later-maturing cultivars, but this interval varies from year to year because of tree vigor and age and water and heat stresses. Applications of large amounts of nitrogen promote vigor and delay maturation. Fruits on older trees with less vigor mature earlier than those on young vigorous trees. Water stress and warm temperatures hasten maturation. Because water stress and the amount of heat units accumulated depend on the weather, which is most variable in the spring, the interval between full bloom and harvest is not a reliable criterion for deciding when to pick cherry, apricot, peach, and plum which ripen early in the season.

In summary, there is no one good criterion for determining the optimum harvest date because the maturation or ripening process of fruits depends on many variables. For these reasons, it is recommended that at least two harvest indices be applied before setting a harvest date.

Abscission and Preharvest Drop

The abscission layer is a specialized tissue whose formation is due to a complex interaction of hormones (Addicott and Lynch, 1951; Crane, 1969). A constant flow of auxins from the distal region of a leaf through the petiole or the fruit via the peduncle to the stem creates a gradient that prevents the formation of an abscission zone. An actively growing fruit rarely abscises because it usually contains more hormone-rich seeds than a fruit that falls. In addition, those that abscise are often located on spurs or stems at positions less advantageous to compete for photosynthates and nutrients. Peduncles of seeded fruits have a steep hormonal gradient because auxins and other growth-regulating compounds originating from the developing endosperm and nucellus within the ovule are being exported to the spur (Luckwill, 1970). The amount concurrently being imported into the fruit from leaves and roots is presumably much less. If a fruit is severed from its peduncle or if a leaf blade is excised, the remaining peduncle or petiole will turn yellow and abscise from the spur because the hormonal source is eliminated. This abscission can be delayed by applying synthetic auxin suspended in lanolin on the cut surface of the peduncle or petiole (Addicott and Lynch, 1951).

The abscission zones of leaves and fruits usually consist of the separation and protective layers (Fig. 6.41). As these organs mature, cells in the separation layer become meristematic, and the resulting parenchyma cells enlarge and become turgid. Pectinase and cellulase activities cause dissolution of the middle lamella so that separation occurs readily. The vascular connections are ruptured, and a protective layer is formed as a result of an exudation of gummy sub-

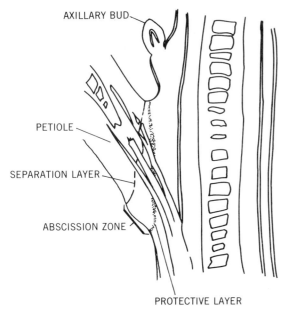

Figure 6.41
Abscission zone of a walnut leaf with a protective layer.

stances. A periderm forms underneath the protective layer so that a corky layer seals the leaf or fruit scar against possible invasion by pathogens.

The point at which fruits separate from their stems varies. At maturity a pear or an apple with attached pedicel separates from the spur. Some stone fruits, for example, peach, nectarine, apricot, and almond, separate between the pericarp and the peduncle; others such as the Japanese plum and sweet cherry that are marketed as fresh dessert fruits retain their pedicels. These are selected cultivars that do not possess an abscission layer between the pedicel and fruit. However, sweet cherry cultivars that readily become detached from the pedicel, such as Vittoria and Bianca di Verona, have been developed with the hope that they can be mechanically harvested. The receptacles of these cultivars form a callus so that the fruit separates readily from the pedicel without "bleeding."

An immature persimmon abscises between the pericarp and the receptacle, leaving the green calyx on the tree. The calyx eventually withers and abscises from the peduncle, which is persistent and remains on the tree. But calyxes of mature persimmons adhere very tightly to the pericarp, whether the fruits are seeded or seedless.

Untimely fruit abscission can spell profit or loss. A light "June drop" means an expensive thinning bill for a peach grower, but to a cherry grower it could mean a large crop of small unmarketable fruit. A heavy preharvest drop is detrimental. Some peach cultivars are notorious for continuously dropping their fruits as they ripen. In other cultivars physiologically ripe but poorly colored fruits fall because of excessive applications of nitrogen fertilizer.

Several synthetic chemicals have been used experimentally to reduce or increase fruit abscission. A common practice to prevent preharvest drop of fruit is to spray trees with naphthaleneacetic acid (NAA), 2,4-dichlorophenoxyacetic acid (2,4-D), or 2,4,5-trichlorophenoxypropionic acid (2,4,5-TP). The use of 2,4-D on Bartlett pears has inhibited the formation of a separation layer, but shears have to be used to harvest the fruit. Because these treated fruits do not fall naturally, growers have occasionally delayed harvest and allowed them to become overmature on the tree.

Treatment with ethylene-generating substances, for example, ethephon, hastens the onset of senescence and abscission. Hence, the use of such compounds as chemical "looseners" has become a common practice for mechanically harvesting walnuts.

Physiological Disorders

Pathogens and insects are not the only factors rendering fruits abnormal and unmarketable. Fruits may have physiological disorders. These abnormalities are brought about by inclement weather, such as excessively high and low temperatures, frost, rain, and high or low humidity. Harvesting at improper times, poor canopy management, use of certain rootstocks, overthinning the crop, and water stress also induce abnormal fruit development. Nutritional deficiencies and excesses also bring about physiological disorders, but they are treated separately (see Mineral Nutrition, Ch. 8).

The following disorders are associated with environmental factors and poor timing of certain cultural practices. Some of them can be avoided or at least alleviated through manipulation of the environment and/or the trees.

TEMPERATURE-RELATED DISORDERS

Pit-Burning of Apricot

The inner mesocarp of ripe apricots becomes brown and gelatinous when air temperatures approach 40° C for a few days. Green, immature apricots do not manifest this symptom when exposed to equally high temperatures for the same length of time. The disorder is attributed to lack of oxygen in the inner mesocarp brought about by the high respiration rates of fruits as they approach the climacteric peak. Experimentally, when ripe apricots are held at normal oxygen level and 40° C, pit-burning becomes pronounced and the fruits deteriorate rapidly. Those held at an equally high temperature but in an oxygen-enriched atmosphere do not exhibit this symptom. The cultivars Blenheim (syn. Royal) and Tilton are susceptible to pit-burning, but there are seedling selections that are resistant. Hence, the disorder may be reduced or eliminated in the future by breeding the heat-resistant character into commercial cultivars.

Blackening of Peach

The mesocarp tissue of ripening peaches becomes grayish black when the temperature exceeds 39° C for a few days. The discoloration is difficult to detect in yellow-fleshed cultivars but is readily visible in white-fleshed cultivars, for example, White Heath and Nectar. The darkening is worse near the pit than in the epidermis and subepidermal cells of the mesocarp, suggesting that the oxygen supply is insufficient adjacent to the endocarp. Apparently, the distance that oxygen must diffuse from the external atmosphere to the inner zones of the mesocarp is too great and the rate of movement too slow to replenish the oxygen used for respiration by the overheated cells.

Internal Browning of Prune

As prunes ripen on the tree, the mesocarp tissue normally changes from a light yellow to amber. When the fruits become overripe, the flesh turns from amber to dark brown and even blackish. The rate of change in coloration is hastened by high temperature. The browning is worse next to the pit and is related to a lack of oxygen in this zone. Discoloration is accentuated when the fruits are handled roughly after they have been harvested because any compression of the flesh injures the cells and causes oxidation of phenolic substances by the enzyme polyphenoloxidase.

Premature Shriveling of Prune

French prunes turn blue and shrivel prematurely when exposed to hot, dry winds. This disorder is attributed to a combination of heat and water stress. The trees are unable to supply the fruit with sufficient water during this period because of the rapid transpiration rate.

Water Core of Apple and Pear

The core zones of apple and pear often acquire a glazed appearance because the cell contents diffuse into the ovarian tissue and fill the intercellular air spaces. The water-soaked appearance may be symptomatic of various stressful conditions. One form of water core is attributed to the temperature gradient, which steepens between the portion of the fruit exposed to solar radiation and the shaded part. Highly pigmented red apples are more susceptible to this disorder than yellow- or green-skinned cultivars because dark skin absorbs more heat than light skin does. The temperature gradient is reversed at night because the part that was exposed to the sun during the day loses heat rapidly from nocturnal radiation. This diurnal alternation in temperature results in a gradient in water activity in one direction during the day and in the opposite direction at night. This process causes cells to lose their ability to retain solutes within the vacuoles. Hence, the sorbitol-rich solutes diffuse into the core zone, giving it a glazed or water-soaked appearance.

Variations in susceptibility to water core exist among apple and pear cultivars. In general, Asian pears are more susceptible to water core than European pears.

Pink Calyx

The calyx of Bartlett pear frequently turns pink when the fruit is exposed to temperatures at or below 20° C for four to six weeks before harvest (Wang et al., 1971). These fruits tend to soften rapidly after harvest; they should be marketed as soon as possible to avoid losses.

Doubling of Sweet Cherry and Other Stone Fruits

Stone fruit tree buds that are exposed to temperature stress or direct solar radiation when they are initiating flowers often form double pistils. If both cherry pistils are pollinated and set, a doubled fruit (Fig. 6.42) develops. If only one pistil sets while the other is merely pollinated and sets parthenocarpically, the fruit is spurred. Sweet cherry flowers with two pistils are also found in orchards located in relatively cool coastal fruit-growing districts, but their percentage set is small compared to those growing where the summers are hot. Apricot and peach buds initiate double pistils if trees are water-stressed under warm growing conditions.

The frequency of doubles and spurred fruits is reduced slightly by cooling trees with water from overhead sprinklers. But increasing the humidity may cause higher incidences of diseases and skin cracking of ripe fruits.

Unfortunately, principal cultivars such as Ear-

Figure 6.42
"Spurred" (upper) and "doubled" (lower) fruits of sweet cherry.

ly Burlat, Bing, Van, and Bigarreau Napoleon (syn. Royal Ann) are susceptible to this disorder. These cultivars have set more than 50 percent doubles. Black Republican, a dark-fleshed cultivar, and Rainier, a white-fleshed one, rarely form doubles. Thus, there is promise that doubling can be eliminated through breeding.

Black Kernel of Persian Walnut

Nuts are borne terminally on spurs so that they are exposed to solar radiation and are vulnerable to sunburning. When nuts are sunburned, the heat is transmitted to the kernel, which shrivels, darkens, and becomes worthless. Spraying nuts (and foliage) with whitewash to reflect solar radiation does not lower the fruit temperature enough to prevent sunburning.

Bud Failure of Almond

The physiological disorder bud failure appears sporadically in the whole tree or parts of a tree after they have been exposed to extremely high temperatures during the summer. Many vegetative buds do not survive the stress and abscise; those that survive either grow vigorously or remain dormant. The sparse growth on some parts of a tree and a concentration of shoot development on another part ultimately give the tree an appearance described as "crazy top" or "witches'-

broom." These symptoms worsen after each heat spell. All cultivars are susceptible to this heat-induced disorder; among the commercially grown cultivars in California, Jordanolo is the most sensitive.

WATER-RELATED DISORDERS

Cracking of Sweet Cherry

When water stands in the cup between the pedicel and a ripe fruit, or hangs from the fruit as a droplet, it moves through the epidermis into the mesocarp. Water penetrates faster than the epidermal cells can expand and accommodate the rapid increase in volume; thus, the fruit cracks (Fig. 6.43). All cultivars are susceptible to cracking. Lambert seems most resistant, whereas some soft, tender-skinned ones are highly susceptible.

Ripe cherries crack faster after being dipped in distilled water than do those treated with calcium salt solutions. Calcium may impede the rate of water uptake and also strengthen the cell wall. Under field conditions, spraying with calcium or with a wax spray to prevent water uptake has not been feasible. Blowing the water from fruits with an air blast sprayer or helicopter is not effective unless the duration of the rain is very short. Any droplet of water left on a ripe fruit will cause it to split.

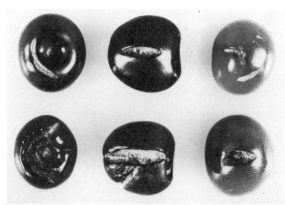

Figure 6.43
Ripe Bing cherries with cracks as a result of rain.

End-Cracking of Prune

Cracking of the distal ends of prunes appears in mid-June on trees that were water-stressed earlier in the season and subsequently irrigated. The mesocarp separates from the endocarp when the fruit regains its full turgidity from the influx of water. The epidermal cells on the distal end of the fruit lose their elastic qualities so that the mesocarp cells separate, usually along the ventral suture.

Side-Cracking of Prune

Side-cracking becomes apparent in mid-July on fruits exposed to direct sunlight, especially if nights are cool and temperatures reach dew point. The epidermal cells lose their elasticity from exposure to sunlight. In the early-morning hours the mesocarp cells become fully turgid because the water potential in the xylem approaches zero. If additional free water is absorbed through the skin, the epicarp and mesocarp crack (Fig. 6.44) as individual cell volume increases.

When immature prunes crack, the mesocarp cells become meristematic and form callus, which is unsightly after the fruits are canned or dried.

Figure 6.44
French prunes with side-cracking. (Photograph courtesy of K. Uriu, Department of Pomology, University of California at Davis)

Physiological Disorders

When ripe prunes crack, sugars, which support mold colonies, and gums are exuded. Not only do molds give an off-flavor to the dried product, but some molds also produce aflatoxins, microbiol poisons that are carcinogenic.

Cracking of Apple Skin

Apple skin cracks, as do the sides and ends of prunes when trees are irrigated after undergoing long periods of moisture stress. Cracking is particularly bad on the side of the fruit exposed to the sun and especially on limbs with sparse foliage. Cracking is attributed to the loss of elasticity by the skin from heat stress and subsequent rapid uptake of water by fruit cells when the tree is irrigated. The problem is reduced or eliminated by (1) irrigating the orchard when the soil moisture content reaches −40 centibars to assure an adequate water supply for transpiration, and (2) pruning and fertilizing the trees to maintain full canopies.

Russeting of Apple and Pear

Russeting is associated with moisture condensation on the surface, fruit thinners, and virus infections. When dew, fog, and spray droplets do not dry rapidly, they induce the formation of the brown corky layers on the epidermis. Newton Pippen apple and Bartlett pear grown in the coastal districts of California display varying degrees of russeting because of the nearly daily morning fog. Covering the fruits with paper bags to avert condensation on the skin or frequent spraying with low dosages of GA has been shown to reduce russeting.

Young apple and pear fruits injured by frost develop russeting, known as "frost rings" (Fig. 6.45). Growth of the cortical tissue is inhibited in this portion of the fruit.

GROWTH-RELATED DISORDERS

Whenever a crop is light as a consequence of spring frost, injudicious use of chemical thinners,

Figure 6.45
"Frost rings" on Golden Delicious (left) and Red Delicious (right) apples.

or alternate bearing, a large leaf to fruit ratio ensues. The enhanced fruit growth rate leads to excessively large fruits having various disorders. These disorders, which are directly or indirectly associated with rapid fruit growth, are described here, together with their symptoms and causal factors.

Splitting of the Endocarp

Peaches and nectarines frequently have split endocarps or pits (Fig. 6.46). This condition is asso-

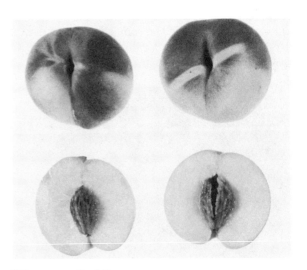

Figure 6.46
Clingstone peaches with an intact endocarp (left) and a split endocarp (right).

ciated with a light crop and is especially prevalent in early-ripening cultivars and those with elongated endocarps rather than spherical ones. Almond, olive, plum, and some sweet cherry cultivars manifest split pits. The endocarp usually fractures along the dorsal side because cells along the ventral suture are merely appressed to each other. When the crop is extremely light, the endocarp may crack in different directions (Fig. 6.47).

Splitting or cracking occurs as the endocarp ceases to grow and its cells begin to lignify and lose their elasticity. Cells of the mesocarp continue to enlarge while the innermost cells adhere tightly to the endocarp. Any cultural operation favoring fruit growth, such as thinning or applying irrigation water during this critical pit-hardening period, causes a sudden surge of carbohydrates and/or water into the mesocarp cells. This sudden increase in osmotic concentration raises the turgor pressure of the mesocarp cells. The natural tendency of parenchyma cells is to become spherical, but because of the lack of space, those next to the pit elongate radially and exert considerable pressure transversely. Under such internal stress between the mesocarp and endocarp, the halves of the endocarp may be pulled apart. If the endocarp is partially lignified when the stress occurs, a fissure may occur between this part and the part that is still nonlignified and plastic.

Figure 6.47
Jordanolo almond pericarp with a cracked endocarp.

Split pits are undesirable for several reasons. (1) If the split extends to the abscission zone of the endocarp, even the peduncle is split. This opens up a pathway by which fungal spores can enter the seed cavity and develop mold colonies, which impart an off-flavor to the fruit. (2) Pit fragments not removed by pitting machines and cannery workers are processed with the fruit. (3) In almonds, callus forms on the inner lining of the endocarp owing to regeneration of cambial activity in numerous vascular bundles. The callus grows against the integument, causing an indentation on the developing kernel. At harvest, when the callus dries, it becomes corky and unsightly. The almond industry keeps these kernels out of the trade because the dry superficial calluses can be mistaken for excrements of larvae or fungal infections. These deformed kernels are sliced for the confectionery trade after the seed coats are removed. Canned or dried prunes with cracked endocarps are impossible to detect and sort out because most of the fruit is processed intact, that is, without pitting.

Some thin endocarps of sweet cherry cultivars, for example, Rainier, split naturally while the fruits are growing, but others shatter when they are bitten or mechanically pitted during the canning process.

The presence of pit fragments in canned and processed products poses a potential hazard to the consumer. The food processing industry is keenly aware of the problem and is constantly striving to eliminate it. The incidence of split pits in peach and nectarine can be reduced by not thinning the crop just prior to the pit-hardening period or Stage II. Thinning should either be done in the middle of Stage I when the endocarp is still plastic or be delayed until the fruits are in Stage II and the pit is inelastic and flinty. Endocarps harden completely to the receptacle tissue within 10 to 14 days after lignification begins in cells at the distal tip, but lignin continues to be deposited in the cells for another 30 days. Orchards should not be irrigated immediately after thinning the crop, especially during the critical early pit-hardening period, for the influx of photosynthates and water will create a sudden increase in turgor pressure of cells.

Gumming

Gums and mucilages, which are complex carbohydrates, are produced by leaves and fruits of various stone fruit species. The quantity of gum that the mesocarp cells accumulate is a heritable characteristic. Peach and almond cultivars that accumulate considerable gum in mesocarp cells of immature fruits have a greater tendency to exude gums than do those accumulating small quantities.

Gum pockets commonly form in the distal end of peach and prune fruits because the vascular bundles end in this region. The gums are hydrophilic; they hydrate readily and create an internal pressure which mesocarp cells are unable to retain within vacuoles. The gums then diffuse through the weak zones created in the mesocarp and epidermis (Fig. 6.48).

When gumming occurs early before pit hardening, peaches usually abscise, but those in which gum is exuded after this period usually persist on the tree until harvest. These are considered culls because gum pockets are unsightly and the gums become mucilaginous.

Gumming is not always associated with large fruits. Almond fruits on trees that are deficient in boron often exude gum. Injuries caused by birds, hailstones, and insects, will also cause fruits to exude gum. Gumming is probably a defensive mechanism to prevent fungal attack. However, in almond, fungal invasion causes gumming so that two or more fruits adhere to each other. The abnormal metabolite created by the fungus or its interaction with the fruit inhibits the formation of an abscission layer, making fruit removal at harvest difficult. Such fruits remain on the tree as "stick-tights," only to be invaded by the Navel Orange worm which overwinters while feeding on the kernel.

Blossom-End Rot of Peach

The desiccation of epidermal and subepidermal cells on the distal end of peach fruits (Fig. 6.49) as they ripen is an early indication of blossom-end rot. All peach cultivars seem susceptible but

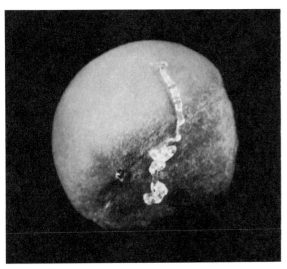

Figure 6.48
Peach fruit with a beady distal gum.

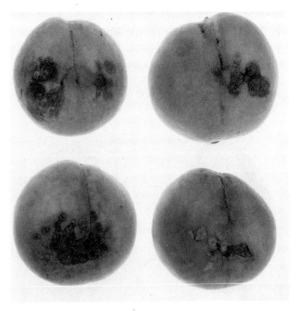

Figure 6.49
Peach fruits with varying amounts of blossom-end breakdown symptoms.

some more than others. In extreme cases, the deterioration spreads over the entire fruit. The fruit looks unappetizing, even though the unaffected portion is edible.

Bitter Pit of Apple

Bitter pit occurs when the leaf to fruit ratio is large because the crop is light through overthinning, frost damage during bloom, poor pollination during inclement weather, and/or alternate bearing. When the crop is light, shoot growth continues into late summer, resulting in an unusually large, thick canopy. The large transpiring area draws the mineral elements that are dissolved in the xylem, including calcium, into the foliage rather than to the fruits. The accumulation of calcium by relatively large fruits is slow. Consequently, a group of epidermal and subepidermal cells, usually near the calyx end, collapses and becomes necrotic (see Plate 8). These necrotic spots can appear on fruits while they are still on the tree, or after harvest while they are stored, or while they are in transit to distant markets.

Large fruit cells with thin, weak walls and limited calcium to regulate membrane permeability are more prone to collapse than small cells with thick walls. These large, low-calcium fruits respire more rapidly and, therefore, have a relatively shorter shelf life than those with adequate calcium.

Bitter pit is somewhat difficult to control, partly because calcium is not mobile. That is, the element does not move readily either from leaves to fruits or from fruits to leaves. Thus, when calcium salt was applied to only one side of an apple, that side did not exhibit bitter pit symptoms, whereas the untreated side did. Because of this relative immobility, spray applications of calcium salts have not always prevented the appearance of bitter pit symptoms since complete coverage of all fruits is not possible. In districts where bitter pit is a serious problem, packing houses dip sound fruits in a calcium salt solution for a few minutes in order to reduce the number of fruits that develop the symptoms.

Cultural practices that tend to reduce the incidence of bitter pit are (1) summer pruning because it decreases the leaf area, thereby diverting less calcium to leaves, and because it eliminates many shoot apices which are strong competitors for calcium, and (2) maintenance of a relatively high level of soil moisture throughout the growing season.

Nearly all cultivars of apples display symptoms of bitter pit if conditions are favorable for its expression, but some cultivars are more susceptible than others.

DISORDERS OF IMMATURE AND OVERMATURE FRUITS

Harvesting fruits prematurely is a common mistake among dessert fruit growers who are overly anxious to get their crops to market while prices are high. Conversely, when the market is saturated and prices are low, growers may delay harvesting and allow their crops to become overmature. In either case, fruits are likely to develop postharvest disorders.

Wilting and Shriveling

Fruits that are harvested prematurely wilt and shrivel, even though they may be cold-stored under high relative humidity. Because the soluble solids content in cells of immature fruits is low, they are less able to prevent water loss than the more mature fruits. Firm-ripe or mature fruits stored at 0° to 1° C at 85 percent relative humidity retain their turgidity and appearance better than do immature fruits, but the shelf life of ripe fruits will be short after storage. The ability to prevent water loss also depends on lenticel density and cuticle thickness. These traits vary among species and cultivars.

If fruits are coated with wax or wrapped in waxed paper or plastic wrap to prevent water loss, the appearance of a wilt symptom can be delayed. Some water lost by transpiration through the cuticle or stomata and lenticels of the fruits is replaced by metabolic water. Metabolic water is given off in the respiration of sugars; the other product of respiration is carbon dioxide. Fruits should not be placed directly in front of refrigeration fans because winds have a desiccating influence, even if the relative humidity is high.

Skin Cracking

Skins of ripe apricot, cherry, peach, and plum often crack when fruits are dipped in a hydrocooler to remove their field heat and immediately packed in cellophane trays. The free water that is absorbed during storage causes the skin to crack.

Scald

Scald is the surface discoloration of fruits that appears during or after a period of storage (Fig. 6.50). It is symptomatic of fruits, especially apple and pear, that are harvested prematurely or when they are overmature. At the incipient stage, the epidermal and subepidermal cells of pomes turn brownish. The pigmentation intensifies with continued storage.

The emanation of some undesirable vapors by scalded fruits is believed to spread the disorder, for sound fruits stored with them become scalded. The onset of scald symptoms can be delayed by (1) scrubbing the atmosphere within the cold-storage facility with activated charcoal which adsorbs volatiles, (2) passing cold air over potassium permanganate which oxidizes organic vapors, (3) dipping fruits in solutions of diphenylamine or mineral oil, or (4) wrapping fruits with mineral oil-impregnated paper.

Internal Browning

Apples that were held for an extended period in cold boxes without ample air exchange or shipped in poorly ventilated holds of transoceanic ships often displayed internal browning. The problem was attributed to the buildup of carbon dioxide. Installation of efficient gas scrubbers and controlled-atmosphere storage on board ships and at shipping points, receiving docks, and wholesale storage facilities has overcome the problem of internal browning.

Figure 6.50
Normal, sound Bartlett pears (upper row) and those with scald symptoms (lower row).

ROOTSTOCK-RELATED DISORDERS

Black and Hard End of European Pear

During World War I when the importation of pear seeds into the United States from France was curtailed, nurseries, especially those on the Pacific Coast of North America, began importing seeds of Asian pears, *Pyrus serotina* (syn. *P. pyrifolia*), from Japan and China. Subsequently, Bartlett pears propagated on these seedling

Figure 6.51
Bartlett pears afflicted with black end disease.

stocks began to yield fruits with calyx ends that were hard and shiny and/or blackened. In extreme cases, the black calyx end became sunken and desiccated (Fig. 6.51).

Davis and Tufts (1931) inarched seedlings of *Pyrus communis* into scaffold limbs of trees bearing these abnormal fruits. When the inarched stocks were large enough to support the limbs, the limbs were severed from the mother tree below the new graft unions (Fig. 6.52). Pears borne in subsequent seasons on these scaffold branches that were now on French pear rootstocks were normal. Thus, the suspicion that a virus was the causal factor was allayed, for normally a virus cannot be eliminated from a scion by changing the rootstock.

Surveys of orchards disclosed that certain trees consistently produced abnormal fruits, some more than others, whereas many trees bore normal fruits. These normal fruits are often found on the same spurs as those with black end symptoms. The number of abnormal fruits produced by a given tree was inversely proportional to the size of the crop, more being produced in years of a light crop and fewer during years of a heavy one. Bartlett, Conference (Fig. 6.53), Superfin, Beurre d'Anjou, Winter Nelis, Doyenne du Comice, Colonel Wilder, Easter Beurre, and Beurre Clairgeau are susceptible to the disorder. Beurre

Figure 6.52
Scaffold branch (right) of a Bartlett pear/*Pyrus serotina* scion–rootstock combination that was inarched with French seedling tree and later severed from the mother tree (left). (From Davis and Tufts, 1931)

Hardy, when topworked on identical Asian rootstocks, does not manifest the symptoms. However, Beurre Hardy fruits on these stocks are not as smooth as those borne on *P. communis* seedlings.

The disorder is tentatively attributed to the production of more roots and root metabolites in years when the crop is light than are produced when the crop is heavy. In a light-crop year, root growth is enhanced and more photosynthates are translocated to roots for growth than when the

Physiological Disorders

Figure 6.53
Conference pears grown on *Pyrus communis* seedling (left) and those grown on *P. serotina* seedlings exhibiting black end symptoms (right).

crop is heavy. The root metabolites that are abnormal to French pears are translocated to the fruit and accumulated in the distal end where the vascular bundles terminate. With the accumulation of more root metabolites per fruit, the number of abnormal fruits on lightly cropped trees increases. Most cultivars are seemingly unable to catabolize these substances, but others such as Beurre Hardy metabolize them and do not show the symptoms.

Yuzuhada (Pomelo Skin Disease) of Asian Pear

When Asian pear cultivars are topworked onto seedlings of *P. serotina*, the entire fruit surface becomes dimpled like the citrus pomelo (Fig. 6.54), from which the name yuzuhada was derived (Hayashi, 1955). Furthermore, the flesh remains hard and granular instead of softening at maturity. Yuzuhada disorder is also called *Ishinashi* or rock-pear in Japan. The disorder occurs to a lesser extent on trees propagated on seedlings of *P. betulaefolia*. The number of fruits exhibiting the symptoms increases if the trees are water-stressed in July. Apparently, the enzymes that degrade the cell walls are inhibited so that these abnormal fruits do not soften at maturity.

Le Conte and Kieffer, hybrid seedlings between *P. communis* and *P. serotina*, are also highly susceptible to this disorder (Fig. 6.55). Stocks that produce hard and black end of Bartlett also produce symptoms of yuzuhada; thus, the two disorders are probably caused by the same factor(s) but are manifested differently. Not all seedlings of *P. serotina* and *P. betulaefolia* induce black end and yuzuhada symptoms, and the percentage of abnormal fruits may not be high on any one seedling stock. These disorders can be easily avoided by propagating European and Asian pear cultivars on *P. communis* seedlings or on clonal pear or quince rootstocks. A compatible interstock is recommended for quince roots if a possibility of graft incompatibility exists.

Figure 6.54
Normal Nijis-seiki syn. Twentieth Century Asian pear (left) and pear afflicted with yuzuhada or pomelo skin disease (right). (Photograph courtesy of S. Hayashi, Tottori University, Japan)

Figure 6.55
Kieffer pears harvested from a tree on *Pyrus serotina* seedling rootstock (left) and from a tree grafted on Old Home, a *P. communis* rootstock clone (right).

Cork Spot of Beurre d'Anjou

Beurre d'Anjou pears with low-calcium content often develop cork spot. Thus, this disorder is comparable to bitter pit of apples. It frequently appears when Beurre d'Anjou is topworked on seedlings of *Pyrus betulaefolia*. The fruit surface becomes dimpled with necrotic sunken areas. In mild cases, the corky condition is not visible in fruits while they are in trees but appears after they are harvested.

References Cited

ABBOTT, D. L. 1984. *The Apple Tree: Physiology and Management.* Grower Books. London.

ADDICOTT, F. T., AND R. S. LYNCH. 1951. Acceleration and retardation of abscission by indoleacetic acid. *Science* 114:688–689.

ASSAF, R., I. LEVIN, AND B. BRAVDO. 1975. Effect of irrigation regimes on trunk and fruit growth rates, quality, and yield of apple trees. *J. Hort. Sci.* 50:481–593.

AVERY, D. J. 1975. Effects of fruits on photosynthetic efficiency. In *Climate and the Orchard.* Ed. H. C. Pereira. Commonwealth Agricultural Bureaux, Farnham Royal Slough, England.

BLAKE, M. A. 1926. The growth of the fruit of the Elberta peach from blossom bud to maturity. *Proc. Amer. Soc. Hort. Sci.* 22:29–39.

BROWN, D. S. 1952. Climate in relation to deciduous fruit production in California. V. The use of temperature records to predict the time of harvest of apricots. *Proc. Amer. Soc. Hort. Sci.* 60:197–203.

BUKOVAC, M. J., AND S. NAKAGAWA. 1968. Gibberellin-induced asymmetric growth of apple fruit. *HortScience* 3:172–174.

CHANDLER, W. H. 1934. The dry matter residue of trees and their products in proportion to leaf area. *Proc. Amer. Soc. Hort. Sci.* 22:39–56.

COOMBE, B. G. 1960. Relationship of growth and development to changes in sugars, auxins, and gibberellins in fruit of seeded and seedless varieties of *Vitis vinifera*. *Plant Physiol.* 35:241–250.

CRANE, J. C. 1969. The role of hormones in fruit set and development. *HortScience* 4:108–111.

CRANE, J. C., AND I. M. AL-SHALAN. 1974. Physical and chemical changes associated with growth of the pistachio nut. *J. Amer. Soc. Hort. Sci.* 99:87–89.

DAVIS, L. D., AND M. M. DAVIS. 1948. Size in canning peaches. The relation between the diameter of cling peaches early in the season and at harvest. *Proc. Amer. Soc. Hort. Sci.* 51:225–230.

DAVIS, L. D., AND W. P. TUFTS. 1931. A study of growth of pear inarches. *Proc. Amer. Soc. Hort. Sci.* 28:485–488.

DENNY, F. E. 1924. Hastening the coloration of lemons. *J. Agric. Res.* 27:757–769.

GRAEBE, J. E., D. T. DENNIS, C. D. UPPER, AND C. A. WEST. 1965. Biosynthesis of gibberellins. I. The biosynthesis of (-)-kaurene, (-)-kaurene-19-oland trans-geranylgeraniol in endosperm nucellus of *Echinocystis macrocarpa* Greene. *J. Biol. Chem.* 240:1848–1854.

GRANT, J. A., AND K. RYUGO. 1984. Influence of within-canopy shading on fruit size, shoot growth, and return bloom in kiwifruit. *J. Amer. Soc. Hort. Sci.* 109:799–802.

GRIGGS, W. H., L. L. CLAYPOOL, AND B. T. IWAKIRI. 1960. Further comparisons of growth, maturity, and quality of seedless and seeded Bartlett pears. *Proc. Amer. Soc. Hort. Sci.* 76:74–84.

HANSEN, P. 1967a. ^{14}C-Studies on apple trees. I. The effect of the fruit on the translocation and distribution of photosynthates. *Physiol. Plant.* 20:382–391.

HANSEN, P. 1967b. ^{14}C-Studies on apple tree. III. The influence of season on storage and mobilization of labelled compounds. *Physiol. Plant.* 20:1103–1111.

HANSEN, P. 1970. ^{14}C- Studies on apple trees. V. Translocation of labelled compounds from leaves to fruit and their conversion within the fruit. *Physiol. Plant.* 23:564–573.

HANSEN, P., AND K. RYUGO. 1979. Translocation and metabolism of carbohydrate fraction of ^{14}C-photosynthates in 'French' Prune, *Prunus domestica* L. *J. Amer. Soc. Hort. Sci.* 104:622–625.

HANSEN, P., K. RYUGO, D. E. RAMOS, AND L. FITCH. 1982. Influence of cropping on Ca, K, Mg, and carbohydrate status of 'French' prune trees grown on potassium limited soils. *J. Amer. Soc. Hort. Sci.* 107:511–515.

HAVIS, L. 1947. Studies in peach thinning. *Proc. Amer. Soc. Hort. Sci.* 49:55–58.

HAYASHI, S. 1955. Studies on "Yuzuhada" disease of fruits of Nijis-seiki pears *(Pyrus serotina).* I. The relation of osmotic pressure of leaves and fruits to the development of "Yuzuhada." *J. Hort. Assn. Japan.* 24:94–102.

HULME, A. C. 1936. Biochemical studies in the nitrogen metabolism of the apple fruit. II. The course followed by certain nitrogen fractions during the development of the fruit on the tree. *Biochem. J.* 30:256–268.

HULME, A. C. 1954. The climacteric rise in respiration in relation to changes in the equilibrium between protein synthesis and breakdown. *J. Exp. Bot.* 5:159–173.

ITO, H., Y. MOTOMURA, Y. KONNO, AND T. HATAYAMA. 1969. Exogenous gibberellin as responsible for the seedless berry development of grapes. I. Physiological studies on the development of seedless Delaware Grapes. *J. Tohoku Agric. Res.* 20:1–18.

KAWASE, M. 1965. Etiolation and rooting in cuttings. *Physiol. Plant.* 18:1066–1076.

KIDD, F., AND C. WEST. 1930. Physiology of fruit. I. *Proc. Roy. Soc.* B106:93–109. London.

KUMASHIRO, K., AND S. TATEISHI. 1966. Effect of soil moisture on tree growth, yield, and fruit quality of Jonathan apples. *J. Japan. Soc. Hort. Sci.* 36:9–20.

LILLELAND, O. 1935. Growth study of the apricot fruit. II. The effect of temperature. *Proc. Amer. Soc. Hort. Sci.* 33:269–279.

LOMBARD, P. B., C. B. CORDY, AND E. HANSEN. 1971. Relation of post-bloom temperature to Bartlett pear maturation. *J. Amer. Soc. Hort. Sci.* 96:799–801.

LOVEYS, B. R., AND P. E. KRIEDEMANN. 1973. Rapid changes in abscisic acid-like inhibitors following alterations in vine leaf water potential. *Plant Physiol.* 28:476–479.

LUCKWILL, L. C. 1953. Studies on fruit development in relation to plant hormones: II. The effect of naphthalene acetic acid on fruit set and fruit development on apples. *J. Hort. Sci.* 37:137–140.

LUCKWILL, L. C. 1970. Control of growth and fruitfulness in apple trees. In: *Physiology of Tree Crops.* Ed. L. C. Luckwill and C. V. Cutting. Academic Press. New York.

MAGNESS, J. R., F. L. OVERLAY, AND W. A. LUCE. 1931. Relation of foliage to fruit size and quality in apples and pear. Wash. Agric. Exp. Sta. Bull. 249.

MARKAKIS, P., AND R. J. EMBS. 1964. Conversions of sugars to organic acids in strawberry fruit. *J. Food. Sci.* 29:629–630.

MATSUI, S., K. RYUGO, AND M. W. KLIEWER. 1985. Growth inhibition of Thompson Seedless and Napa Gamay berries by heat stress and its partial reversibility by applications of growth regulators. *Amer. J. Enol-Vit.* 37:67–71.

MOLISCH, H. 1921. Pflanzen-physiologies al Theorie Gartnerei. 4th Ed. Gustav Fischer. Jena, Germany.

NAKAGAWA, S., M. J. BUKOVAC, N. HIRATA, AND H. KUROOKA. 1968. Morphological studies of gibberellin-induced parthenocarpic and asymmetric growth in apple and pear fruits. *J. Japan. Soc. Hort. Sci.* 37:9–19.

NAKAMURA, M. 1967. Physiological and ecological studies on the calyx of the Japanese persimmon fruit. *Res. Bul.* #23, Gifu University, Gifu, Japan.

NELJUBOW, D. 1901. Uber die horizontale Nutation der Stengel von *Pisum sativum* und einigen anderen Pflanzen. *Beih. Bot. Zentralbe.* 10:128–138.

NITSCH, J. P. 1950. Growth and morphogenesis of the strawberry as related to auxin. *Amer. J. Bot.* 37:211–215.

OKUSE, I., AND K. RYUGO. 1981. Compositional changes in the developing 'Hayward' kiwifruit in California. *J. Amer. Soc. Hort. Sci.* 105:642–644.

OLIEN, W. C., AND M. J. BUKOVAC. 1982. Ethephon-induced gummosis in sour cherry (*Prunus cerasus* L.). I. Effect on xylem function and shoot water status. *Plant Physiol.* 70:547–555.

PATTEN, K. D., AND E. L. PROEBSTING. 1986. Effect of different artificial shading times and natural light intensities on the fruit quality of 'Bing' sweet cherry. *J. Amer. Soc. Hort. Sci.* 111:360–363.

PLINIUS, SECUNDUS C. (The Elder). *Natural History*, Vol V: Book XVII. Translated by H. Rackham. 1950. Harvard University Press. Cambridge, Mass.

POWELL, L., AND C. PRATT. 1966. Growth promoting substances in the developing fruit of peach (*Prunus persica* Batsch.) *J. Hort. Sci.* 41:331–348.

RIBEREAU-GAYON, G. 1968. Étude des mecanismes de syntheses et de transformation de l'acide malique, de l'acide tartrique et de l'acide citrique chez *Vitis vinifera* L. *Phytochem.* 7:1471–1482.

ROGERS, B. L., AND L. P. BATJER. 1954. Seasonal trends of six nutrient elements in the flesh of Winesap and Delicious apple fruits. *Proc. Amer. Soc. Hort. Sci.* 63:67–73.

ROGERS, B. L., AND A. H. THOMPSON. 1983. Effects of dilute and concentrated sprays of NAA and carbaryl in combination with daminozide and pesticides on fruit size and return bloom of 'Starkrimson Delicious' apple. *HortScience* 18:61–63

ROSS, N. W. 1952. The relation of number and size of cells in the flesh of the peach to the size of the fruit at time of pit hardening. Master of Science thesis, University of California at Davis.

RYUGO, K. 1964. Relationship between malic and citric acids and titratable acidity in selected peach and nectarine clones. *Proc. Amer. Soc. Hort. Sci.* 85:154–160.

RYUGO, K. 1969. Seasonal trends of titratable acids, tannins and polyphenolic compounds, and cell wall constituents in oriental pear fruit (*Pyrus serotina*, Rehd.). *Agric. Food Chem.* 17:43–47.

RYUGO, K. 1976. Gibberellin-like substances in the endosperm-nucellus tissues of the developing almond *Prunus amygdalus* Batsch. cv. Jordanolo. *J. Amer. Soc. Hort. Sci.* 101:565–568.

RYUGO, K., AND L. D. DAVIS. 1958. Seasonal changes in acid content of fruits and leaves of selected peach and nectarine clones. *Proc. Amer. Soc. Hort. Sci.* 72:106–112.

RYUGO, K., AND L. D. DAVIS. 1959. The effect of the time of ripening on the starch content of bearing peach branches. *Proc. Amer. Soc. Hort. Sci.* 130–133.

RYUGO, K., B. MARANGONI, AND D. E. RAMOS. 1980. Light intensity and fruiting effects on carbohydrate contents, spur development, and return bloom of 'Hartley' walnut. *J. Amer. Soc. Hort. Sci.* 105:223–227.

RYUGO, K., N. NII, M. IWATA, AND R. M. CARLSON. 1977. Effect of fruiting on carbohydrate and mineral composition of stems and leaves of French prunes. *J. Amer. Soc. Hort. Sci.* 102:813–816.

SPARKS, D. 1977. Effects of fruiting on scorch, premature defoliation, and nutrient status of 'Chickasaw' pecan leaves. *J. Amer. Soc. Hort. Sci.* 104:195–199.

TAKETA, F., K. RYUGO, AND J. C. CRANE. 1980. Translocation and distribution of ^{14}C-photosynthates in bearing and nonbearing pistachio branches. *J. Amer. Soc. Hort. Sci.* 105:642–644.

WALSH, C. S., H. J. SWARTZ, AND L. J. EDGERTON. 1979. Ethylene evolution in apple following post-bloom thinning sprays. *HortScience* 14:704–706.

WANG, C. Y., W. M. MELLENTHIN, AND E. HANSEN. 1971. Effect of temperature on development of premature ripening in Bartlett pears. *J. Amer. Soc. Hort. Sci.* 96:122–126.

WEINBERGER, J. H. 1932. The relation of leaf area to size and quality of peaches. *Proc. Amer. Soc. Hort. Sci.* 28:18–22.

WILLIAMS, K. M. 1979. Seasonal gibberellin- and cytokinin-like substances in developing sweet cherry fruit. Master of Science thesis, University of California at Davis.

WILLIAMS, M. W. 1981. Response of apple trees to aminoethoxyvinylglycine (AVG) with emphasis on apical dominance, fruit set, and mechanism of action of fruit thinning chemicals. *Act. Hort.* 120:137–141.

WILLIAMS, MAX W., AND L. P. BATJER. 1964. Site and mode of action of 1-naphthyl *N*-methylcarbamate (Sevin) in thinning apples. *Proc. Amer. Soc. Hort. Sci.* 85:1–10.

WINKLER, A. J. 1948. Maturity tests for table grapes—the relation of heat summation to time of maturing and palatability. *Proc. Amer. Soc. Hort. Sci.* 51:295–298.

WOOD, B. W., AND J. A. PAYNE. 1983. Flowering potential of pecan. *HortScience* 18:326–328.

YAMAKI, S. 1983. Biochemical changes of cell wall elements with fruit development and ripening, and the harvesting period in Japanese pear fruit (*Pyrus serotina* Rehder var. *culta* Rehder). *Japan. J. Agric. Res. Quarterly* 17:134–140.

ZOCCA, A., AND K. RYUGO. 1975. Changes in polyphenoloxidase activity and substrate levels in maturing 'Golden Delicious' apple and other cultivars. *HortScience* 10:586–587.

PART II

Orchard Management and Nursery Practices

Whenever possible, cultural practices in Part II are discussed with respect to principles covered in Part I. Some practices, such as root pruning to promote flowering and the need for pressure between scion and stock to induce differentiation of vascular elements, evolved empirically and have no adequate scientific explanation. Some technical aspects of fruit culture, for example, mineral analysis and cold hardiness, are included in Part II because such knowledge is essential to sound orchard management.

CHAPTER 7

Establishing and Managing an Orchard

Site Selection and Preparation

REGIONAL AND LOCAL CONSIDERATIONS

Seasonal Temperatures

Chilling Temperatures To Satisfy Rest Requirements. Temperate zone fruit species need to be exposed to chilling temperatures during the winter to satisfy their rest requirement and be able to resume normal growth the following spring. The amount of chilling required varies from species to species and even within a given species (see Ch. 2, Table 2).

Minimum Winter Temperatures. Although certain species can tolerate considerable cold, for example, apple, American plum, pear, and cherry, others suffer winter chilling injury if they are exposed to temperatures below $-22°$ C. The ability to survive minimum winter temperatures is but one factor that limits where fruits might be grown.

Frost-Free Days and Cumulative Heat Units. The number of days between the last frost in the spring and the first freeze in the autumn and the number of heat units to which fruits are exposed (see Temperature, Ch. 6) limit cultivation of fruit trees. Open flowers and young developing fruitlets of temperate zone fruit tree species are sensitive to spring frosts, and some mature fruits and vigorously elongating shoots do not tolerate early fall frosts.

With an increase in latitude and elevation, the number of frost-free days decreases because the temperatures are colder. Thus, late spring frosts are prevalent and fall frosts occur earlier. Late-maturing cultivars cannot be grown commercially in such areas. In the fruit-growing districts of the Central Valley of California, Elberta peach blooms about March 7 and ripens in early August. In upper New York and British Columbia, the same cultivar blooms about a month later and ripens in September, about the time of the first fall frost.

The need for a long, warm growing season limits pecan cultivation to the southern latitudes of the United States and the growing of Granny Smith apples to areas south of the Alps in Europe.

At higher elevations in Taiwan, Indonesia, Israel, and central Africa, the temperatures are cool enough in winter to satisfy the chilling requirement and sufficiently warm during the summer to grow peaches and apples.

Coastal Influence on Temperature. Oceans and large lakes tend to ameliorate the temperature of the adjacent landmass such as Denmark, the Niagara peninsula in New York, the western shorelines of Michigan, and the Pacific Coast states. Large bodies of water serve as good heat exchangers. They give off heat as the water cools and/or freezes in the late fall and winter and absorb heat in the spring and summer, keeping the region relatively cool. Cool temperatures during spring delay budbreak and anthesis and allow trees to escape frost damage.

Edaphic Considerations

Soils. Fruit trees fare best in deep, well-aerated sandy loam soils. Thus, the world's best orchards are those established on (1) deltas at the mouths of rivers and silty soils along river banks, (2) deep alluvial fans formed at the mouths of canyons opening into broad valleys, and (3) glacial moraines that are rocky but have excellent drainage.

For optimum performance of most temperate zone deciduous fruit tree species, the pH of soils should be slightly acidic, between 6.5 and 5.5. At pH above 7.5 or below 4.5, many essential elements, for example, copper, iron, and zinc ions, will either form complex ions such as a zincate ion or precipitate as salts with very low solubilities, such as iron phosphates. Roots of fruit trees growing in soils with extreme acidity or alkalinity are unable to extract sufficient amounts of these elements because their concentrations in the soil solution are too low. Consequently, the trees develop foliar deficiency symptoms and grow poorly. Kiwifruit vines will exhibit chlorosis in soils having a pH of 7.3. Peaches grafted on peach rootstocks become chlorotic if the soil pH is greater than 7.5.

Height of Water Table. The height of the water table dictates which species might grow profitably on the site. In certain parts of the Po Valley of Italy and in narrow valleys of Chile, the water table is less than a meter from the soil surface during the summer months. In these areas, drainage ditches are essential for maintaining the water table below the root zone. Fortunately, the soils in these areas are porous, and water drains laterally into the ditches so that the underground water does not become stagnant and depleted of oxygen. However, open drainage ditches prevent cross-cultivation and hamper many other farming operations. If it is physically and economically feasible, permanent drainage tiles should be installed underground to remove excess water and to lower the water table. Drainage water containing salts should not be placed in an evaporation pond where salts such as selenium could and have become concentrated to a toxic level.

Annual Precipitation and Underground Water Supply

Some of the world's best fruit-growing districts are located where trees are irrigated during the summer months. Adequate annual precipitation of rain and/or snow is essential to replenish irrigation water taken from rivers and canals or pumped from wells. In areas where irrigation is routinely practiced or where summer rainfall supplies the needs of trees, occasional droughts are disastrous.

Topography

A steep terrain makes orcharding difficult and expensive; mechanization is not possible and soil erosion is always a constant danger, especially if the planting is irrigated. If the land is not steep, the following questions need to be answered. What should be planted on the north slope which is usually cooler in the spring than the south slope? Should the land be terraced? If the orchard is to be irrigated, will it be by contour basins, sprinklers, or drip lines? Are there low spots where cold air might gather in the spring to pose a frost hazard?

If the land is terraced, the edges should be raised to prevent erosion. The use of sod cover with a thick fibrous root system will hold the topsoil and prevent erosion and formation of gullies.

If the land is contoured for planting, the planting distance and direction of the rows may be determined by the lay of the land. It may not be possible to orient hedgerows in the north–south direction to take advantage of sunlight. If there are low spots where cold air tends to accumulate, species and cultivars that have cold tolerance, require a lengthy chilling period, and bloom after the danger of frost is past should be chosen.

History of the Planting Site

A documented crop history of the planting site would be helpful in making farming decisions. For instance, if crops such as tomato and bean that favor a buildup of *Verticillium* spp. have been grown there and almond trees are to be planted, the land may need to be fumigated, for the organism readily attacks almond trees. If the site had been an orchard, a change in tree species or at least in rootstock is recommended to solve the replant problem.

Replant or Soil Sickness Problem. The replant or soil sickness problem is a soil–plant phenomenon. When the same rootstock species is planted in succession in the same orchard, the second generation of trees grows poorly. Replanted trees are usually stunted for the first growing season, but in some areas the trees fare poorly for several seasons. The problem is not limited to fruit trees alone but is common in other agronomic and ornamental crops such as alfalfa and the rose.

Constituents in the root exudate are suspected as a factor in the phenomenon. When excess irrigation water collected from one potted plant is applied to a second one, its growth is often inhibited. But the replant problem in orchards can be ameliorated or avoided by soil fumigation (see the next section, Preparation of the Site). Thus, the problem cannot be ascribed entirely to toxic residues.

One means of overcoming the replant problem is to plant trees grafted on a different rootstock species, a form of crop rotation. For example, if a grower removes a peach orchard of Elberta/Nemaguard and wishes to replant with the same scion cultivar, he or she could plant the Elberta/apricot combination.

PREPARATION OF THE SITE

Leveling. Leveling is essential if the planting is to be irrigated by furrows or by basins. One danger of leveling naturally uneven land is that the best topsoil is usually scraped to fill in low spots. Trees in the scraped area do not grow well because of the low fertility of the soil, and trees in the filled-in areas often suffer from a high water table during the winter months.

Chiseling and Back-Hoeing. Water penetration downward is generally slow in these landfills because they are often former creek beds or drainage channels. Before these areas are filled with soil, they should be ripped with a deep chisel in order to improve drainage. Similarly, areas known to have a hardpan or plow sole, which impairs water penetration into the subsoil, should be chiseled in two directions. A hardpan is a layer of soil compacted through the natural concretion of soil particles; a plow sole is a layer of earth at the bottom of a furrow compacted by repeated plowing to the same depth, especially when the soil is wet. Back-hoeing individual tree-planting sites mixes the soil during digging and refilling operations so that the profile becomes more uniform; any hardpan or plow sole that may have been present is now destroyed, reducing a potential problem with perched water tables.

Preplanting Fumigation. If a site had previously been planted with fruit trees, the grower should have the soil tested for the presence of nematodes and other microorganisms harmful to trees. If the analysis shows the existence of harmful microorganisms such as root-knot nematodes or oak root fungus, the grower should have the land fumigated by a licensed pesticide applicator.

Note: Be sure to post the property with adequate warning signs because soil fumigants such as methyl bromide and chloropicrin are highly toxic.

Soil–Plant–Water Relationships

IRRIGATION PRINCIPLES

The water-holding capacity of a soil is not as important a criterion as available water in the soil–water–plant relationship (Veihmeyer and Hendrickson, 1950). Available water is the amount of water held by soils that is extractable by plant roots or the difference between *field capacity* (FC) and *permanent wilting percentage* (PWP).

Field capacity is the amount of water held by soils against the pull of gravity measured several days after an irrigation. As water is extracted by roots, the soil becomes drier and eventually reaches a moisture potential at which point roots are no longer able to absorb sufficient water and the plant wilts. If the plant remains wilted until reirrigated, the soil is at its permanent wilting percentage. The soil moisture potential at PWP is approximately -14 bars.

Field capacity and permanent wilting percentage are soil characteristics, which are expressed as a percent of oven-dried soil. Clay soils have the largest field capacity, whereas sandy soils have the smallest (Fig. 7.1). Moisture depletion curves often show that field capacities decrease with increasing soil depths in the same orchard (Fig. 7.2). The finer soil texture in the upper stratum or topsoil helps it to retain water. In most soil profiles, the texture becomes coarser with depth. Furthermore, the upper stratum of the soil con-

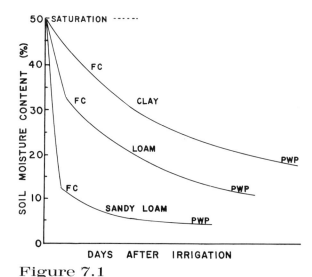

Figure 7.1

Patterns of moisture extraction by trees growing in three different soil types. FC and PWP are field capacity and permanent wilting percentage, respectively.

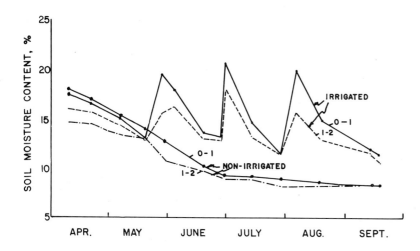

Figure 7.2
Moisture extraction patterns at soil depths of 0 to 1 meter and between 1 and 2 meters in a prune orchard irrigated three times compared to those in a nonirrigated one. (From Veihmeyer and Hendrickson, 1950)

tains more organic materials, which tend to retain water, than do soils at the lower strata.

It is difficult to determine the permanent wilting percentage for soils containing considerable clay because (1) roots are unable to explore the entire soil mass; (2) water penetration rate is so slow that the anaerobic condition injures or kills the fine roots; and (3) deep cracks form as moisture is extracted so that some water is lost by evaporation from surfaces of these fissures as well as by transpiration via plants.

Shoot and fruit growths are impaired long before soil moisture tension approaches PWP, so that water is not equally and readily available to roots between FC and PWP. Water stress within a plant increases as the soil becomes dry. The moisture potential of orchard soils should be about −0.5 bar for maximum shoot and fruit growth. To compensate for the time required for the water to drain to the root zone, growers should irrigate when the soil moisture potential reaches −0.4 bar (Univ. Calif. Leaflet 21212, 1981; Fig. 7.3).

Soil texture or structure is also a function of the exchangeable cations present on the colloidal clay particles. Soils containing a large amount of exchangeable calcium ions tend to form aggregates or "crumbs." Water drains rapidly through these aggregates; this action pulls air into the soil, improving aeration of the root zone. Contrarily, soils that have large amounts of exchangeable sodium ions tend to disperse or deflocculate, so that water penetration is impeded (Univ. Calif. Leaflet 2280, 1978).

How the water is delivered to the root zone is not as critical to the plant as how much water, when, and how often it is applied. Of the various irrigation systems devised, there is no one best way; each has its advantages and disadvantages.

IRRIGATION SYSTEMS

Irrigation systems are classified as (1) gravity flow or surface, (2) sprinkler, and (3) drip or trickle. Gravity flow systems of various designs have been used since human beings began to cultivate plants. Water catchment basins and irrigation systems dating back to the pre-Christian period in the Middle East and the pre-Columbian era in South America have been uncovered by archaeologists, indicating that prehistoric people recognized the water requirements of plants. From ancient times, Asians cultivated rice in paddies on hillside terraces. Rice paddies were flooded, and the surplus water was allowed to descend by gravity flow from higher planting sites to lower sites. Roman aqueducts and stone-held terraces constructed in the fourth century are still evident

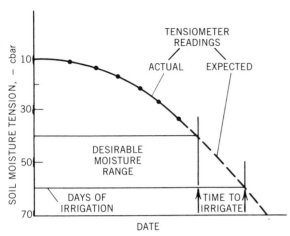

Figure 7.3
Irrigation schedule based on the moisture extraction rate, using the tensionometer. The orchard is irrigated each time the soil moisture tension is expected to reach minus 40 centibars. (Adapted from Univ. Calif. Leaflet 21212, 1981 Fig. 27)

in various parts of southern Europe. These remains attest to the important role that water played in the expansion of the Roman Empire.

Gravity Flow Systems

The gravity flow system is widely used where the terrain is relatively level. It is the cheapest system to install but may be more expensive to maintain than a sprinkler or drip system. The gravity flow system uses square and contour basins or small, narrow ditches or furrows. Water is delivered into basins from a large ditch by syphon tubes and into furrows through gated aluminum pipes.

Basin Method. In a basin system levees or ridges must be formed and then taken down after one or two irrigations, for many soils crack and shrink as they dry. These cracks cause breakages in the levees, so that water seeps into adjacent basins instead of flooding the intended one. Ridges must also be taken down and the land leveled again to allow equipment access for cultivation, spraying, and harvesting. Some of the disadvantages are (1) the cost of operating the tractor to erect and take down the ridges, (2) soil compaction by constant use of heavy equipment, and (3) exclusion of a permanent sod.

The basin system requires a large volume or head of water. The amount of water to apply depends on soil type, basin size, ridge height, and the distance the water must flow. If a square basin is filled rapidly, the ridges must be high enough to hold about 20 centimeters (8 inches) of water. This amount of water will moisten an average sandy loam soil to a depth of about 1.5 meters (5 feet).

Furrow or Strip Irrigation. The land must be leveled and graded for furrow or strip irrigation. For strip irrigation, planting trees on berms, narrow, raised beds, or on levees, so that tree crowns are above the high-water level, provides protection from crown rot organisms. With furrow irrigation, the number of furrows and the distance between them within tree rows depend on soil texture. Furrows must be spaced closer in sandy soils because water does not move as far laterally as it does in heavy clay soils (Fig. 7.4).

Furrows are filled by siphons or gated pipes (Fig. 7.5). To ensure that the soil under the entire length of the furrows is moistened to a uniform depth, growers (1) control the flow of the stream or (2) adjust the length of the furrows so that the water reaches the lower end between 20 to 25 percent of the irrigation time. That is, if the water reaches the far end in one hour, it should be shut off after running four to five hours.

Sprinkler Irrigation

Sprinkler systems are used in hilly areas where leveling or terracing is not feasible. Soil erosion, which is always a management problem in hilly areas, can be minimized under this system. Moveable sprinkler systems have heads set in rigid aluminum pipes equipped with quick connectors or spaced on flexible plastic hoses which can be dragged through orchards. Fixed sprinkler heads with underground water delivery systems are

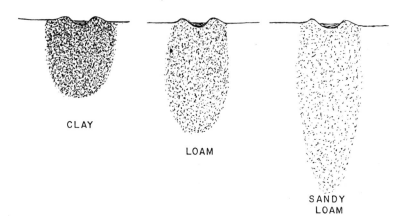

Figure 7.4
Soil profiles depicting the width and depth of wetness of clay, loamy, and sandy loam soils under furrow irrigation, each receiving the same amount of water.

more expensive to install initially, but their operation requires less manual labor than do moveable systems. A disadvantage of fixed heads is their vulnerability to breakage by passing equipment.

Drip or Trickle System

Drip irrigation systems are used in both level and hilly terrains. The principle of drip irrigation is to moisten the portion of the soil mass occupied by most of the tree roots. Roots growing outside of this wet zone will sparingly absorb soil moisture remaining from winter precipitation. Ideally, a drip system delivers just enough water daily to replenish that lost by transpiration and evaporation from the soil surface. Hence, drip or trickle systems are engineered to deliver small amounts of water at frequent intervals (Fig. 7.6).

The amount of water to apply per irrigation is estimated by considering the amount that evaporates during the day from a given surface area. This value is determined from a U.S. Weather Bureau Class A pan, yielding the evaporation rate, EP. The crop coefficient is the ratio of the evapotranspiration rate to the evaporation rate, ET/EP. This constant is difficult to derive because it depends on many variables, such as ambient temperature and humidity, wind velocity, the percentage of ground area shaded by the growing plant, the age of the plant, and the openness and height of the canopy. The ratio stabilizes

Figure 7.5
Gated aluminum pipes used to deliver water into separate furrows in a prune orchard.

Figure 7.6
Drip irrigation system with emitter suspended from a trellis wire in a kiwifruit planting.

Soil–Plant–Water Relationships

for a mature orchard but varies during the growing season as the canopy closes over and the weather changes. ET/EP ratios under California conditions vary from a low of 0.45 in March to a maximum of 0.75 during July and August. EP values for several consecutive days are published in newspapers or are broadcasted daily in large farm communities.

The number of hours a drip irrigation system must operate per day is computed by substituting the evapotranspiration rate, ET, for the particular part of the season (Fig. 7.7) in the following equation:

$$\text{Hours/day} = \frac{\text{tree spacing} \times 0.6234 \times \text{ET} \times \text{crop coefficient}}{\text{number of emitters} \times \text{flow rate} \times \text{irrigation efficiency}}$$

Tree spacing is in square feet, and the flow rate is in gallons per hour; the constant, 0.6234, converts square feet-inch into gallons. (Substituting equivalent metric values into the equation yields the same answer.) When ET is used, the crop coefficient is unity for an orchard with a full canopy or for one in which the canopy covers 80 percent of the orchard floor and a sod covers 20 percent.

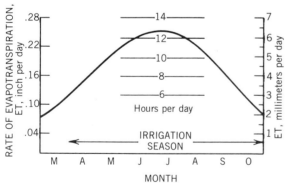

Figure 7.7
The mean evapotranspiration rates, ET, in almond orchards of the Sacramento Valley during an irrigation season on which the hours of operating a drip system are based. (See text for computations.)

Irrigation efficiency is unity for a drip system that is properly designed and well maintained (Univ. Calif. Leaflet 21259, 1981).

A drip system conserves water when the trees are small because it delivers only what is needed. But as the trees grow and the leaf area increases, water usage becomes equal to that of other systems. This eventual increase in water usage should be considered in planning the delivery system.

Precautionary Instruction for Sprinkler and Drip Systems

With the sprinkler and drip systems, the rate at which water is applied must be slower than the infiltration rate. This prevents puddling of water about the trunk; avoiding free water diminishes the possible invasion of trees by microorganisms. These systems require filters at the water inlet, which should be checked regularly to see that nozzles and emitters are not clogged with debris. Clogging of the system causes water stress and eventually death of trees.

Over-the-tree sprinkler systems may be used to apply fertilizers and pesticides and for frost protection. But care must be exercised in the selection of materials, for some chemicals corrode metals in the sprinkler head; others react with plastic pipes and connections. Hence, delivery lines should be flushed after each chemical application. A minimal amount of water should be used after a pesticide application so that the freshly applied material is not removed from plant parts. All pipelines should be flushed with a dilute solution of sodium hypochlorite (household bleach) at the beginning of the season to eliminate any microorganisms and sediment that might have accumulated. At the end of the irrigation season the system should be emptied of water to avoid damage from freezing.

Some Hazards of Perennially Irrigated and Cultivated Soils

Depending on the salt content of the water, plants may absorb water faster than the dissolved minerals. In areas where flood irrigation is practiced or where summer rains are prevalent, salt residues are leached beyond the root zone. With a sprinkler or drip irrigation system that delivers a given volume of water at each irrigation cycle, salts could accumulate at the junction of the wet and dry zones. This salinity problem can be remedied by flooding the area periodically to dilute the salt concentration and to leach the salts below the root zone.

The structure of soil under intensive cultivation with heavy equipment is inevitably broken down. Even a sandy soil can be compacted by cultivating or traversing it while the soil is still moist. A dense soil inhibits root growth, resists water penetration, and becomes depleted of oxygen. Several cultural practices minimize structural breakdown: (1) cultivate the soil when it is neither too moist nor too dry; (2) avoid tillage as much as possible, especially after loosening the soil; (3) apply herbicides around tree trunks; (4) maintain a sod or cover crop with long, fibrous tap roots to open up pore spaces among soil aggregates (see Sod Culture later in this chapter); and (5) turn under crop residues whenever possible to build up the soil organic matter.

Orchard Design

PLANTING SYSTEMS

During the planning stage of an orchard, the following factors should be considered in selecting a planting system: (1) the need for cross-pollinizers; (2) the ultimate size of trees, with a schedule for removing designated temporary trees to avoid overcrowding; (3) the direction of the flow of irrigation water; (4) the placement of permanent sprinkler heads and drip irrigation lines; and (5) whether the crops will be harvested mechanically or manually. Some common planting systems are square, rectangular, quincunx, hexagonal, alternate (Fig. 7.8), and single- and multiple-row hedgerows (Fig. 7.9).

Within each system, planting distances depend on the final size of mature trees, which in turn is a function of the growth potential of the scion–rootstock combination of the species. A hectare of land will accommodate only 55 trees of the terminal bearing walnut cultivar Franquette, spaced 13.5 meters (45 feet) apart, whereas the same area can be planted with 445 trees of the lateral bearing cultivar Chico, spaced 6.6×3.3 meters (22×11 feet) apart and not be crowded. Spur-type apples on dwarfing rootstocks can be planted as densely as 2800 trees per hectare (1130 trees per acre) (Wertheim, 1985).

No matter what planting system is used, a good contour map should be made indicating the locations of pollinizers and irrigation lines, if appropriate, and changes in soil type and depth.

Square and Rectangular

Square and rectangular planting systems allow cross-cultivation. The land must be nearly level if the orchard is to be irrigated by furrows or basins. If the trees become too large and begin to cast shadows on one another, alternate rows in both directions can be removed in order to provide space for the canopies of remaining trees to enlarge. If pollinizers are necessary, they should be planted in permanent tree positions.

Quincunx

In the quincunx system, four trees are planted on the square and one in the middle, leaving a diagonal avenue for cultivating, spraying, and harvesting the crop. If pollinizers are needed, every third tree in every third diagonal row should be so designated. If the trees become crowded, the middle tree can be headed back for a few years and eventually removed. This results in a nine-tree square planting, with a pollinizer in the middle.

Hexagonal

The purpose of the hexagonal system is to have all trees equidistant from each other. This planting scheme does not lend itself to systematic tree removal if and when trees begin to crowd and shade each other. Eliminating trees in any one direction destroys the initial objective of having trees

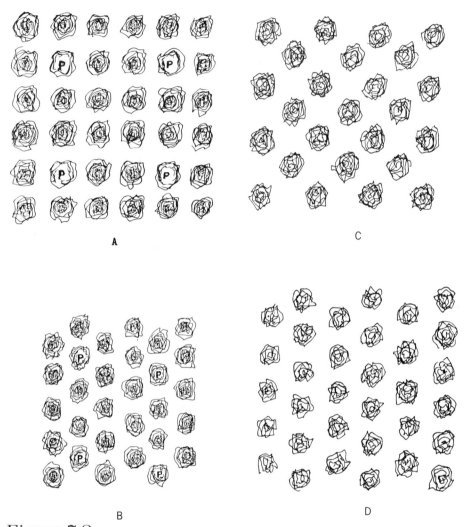

Figure 7.8
Planting systems: (A) square, (B) quincunx, (C) hexagonal, and (D) alternate. The letter P denotes where pollinizers might be placed.

equally spaced and complicates the placement of pollinizers and other cultural operations. Thus, with this system, scion–rootstock combinations whose size can be controlled for a long time should be planted.

Alternate

In an alternate scheme, trees are planted on the square or rectangle with an extra tree within the row. This pattern is alternated with another set of trees planted at the same distances but offset by half the distance between trees. Initially, the scheme appears to be a hexagonal planting. If trees become crowded and their size cannot be controlled, trees in alternate rows can be removed. Tree density is reduced to half; the planting system is altered to a rectangular or hedgerow system. If there is further crowding and alternate trees within rows are eliminated, the planting will become a square one.

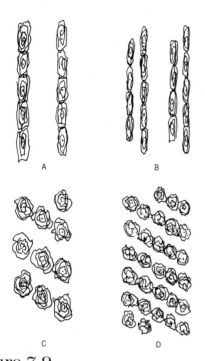

Figure 7.9
Planting schemes for (A) a single-row hedgerow and (B) a double-row hedgerow; and (C) three-row and (D) five-row high-density plantings with picking aisles.

Hedgerow System

In the hedgerow system, the distance between trees in a row is one-half to one-third the distance between rows. In the foothills of the Alps near Bolzano, Italy, where light is intense, apple trees on dwarfing stocks are planted in double offset rows. That is, two rows are planted 1.3 meters apart, separated by an alleyway just wide enough to allow passage of equipment. If the topography permits, the rows are planted in the north–south orientation so that the trees are exposed to morning and afternoon sunlight. The tree height is kept one meter higher than the row is wide so that shade from one row is cast to the next row only in the early morning and late afternoon. At lower latitudes, row orientation is less important because the path of the sun is such that both the north and south sides of trees receive adequate light during the summer. The individual trees in the hedgerow systems are trained to various free-standing or trellis systems (see Training Systems, Ch. 9). High-density plantings with as many as 2800 apple trees per hectare are becoming popular because (1) excellent cultivars on size-controlling stocks are available, (2) cultural practices can be mechanized, and (3) early production brings quick returns on investments. In these planting systems, alleyways just wide enough to accommodate spray equipment and harvesting trailers are maintained between double or triple rows of trees. Five-row plantings with a 0.5-meter lateral pathway are being tried, but unless the spray application is thorough, the middle trees may not receive good coverage even if trees are trained to a slender spindle (Wertheim, 1985). Trees in high-density planting systems are designed to maintain a good light environment in the canopy and to promote the formation of flower buds and fruit color.

Contour Planting

If the land is hilly and leveling is not feasible, the land should be surveyed and terraces constructed following the natural contour as much as possible. The edges of the terraces should be slightly higher than the side against the hill so that the irrigation water will remain in the terraces. The grower should take into consideration (1) the source of irrigation water and how it will be applied; (2) the width of the avenue needed for equipment to mow the sod, to apply fungicides and herbicides, and to carry out other cultural practices; (3) drainage of water from one terrace to the one below; and (4) the method of harvesting the crop. The topography and shapes of the terraces will dictate the planting distances between trees. As irregular as the planting scheme may be, trees should be positioned near the edge of the terrace so that irrigation or rain water will not stand around tree trunks. Trees are irrigated by using sprinklers or by filling the highest terrace and allowing the surplus water to drain to successively lower terraces. Sod culture should help minimize problems of soil erosion and water runoff, which are associated with such a terrain.

TREE PLANTING

Planting a tree is a simple task, but if it is not done properly, the tree may grow poorly or not at all. Trees in other than the contour system are planted on a berm in a straight line by stretching a wire and staking out tree positions. A planting board (Fig. 7.10) is used to locate the exact position of the tree. A hole, slightly larger and deeper than necessary to accommodate the roots, is dug either manually or mechanically.

The soil should not be so wet that the inside of the hole is glazed by the mechanical drill blade, nor should the soil stick to the shovel; the soil should be friable. Some topsoil is returned to the pit and mounded. The trunk is placed in the middle notch of the planting board and held about 5 centimeters higher than its former level in the nursery. After the roots are spread over the mound in the pit, the hole is gradually filled with more topsoil. A shovel handle can be used to force the soil between roots and to eliminate large air pockets. Roots are covered completely with soil and a small basin is made; the tree is watered immediately to settle the soil around the roots.

A more rapid means of preparing an orchard for planting is to use a subsoiler to chisel the ground along a stretched wire in two directions, forming a grid. A hole is excavated at the intersection and a tree is planted. Another method is to use a mechanical transplanter that spaces trees at desired intervals in a furrow made by a plow and covers the roots immediately with a set of disks.

A week or ten days after planting the tree, additional soil is mounded about the trunk slightly higher than the surrounding soil surface to promote runoff of water in case of rain. This precaution is taken to prevent free water from standing about the trunk, which leads to an anaerobic condition and predisposes the tree to attack by the crown rot organism *Phytophthora* or other soil organisms.

In areas where rabbits and other rodents are likely to damage newly planted trees by eating their buds and bark, basal shoots on the trunk should be removed. A tree guard—a stiff cardboard or plastic sleeve—is slipped around the trunk to deter predators. Tree guards should be loose and slotted at the bottom to allow air movement and to prevent rain or irrigation water from being trapped. If rodents are not a problem but sunburning is, the trunks should be painted with white latex paint or whitewash (calcium carbonate mixed with some adhesive).

POLLINIZERS, THEIR SELECTION AND PLACEMENT

Bases for Selection

Pollinizers fall into several categories by their specific traits and cultural requirements. They may be (1) wind- or insect-pollinated; (2) dioecious, as in kiwifruit and pistachio, which need male plants to supply pollen, or monoecious; (3) wholly or partially *self-unfruitful*, as in sweet cherries, almonds, chestnuts, and some apple cultivars; and (4) dichogamous but *self-fruitful*.

Besides these genetic traits, several other factors should be considered in selecting pollinizers.

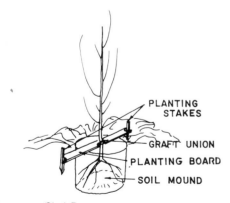

Figure 7.10
Planting a grafted tree using a planting board and stakes. Notice that the roots are spread over a mound before covering with soil.

They should (1) be cross-compatible with the main cultivar being grown; (2) form an abundance of flowers annually that bloom simultaneously with the main cultivar; and (3) if all other factors are equal, bear a marketable crop. Although their fruits may be worthless, crab apples, owing to their cross-compatibility, profuse blossoms, and abundance of pollen, are interplanted as pollinizers for apples.

Additional factors related to cultural practices and marketing dictate the choice of pollinizers.

Harvesting and Marketing of Mixed Cultivars. If the market will accept mixed cultivars, the harvested fruit of the main and pollinizer varieties should be similar in appearance and mature at the same time. If the fruits of the main and the pollinizer cultivars do not mature at the same time, the two should be planted in separate rows to maximize the potential for cross-pollination while facilitating pest control and harvesting. Irrigating such an orchard poses a problem because water must be applied well ahead of the harvest period of the first cultivar so that the soil will be dry and yet not cause the second cultivar to undergo water stress before the next irrigation.

If the two cultivars require applications of special pesticide or growth regulators, as apples do, care should be taken to minimize cross-contamination and ensure that the residue level is below legal tolerance limits.

Means of Harvesting. If the main cultivar matures at a different time from the pollinizer and the crop is to be mechanically harvested as in nuts, it is preferable to plant the main cultivar and pollinizers in separate rows, so that the crops can be picked up separately as they mature. The number of rows of each cultivar depends on the species (see the pollinizer ratios given later). If the harvest is handpicked by a person on a ladder, and the cultivars are easily distinguishable, trees of the main and pollinizer cultivars can be alternated and there will be little likelihood of mix-ups.

The Direction of the Prevailing Winds. The direction of the prevailing winds is important in wind-pollinated crops, such as the walnut, chestnut, pecan, and filbert. If winds are from either the north or the south, the pollinizer rows are oriented in the east–west direction.

Ratio of Pollinizers to the Main Crop and Their Placement

If the value of the pollinizer crop and its maintenance costs are equal to that of the main one, their ratio can be 1:1. This is often not the case; thus, the ratio of the pollinizer to the main crop is minimized but is kept high enough to maximize the yield. The ratio of male to female kiwifruit and pistachio plants is usually 1:8; in some pistachio plantings the ratios are 1:11 or 1:15. Similar ratios are encountered in sweet cherry orchards where Black Tartarian and Chapman are used as pollinizers for Bing trees. These pollinizers are the first cherries to ripen, about ten days before Bing. The fruits are tasty, but their pulp is soft and their shelf life short. They can be shipped to distant markets by airplane but only a short distance by truck or railcar. Their sales diminish rapidly after firm-fleshed cultivars begin to arrive on the market. For these reasons, the number of pollinizers should be kept at a minimum.

For marketing purposes, almonds are mechanically harvested separately by cultivars. Thus, three to four rows of Nonpareil, the most important commercial cultivar in California, are interplanted with a row of early-blooming Ne Plus Ultra or Peerless on one side and a row of Mission or Thompson, which blooms after Nonpareil, on the other. This planting scheme assures a maximum overlap of pollinizers with Nonpareil.

Walnuts are harvested mechanically, processed, and graded separately by cultivars unless they are indistinguishable. Commonly, eight to twelve rows of the main cultivar are planted between two rows of pollinizers. Such a planting allows cultivars to be harvested as they mature with the least possibility of a mix-up.

Vegetation Management

CLEAN CULTIVATION

In a clean-cultivated orchard or vineyard, the ground is kept bare of weeds so that they (1) do not compete with trees and vines for nutrients and water; and (2) do not predispose trunks to invasions by soil organisms that thrive in a humid environment created by weeds. After each irrigation, the planting must normally be cultivated to eliminate the newly germinated weed seedlings and to prepare the land for the next irrigation. This means six or more diskings during the growing season under California conditions. Weeds growing adjacent to trunks, especially of young trees, should be hoed or sprayed with a herbicide. The material should not be allowed to drift onto the foliage or penetrate the thin bark of small, young trees. After their bark thickens, certain herbicides can be used safely.

The practice of pulverizing and tamping the soil after each cultivation does not reduce the rate at which water evaporates from the soil surface. It may create dust which covers the leaves, reducing their photosynthetic activity and predisposing them to an invasion by mites. The fruits may also become laden with dust.

SOD CULTURE

The use of sod is recommended in areas where water is relatively abundant and inexpensive, and in hilly areas where soils are shallow and subject to erosion. The leaves of the sod cover intercept solar radiation and keep the soil cool, thus enabling the tree roots to grow near the surface and to explore the entire soil mass. The roots of the cover crop hold the soil in place and hasten the infiltration rate. But the presence of a thatch made up of the clippings that remain after the sod is mowed reduces the sealing of the soil surface by clay particles, a problem known as puddling.

Besides mowing, an orchard with a sod cover must be irrigated and fertilized more frequently than a clean-cultivated orchard. However, mowing can be done with a light, wheeled tractor rather than with the heavy track-laying equipment that is often required for plowing and disking. Thus, sod culture reduces the probability of compacting soil, keeps the surrounding area cooler, and lessens the dust problem as compared to orchards under clean cultivation.

In addition to the extra water and fertilizer required, sod has other disadvantages. (1) Frost damage may be higher under sod culture than clean cultivation because the sod cover intercepts solar radiation during the day and prevents the soil from absorbing heat. (2) Humidity builds up, increasing foliar and fruit diseases and crown rot of trunks. (3) Certain sod species may harbor insect pests, injurious to the tree and its crop. (4) Sod may interfere with mechanical harvesting of nut crops from the orchard floor unless vegetation is cut short. (5) If the cover crop is not mowed during the period the trees are in bloom, bees may be attracted to the blossoms of the ground cover rather than those of the fruit trees. (6) If the shade of the tree canopies begins to cover the entire orchard floor, maintaining a permanent sod may become difficult.

Vegetation should be kept away from tree trunks to prevent possible infection by crown-rotting organisms and to eliminate cover for voles and field mice. Maintaining weed-free strips, 1.0 to 1.5 meters wide, down the tree rows reduces these hazards and renders the sprinklers or drip lines visible so that they will not be damaged.

Protection Against Cold Injury

WINTER INJURY AND ACQUIRED COLD HARDINESS

Winter injury occurs when the temperature drops below freezing and various parts of the tree have not become acclimated or acquired cold hardiness. There are three stages of acclimation

(Weiser, 1970). In the fall when the days become short and the temperatures gradually decrease, plants enter the first stage. Shoot growth ceases, and the tissues mature or harden. Any cultural practice that favors invigoration, such as fertilization with large amounts of nitrogen, delays the onset of this stage. As nights become frosty and cold temperatures prevail, plants enter the second stage of acclimation, which is the basic level of hardiness, and become able to withstand normal freezing temperatures. Plants do not acquire this level of hardiness if they are defoliated early, or mature the crop late in the season, or bear an excessively heavy crop. Plants enter the third stage of acclimation or hardiness after the tissues become frozen. Cells are now able to withstand extremely cold temperatures because water becomes bound to the macromolecules in the cytoplasm, including starch grains, and resists freezing. Some species growing in the boreal region of Canada withstand temperatures of about $-45°$ C $(-50°$ F). At maximum hardiness, dormant buds of most temperate zone fruit trees such as the apple, sweet cherry, and peach can withstand temperatures of $-29°$ C $(-20°$ F) (Proebsting, 1970). Pear, pecan, and walnut are less hardy; the vegetative tissues of these species may withstand $-20°$ C.

If the temperature should rise and the tissues thaw during the third stage, the level of hardiness of tissues reverts to the second stage. Tissues must then undergo the hardening stage again to withstand extreme cold.

As buds swell in late spring, they lose their hardiness and now can tolerate only $-18°$ C $(0°$ F). At pink bud stage when the petals begin to show, they become vulnerable to temperatures of $-3.3°$ C $(26°$ F).

The extent of damage incurred by freezing depends on the rate at which the temperature decreases, the minimum temperature, and its duration. If the temperature drops gradually over a course of hours, tissues become acclimated because water is exported from the vacuoles to the cell wall and intercellular air spaces. The osmotic concentration of the vacuole increases so that the freezing point of the vacuolar content is depressed. An ice nucleus may form on the surface of floral parts or a leaf through physical agitation or often by flagellate, ice-nucleating bacteria (see later). The water in the tissues is often supercooled from -3 to $-6°$ C before it freezes. Once ice forms on the surface, freezing spreads rapidly to the interior tissues because water is contiguous. The water in the cell wall, however, being low in solutes, freezes before that in vacuoles. Ice formed in these cell wall spaces does little or no damage because the cytoplasm is not injured and its functions are not disrupted. When the temperature drops rapidly, cells are not able to export water quickly enough and freezing occurs. Ice forms in the vacuole because the osmotic concentration is relatively low. The sharp edges of the ice crystals rupture the cell membranes, disorganizing the cell and killing it.

Freezing of the supercooled liquid on the plant surfaces, in intercellular spaces, and in cell walls causes the first exotherm, a rise in temperature. As the tissue temperature continues to decrease and additional water freezes, more heat is evolved so that the rate of cooling is delayed. A second exotherm occurs between $-25°$ and $-30°$ C in stem tissue and is associated with freezing of water in the protoplasm.

FACTORS FAVORING FROST

Certain weather conditions and topography are conducive to the formation of frost. The frequency of spring frosts and, to a lesser extent, fall frosts limits the areas where fruits might be commercially grown.

Weather Conditions. Dew point is the temperature at which water vapor condenses and forms dew as the atmosphere cools. Dew point is higher when the humidity is high. White frost forms when the temperature drops to $0°$ C and the dew freezes. Black frost occurs if the air is dry and the dew point is below freezing; no visible ice will form, but plant cells could still be damaged.

A cold, clear night following a cold day is an

ideal situation for forming frost because the earth absorbs little heat during the day. The small amount of stored heat is radiated back to the dark sky beginning at nightfall, and continuing until sunrise. Thus, minimum temperature is usually reached just before daybreak.

Since all living bodies, as well as inanimate ones, radiate heat to a colder surrounding, and conversely, cold bodies absorb heat from warmer ones, floral parts will radiate heat to the sky at night and simultaneously absorb heat radiating from the soil. Thus, the pistils of downward facing flowers are less likely to be damaged than those facing upward.

Small flower parts and shoot tips radiate heat quickly so that their temperatures may be lower than the ambient air. If there is a slight breeze, the warmer ambient air contacting the plant parts provides some degree of protection. During this period the pistil and its enclosed ovules are extremely sensitive to cold because they have small heat capacities. The cells of these organs are dividing and metabolizing rapidly. Thus, their soluble solids content is low and their moisture content high. At this stage the ovule is killed at $-1°$ to $-2.2°$ C ($30°$ to $28°$ F).

Topographical Features. Cold air is denser than warm air so that it tends to flow downward, filling shallow depressions and small valleys. Low branches and brush along fence lines that impede air movement through an orchard should be removed to provide good air drainage. In areas behind earth-filled, elevated highways and depressions or pockets where cold air collects, cold-tolerant species that bloom late in the spring may be planted.

Some geographical features that temper the climate to make fruit growing possible are the Pacific and Atlantic oceans, the Great Lakes, the Sierra Nevada and Cascade mountain ranges, and the Cordillera of the Andes. Large bodies of water store considerable heat so that the air temperature on the adjacent landmass remains relatively stable. High mountain ranges tend to block the movement of prevailing winds, thus creating microclimates favorable for cultivation of fruit trees.

MEANS OF REDUCING FROST DAMAGE

If a large mass of polar air moves into fruit-growing regions of the Northern or Southern Hemisphere, there is no way to avoid frost damage or winter injury. If conditions for frost are local, however, damage may be reduced in several ways, by (1) irrigating, (2) orchard heaters, (3) wind machines (Gerber, 1970), (4) promoting heat absorption by the soil through clean cultivation, (5) providing good air drainage, and (6) preventing bacterial ice nucleation with antibiotics and antagonistic bacteria (Lindow et al. 1978). The extent of frost damage may be diminished by delaying budbreak with growth retardants and evaporative cooling.

Irrigation

Basin and Furrow Irrigation. Since water gives off heat when it cools or freezes, surface-irrigating the orchard when a frost warning is issued is a simple way of avoiding damage. This method has several disadvantages. (1) If the temperature falls too low, the irrigation water will freeze. (2) If the freezing temperature persists for several days and irrigation is continued, the soil becomes waterlogged and the trees may die. (3) Should the trees survive, other cultural operations such as the application of bloom sprays will be hampered by the wet soil.

Overhead Sprinkle Irrigation. Sprinklers have been used to reduce frost damage. The principle underlying their use is that water releases 80 calories per gram (heat of crystallization) upon freezing. A mixture of water and ice (slush) remains at $0°$ C. At this temperature, flowers and leaves covered with slush do not freeze because the soluble solids depress the freezing point of the vacuolar solution. Enough water must be delivered to cover the plants with slush while the air temperature is below freezing. Thus, if the temperature drops to −3.3° C (26° F), an application of 2.5 millimeters of water (0.1 inch) per hour is needed to protect plants. If the freeze is accompanied by a slight breeze, more water is necessary to compensate for additional chilling created by evaporative cooling.

Some disadvantages of sprinklers are similar to those for surface-irrigating the orchard. If sprinkling is continued during extended freezing weather, the soil will become saturated. Such flooding may have long-lasting effects detrimental to plants, but heavy formation of ice will immediately break limbs. Sprinklers are fixed as to their volume and speed of rotation so that they may not be able to deliver the volume of water required. If the speed of rotation is slow, the slush turns to solid ice and its temperature will drop. Chilling damage may occur before additional water is added by the sprinkler to recreate a slush and elevate the temperature (Fig. 7.11). Clogging or icing over of nozzles is a common problem; water lines and nozzles should be checked frequently to make sure that the system is functioning.

Heaters

Orchard Stack Heaters. Portable heaters (Fig. 7.12) which burn fuel oil radiate heat to the

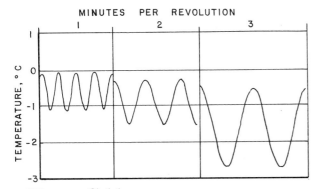

Figure 7.11
Frequency and magnitude of temperature fluctuations in plant tissues as a function of the rate of water delivery by sprinklers.

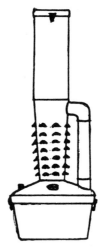

Figure 7.12
An orchard heater with a return stack through which unburned gases are recycled, thus increasing its efficiency.

surrounding trees. Between 10 to 20 percent of the heat is radiant heat, which travels in a straight line. The efficiency of the heat decreases proportionally to the square of the distance; 80 to 90 percent of the heat generated rises above the tree top by convection current and is dissipated to the atmosphere. A return-stack heater recycles unburned gas. For return-stack heaters to be effective, one burning 1 gallon per hour is required for two trees or about 100 per hectare (Gerber, 1970). Under most conditions, this number of heaters has the capacity to raise the ambient temperature about 3° C, and, if there is neither wind nor an intense low inversion, the temperature can rise as much as 10° C.

More heaters should be placed at the periphery of an orchard. After the frost alarm goes off, heaters along the periphery are lit first. This assures uniformity of heat distribution. If heaters in the middle of the orchard are lit first, the convection currents will initiate a chimney effect and draw cold air from the surrounding areas into the center of the orchard.

The high cost of heaters, oil, and storage tanks and the need to have a standby crew are two disadvantages of heaters. When the frost hazard is over, the heaters must be collected, emptied, and stored in a safe place.

Liquid Propane, Natural Gas, and Oil Burners. Sophisticated burner systems utilize liquid propane, natural gas, or oil supplied through underground or surface pipes to fixed individual outlets (mantles) positioned in the orchard. Although the initial installation costs are high, the burners can be readily turned on and off; the subsequent savings in wages and fuel may make them cost-effective. In most fruit-growing areas where spring frosts rarely destroy the entire crop, such a large capital outlay may not be a profitable investment, especially in the face of the increasing cost of fuel and interest on bank loans and the decreasing margin of profit. In areas where spring frosts occur regularly, such an investment will raise the cost of production so that unless other savings can be achieved, the orchard may not be competitive with those located in frost-free areas.

Wind Machines

Wind machines are usually set high over the trees. Their use is predicated on the presence of a layer of warm air aloft; the wind machine mixes this warm air with the cold air near the ground. The warm air aloft was heated during the day. As night falls, the air above the surface of the soil, which was warm during the day, now becomes cold as its heat is exchanged with the soil. As the heat exchange continues, the stratum of cold air adjacent to the soil increases in height, the coldest and densest air being next to the ground. This stratum is in contact with the layer of warm air aloft, which has a temperature gradient; it becomes cooler with increasing elevation. Hence, a reversal in temperature gradient exists between the air warmed during the day and that cooled at night. The height at which the gradient becomes reversed is called the inversion, and the warm stratum above it the inversion layer.

Wind machines are set on the windward side of the orchard to take advantage of the prevailing wind. When more than one machine is used, the effect is more than additive because the possibility of mixing larger volumes of air increases. The effectiveness of wind machines mounted on towers depends on their height and horsepower. Their efficiency lessens the higher the inversion unless the temperature difference between the lower cold air and upper warm air is at least 8° C. Wind machines are more effective if they are used in conjunction with orchard heaters because they can disperse convection heat created by the burners.

Large portable wind machines with built-in heaters mounted on trailers or sleds are available. The principle of these machines is based on generating and dispersing convection heat. Because they are operated under the trees, they do not efficiently mix cold air with that in the inversion layer. However, they can be moved to higher grounds, to the windward side, or to low spots where they might be used very effectively. Helicopters have been used to serve the same purpose.

Cultivation

Sods and weeds intercept and absorb solar radiation and convert it to photosynthates. In this way they prevent the soil from heating. Their transpiration also has a cooling effect. Hence, turning under a sod and tamping the soil allow radiant heat from the sun to warm the soil mass. This heat is radiated back to the sky at night. In this process, buds absorb some heat and are provided some protection from frost.

Inhibition of Ice-Nucleation-Active (INA) Bacteria

Naturally occurring strains of the epiphytic bacteria *Pseudomonas syringae*, *P. fluroescens*, and *Erwinia herbicola* promote ice nucleation. These bacteria possess a unique protein that serves as a focal point about which ice crystallizes. Spraying with antibiotics eliminates these normally pathogenic organisms and helps to prevent ice from forming on plant tissues. Bacterial antagonists to these ice-nucleation-active (INA) bacteria have been discovered and are now being cloned by genetic engineers. When these antagonistic bacteria are experimentally sprayed on plants, ice nucleation by the INA bacteria is suppressed. The vacuolar contents of herbaceous plant cells have been supercooled from $-6°$ to $-8°$ C without freeze damage when ice nucleation by bacteria is prevented.

Delaying Budbreak

By the Use of Growth Retardants.
Growth retardants such as daminozide and an ethylene-generating compound, ethephon, have been experimentally applied at relatively high concentration in the fall to delay budbreak in the spring. These compounds have delayed anthesis by four to ten days, depending on the dosage and species. Excessively high dosages of these chemicals have reduced fruit set. This method may be used on early-blooming pollinizer cultivars not only to avoid frost damage but also to obtain better overlap with the main cultivars of sweet cherry and almond.

The modes of action of these materials in delaying budbreak may be different. During the rest period, gibberellin content in buds increases; the delay in budbreak is attributed to the interruption of gibberellin synthesis by daminozide. Ethephon causes defoliation, and the ethylene generated from it alters membrane permeability. Ethylene is believed to block gibberellin at the site of its action. Hence, ethephon may operate in many modes.

By Evaporative Cooling of Buds.
Misting of trees before buds begin to swell in the spring has been experimentally successful in delaying the onset of bloom. Sprinklers with a fine spray nozzle are set to mist trees when the temperature approaches $15°$ C ($60°$ F). Evaporative cooling from misting causes buds to remain dormant about two weeks longer than those on unmisted trees. For misting to function ideally, good-quality water and a slight breeze to hasten evaporation are necessary. Misting trees with water containing dissolved minerals leaves an encrustation of salt, which will injure shoots and buds. Trees must be drenched periodically to avoid salt accumulation. Even in areas where the water quality is good, this delay in budbreak increases fruit set and reduces harvest size. The quality of fruits from misted trees may not be as good as that of fruits from untreated trees. Thus, delaying budbreak by misting has a different effect from that imposed by cold temperatures.

CARE OF FREEZE-DAMAGED TREES

After a frost or freeze has occurred, the trunk and crotch of trees should be checked because of their vulnerability to damage. Any cracks in the bark should be treated with a fungicide and covered with grafting wax to reduce entry by wood-rotting organisms. Pruning should be delayed until growth begins in the spring so that damage can be better assessed. At that time dead branches should be removed and large cuts covered with grafting wax.

References Cited

GERBER, J. F. 1970. Crop protection by heating, wind machines, and overhead irrigation. *HortScience* 5:428–431.

KAWASE, M. 1966. Etiolation and rooting in cuttings. *Physiol. Plant.* 18:1066–1076.

LENZ, FRITZ. 1985. Water consumption of apple trees for high production. *Compact Fruit Tree* 18:21–26. East Lansing, Mich.

LINDOW, S. E., D. C. ARNY, C. D. UPPER, AND W. R. BARCHET. 1978. The role of bacterial ice nuclei in frost injury to sensitive plants, pp. 249–263. In *Plant Cold Hardiness and Freezing Stress*. Eds. P. H. Li and A. Sakai. Academic Press. New York.

PROEBSTING, E. L., JR. 1970. Relation of fall and winter temperatures to flower bud behavior and wood hardiness of deciduous trees. *HortScience* 5:422–424.

UNIVERSITY OF CALIFORNIA, Div. of Agric. and Nat. Resources. 1978. Leaflet 2280, Soil Physical Environment and How It Affects Plant Growth.

UNIVERSITY OF CALIFORNIA, Div. of Agric. and Nat. Resources. 1981. Leaflet 21212. Irrigating Deciduous Orchards.

UNIVERSITY OF CALIFORNIA, Div. of Agric. and Nat. Resources. 1981. Leaflet 21259. Drip Irrigation Management.

VEIHMEYER, F. J., AND A. H. HENDRICKSON. 1950. Soil moisture in relation to plant growth. *Ann. Rev. Plant Physiol.* 1:285–304.

WEISER, C. J. 1970. Cold resistance and acclimation in woody plants. *HortScience* 5:403–410.

WERTHEIM, S. J. 1985. New developments in Dutch apple production. *Compcat Fruit Tree* 18:1–12. East Lansing, Mich.

CHAPTER 8

Soil and Soil Fertility

Soil Classification

Soils are classified according to (1) their origin, (2) mineral or organic content, (3) ratio of clay, silt, and sand particles, (4) the soil reaction—pH, (5) water-holding capacity, and (6) cation exchange capacity (CEC).

Origin of Soils. A sedentary soil is one that formed in place, for example, volcanic ash, decomposed granitic deposit, or organic peat soils. Transported soils are formed in one place and carried to another by rivers, glaciers, and winds. Sedimentary layers of silt, clay, and sand have been deposited by rivers overflowing their banks; glaciers have left behind moraines, and winds have created dunes.

Mineral or Inorganic and Organic Soils. Surface soils are generally more fertile than subsoils because on them has accumulated dead organic residue from decomposing vegetative and animal matter. The organic matter breaks down rapidly, however, especially in an arid climate, so that the amount rarely reaches 5 percent. Such soils consisting predominantly of inorganic constituents are termed mineral or inorganic soils.

In boggy and marshy areas, organic matter accumulates in considerably larger quantities. Such soils containing as much as 80 percent organic matter are called peat deposits or muck soils.

Ratio of Particle Sizes. Mineral soils are classified according to the percentage ratio of clay, silt, and sand (Fig. 8.1). Soil particles range in diameter from clay, 0.002 millimeter; to silt, 0.002 to 0.02 millimeter; to sand, 0.02 to 2.0 millimeters. Particles larger than 2 millimeters in diameter, such as gravel and stones, may constitute part of the soil mass, afford anchorage to trees, and hasten downward drainage of water, but they do not participate in plant nutrition. Soils containing such large particles are often referred to as a gravelly soil.

Soil Reaction—pH. Soils are categorized as being acidic, neutral, or alkaline according to their hydrogen ion concentration or its negative logarithm—pH. The pH is determined on a soil slurry made up of equal volumes of soil and water with a pH meter.

Water-Holding Capacity. Trees fare best in sandy loam soils because their overall characteristics are optimal. A sandy loam soil has a smaller water-holding capacity than loamy clay soils, but the sandy loam has a faster water infiltration rate than the loamy clay. Soils with large amounts of clay are relatively dense so that roots do not penetrate and explore them extensively. Therefore, roots are unable to extract water and nutrients efficiently, even though they are present in adequate amounts. Sandy soils have better tilth than clay soils but are usually less fertile because their cation and anion exchange capacities are smaller. Consequently, sandy soils require more frequent irrigations and fertilizer applications than loamy soils.

Cation Exchange Capacity (CEC). Clay particles and colloidal organic matter or humus are normally negatively charged so that they readily absorb cations. Decomposing organic matter generates hydrogen ions that replace cations such as potassium and calcium, releasing them to the soil solution. Conversely, sand particles, because of their relatively large size and low electrostatic charge, attract small amounts of cations. The ability of soils to exchange cations, which is a measure of soil fertility, is then a function of the amount of clay and organic matter they contain.

Fertilizers

MANURING

Manures are organic materials originating as animal wastes or vegetation that are incorporated into soil to enrich it. That manures improve the fertility of soils has been known since pre-Roman periods. The practice is still common in the United States, in northern Europe where dairy and beef cattle industries and hog farms thrive, and in underdeveloped countries where chemical fertilizers are unavailable or expensive. When available, manures from dairy feed lots and poultry farms are cheap sources of nitrogen, potassium, and phosphorus (Table 5). Microbial ac-

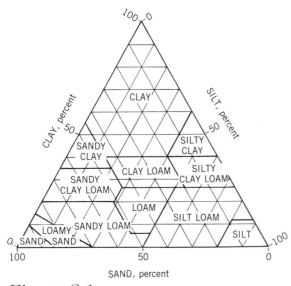

Figure 8.1
Classification of soils by percentage composition of clay, silt, and sand. (From USDA)

Table 5. *Percentage of Composition of Fresh Solid Animal Excrement*

Species	Water	Nitrogen	Phosphorus	Potassium
Horse	75	0.55	0.13	0.33
Cow	85	0.40	0.08	0.08
Sheep	60	0.75	0.22	0.37
Swine	80	0.55	0.22	0.33
Hen	55	1.00	0.35	0.33

From Van Slyke, 1932.

tivity in fresh manure generates enough heat to kill young seedlings. Therefore, if fresh manure is incorporated into the soil, seeding or transplanting should be delayed until the manure is well rotted. Otherwise, manure must be applied lightly or composted prior to use.

Blood from slaughterhouses, sediment from sewage treatment facilities, and meals from seed mills, the residues after oil has been extracted, are also good sources of organic nitrogen (Table 6). Sterilizing manures and other organic fertilizers in order to eliminate undesirable seeds, insects, and microbes increases their cost. Continued use of these organic matters has disadvantages under some conditions. In areas where drainage is poor or irrigation water is applied lightly, large applications of chicken manure can cause sodium to accumulate to toxic levels. Some organic constituents in manures tend to bind or chelate certain essential elements, such as zinc, making them unavailable to plants.

CHEMICAL FERTILIZERS

Manuring is not commonly practiced if inexpensive chemical fertilizers are readily available, even though adding manures and turning under green cover crops improve the texture and water-holding capacity of soils. But organic matter rapidly decomposes in warm, arid areas so that its beneficial effects are not long lasting. The main advantage of chemical fertilizers over organic fertilizers of unknown and varying compositions is that an exact amount of any essential element can be supplied to maintain trees in a healthy state. In many fruit-growing districts, one or more of these essential elements are low or deficient in the orchard soil.

Table 6. *Composition of Animal and Vegetable Wastes*

Material	Source	Nitrogen, percent
Dried blood	Meat packing houses	8–12
Sewage sludge	Waste disposal plants	5–6
Meals from Cottonseed Castor bean Cocoa cake	Mills	4–9

From Lyon and Buckman, 1949.

Mineral Nutrition

BRIEF HISTORY OF ESSENTIAL ELEMENTS

An article by Jan Baptista van Helmont, published by his son in Amsterdam in 1652, reported that a potted willow cutting, which van Helmont had irrigated with rainwater, gained 169 pounds (76.7 kg) in five years. The oven-dried soil in the pot that initially weighed 200 pounds (90.9 kg) lost 2 ounces (56 g). Van Helmont attributed the gain in weight of the willow tree to water and ignored the small decrease in the weight of the soil mass.

About 200 years after van Helmont's article, Justus von Liebig, who is considered the father of modern plant nutrition, analyzed plant tissues and soils. In 1840 he reported that plants utilized ammonia from humus as a source of nitrogen, but neither he nor his contemporaries could account for the abnormal growth of leguminous species in the absence of nitrogenous fertilizers. He became aware of the amount of mineral nutrients that crops were extracting from soils, and he emphasized the need to replace them with appropriate fertilizers.

In 1860 Julius von Sachs used a system of growing plants suspended in nutrient solutions, now known as hydroponics, instead of the sand culture system. Within a few years, the essentiality of most of the following major elements was established: nitrogen, phosphorus, potassium, sulfur, magnesium, and calcium. By this time researchers knew that plants derived carbon, hydrogen, and oxygen from air and water. With the introduction of new chemical techniques, highly refined glassware, and improved analytical instruments, nutritionists added the following micronutrients: iron, manganese, zinc, copper, boron, molybdenum, and chlorine. More recently, evidence that nickel and cobalt may be essential has been uncovered. The terms macronutrients and micronutrients denote the relative quantities of elements needed by plants to remain healthy.

Professor William Henry Chandler's discovery of zinc as an essential element for normal plant development is related here as a tribute to an eminent horticulturist, teacher, and author. He observed that pear trees growing in old corral sites and areas near ancient Indian burial grounds had small chlorotic or yellowing leaves. The disorder known as corral disease was attributed to organic residues and/or soil compaction. Suspecting that the trees were suffering from iron deficiency, Chandler sprayed the trees with solutions of iron salts but found his results to be erratic. Trees treated on some days recovered and developed normal leaves, but others remained chlorotic. On studying his records, he found that symptoms disappeared from trees that were treated on days when zinc-coated iron pails were used to dissolve the iron salts but not when wooden buckets were used. Subsequent trials with zinc salts bore out his suspicion that electrolysis in the metal pails had occurred, dissolving the zinc coating.

Plant nutritionists have established minimal and optimal ranges of mineral elements in various tissues as a means of assessing the fertilizer needs of various species. Shear, Crane, and Myers (1953) demonstrated that symptoms of a deficiency or of an excess of an element are a function of an imbalance among the essential elements rather than of their absolute concentration. They found that if elements were kept at a given equilibrium in young tung trees, no symptoms of deficiency or excess were noted. If the level of one or more of the elements was altered, however, abnormalities soon appeared.

Mineral nutrients are not considered as food by plant scientists, who reserve the term *food* for carbohydrates, fats, and proteins. With improved analytical methods, nutritionists and biochemists are finding that these elements play a great number of roles. Some nutrients exist in extremely minute quantities, but the range between deficiency and toxicity is small. For example, leaves of a healthy fruit tree contain between 30 and 45 parts per million of boron, but they show symptoms of toxicity at 100 ppm.

MACRONUTRIENTS AND THEIR ROLES

Nitrogen. Nitrogen makes up nearly 80 percent of the earth's atmosphere, but it is the one element that is deficient in most agricultural soils. Without annual applications of nitrogenous fertilizers, deficiency symptoms soon appear because temperate zone deciduous fruit trees are unable to fix gaseous nitrogen. They do not have symbiotic nitrogen-fixing bacteria on their roots to accomplish this process. But even species that have the ability to fix gaseous nitrogen respond to additional fertilizer. All crops (and prunings if they are burned) remove much nitrogen from the soil.

Nitrogen is absorbed primarily from the soil as ammonium or nitrate ions. These ions originate from chemical fertilizers such as ammonium nitrate or sulfate, or calcium or potassium nitrate, or sodium potassium nitrate (Chilean nitrate). Urea, a common fertilizer, and decaying organic matter are first degraded to the usable form and then absorbed by roots.

Most of the reduced ammonia in the root is combined with carbohydrates to form amino acids, the building blocks of proteins. These simple amino acids are then metabolized into more complex compounds such as purines and pyridines; vitamins, e.g., biotin, nicotinic acid, riboflavin; hormones, e.g., indoleacetic acid and zeatin and its derivatives; and coenzymes for various enzymes, e.g., nicotinamide adenine dinucleotide (NAD+). The chlorophyll molecule has four atoms of nitrogen. Some very toxic compounds contain nitrogen, e.g., alkaloids and cyanogenetic glucosides. A common metabolite in *Prunus* spp. is the cyanogeneticglucoside amygdalin, which yields hydrocyanic acid (HCN) upon hydrolysis.

Phosphorus. Phosphorus is a constituent of numerous carbohydrates and nitrogenous compounds. Energy-rich phosphorylated sugars appear early in the photosynthetic process. Phosphorylation of adenosine yields adenosine monophosphate, diphosphate, and triphosphate (AMP, ADP, and ATP, respectively), by which plants store energy to fuel other chemical processes. Phytic acid, inositol hexaphosphate, which is concentrated in the chloroplast, is considered to be an important source of phosphate. Phytin is its calcium magnesium salt.

Sulfur. Three amino acids—cystine, cysteine, and methionine—contain sulfur. They are found in some proteins and serve as precursors of biotin, thiamin, and coenzyme A. Coenzyme A is required for the enzymatic acetylation of aromatic amines. Methionine is the starting point in ethylene biosynthesis. Two cysteine residues bind cytochrome c to a protein through a thioether linkage.

Potassium. Potassium is found in large quantities and is omnipresent in tissues but always in the ionic form. It somehow participates in carbohydrate transport, accompanying photosynthates from leaves to fruits and other sinks. As the prune fruit ripens, it accumulates dry matter while maintaining a level of about 1 percent potassium. The element is associated with the movement of sugars in and out of stomatal guard cells, thereby controlling stomatal closure and opening.

Root cells have the ability to accumulate potassium to a high concentration and to discriminate against sodium, possibly as a defense mechanism against toxicity from excess sodium. Analyses of plant cells reveal that total acidity consistently exceeds that accounted for by titration, indicating that potassium and other cations exist as salts of organic acids. They thereby buffer against rapid changes in the pH of cell sap. Potassium and other monovalent cations increase membrane permeability, whereas divalent and trivalent cations have the opposite effect.

Calcium. Calcium exists as part of the structural tissue that is deposited as calcium pectate in the middle lamella. Some membranes contain calcium which governs their permeability. Many species have calcium oxalate crystals of different sizes and shapes; they occur as needles or raphides, as prisms, or as druses—globular clusters of crystals—in specialized cells depending on the species. Apple spurs contain large calcium oxalate crystals, whereas kiwifruits possess many fine raphides. How this quantity of calcium is mobilized by organs that are not equipped to transpire rapidly is a moot question.

The formation of these salts is thought to be a detoxifying mechanism which eliminates the poisonous oxalate ions or which neutralizes excess acidity resulting from organic acid synthesis. Calcium complexes with citrate and malate ions, but whether these complexes are translocated in the phloem is debatable. Calcium citrate and malate are weak acids which have a buffering capacity.

Magnesium. The element magnesium holds the central position in the chlorophyll molecule and serves as an enzyme activator and buffer in aqueous media.

MICRONUTRIENTS AND THEIR ROLES

Copper, iron, and zinc act as coenzymes and parts of prosthetic groups. A prosthetic group is the nonprotein moiety of an enzyme without which the enzyme cannot function.

Iron. Iron porphyrin is the prosthetic group in cytochromes; the element serves as electron donor and receptor in the respiratory process. The terminal oxidases of the cytochrome system reduce molecular oxygen to water. Cytochrome c is attached to its protein moiety by two thioether linkages of cysteine.

Zinc. Zinc is an integral part of carbonic anhydrase, an enzyme that is somehow associated with the synthesis of indoleacetic acid.

Copper. The element copper is the prosthetic moiety of ascorbic acid oxidase.

Manganese. The element manganese plays roles similar to magnesium. It activates enzymes and regulates membrane permeability. When enzymes are isolated from plant tissues and tested *in vitro*, the addition of magnesium, manganese, cobalt, or zinc ions often activates and/or catalyzes the enzymatic reactions. Many poisons, especially carbon monoxide, hydrocyanic acid, and sulfur-containing compounds, have a strong affinity for these metals and, therefore, will inactivate or inhibit enzyme activities.

Boron. Boron improves both flower and fruit set and pollen tube growth. The improvement is attributed to the formation of sugar–boron complexes, which are translocated through protoplasmic membranes more readily than nonborated sugars.

Molybdenum. Molybdenum is a component of nitrate reductase, the enzyme which reduces nitrate ions to nitrite in the roots and leaves. Nitrite is toxic at low concentrations, but it is immediately reduced to ammonium, especially in the presence of light, and is fixed to some carbohydrate to form an amino acid.

Mineral Deficiencies and Excesses

Since essential elements are, by definition, those needed by plants to remain healthy, a deficiency of any one or more of them or an imbalance among them is manifested by abnormal vegetative growth. Consider that a 40-metric-ton (44 short ton) crop of peaches removes about 48 kilograms (kg), or 106 pounds (lb), of nitrogen, 120 kg (264 lb) of potassium, and 9.6 kg (22 lb) of phosphorus per hectare. We can understand why von Liebig was so concerned about the consequence of cropping and the need to replace essential nutrients with fertilizers.

SYMPTOMS AND TREATMENTS

Color plates referred to in descriptions of foliar deficiency and excess symptoms are those of prunes; other fruit and nut crops have similar symptoms.

Nitrogen. A general yellowing of leaves (see Plate 5) is often an incipient symptom of nitrogen deficiency and can be attributable to retardation of chlorophyll synthesis. A greater shortage weakens shoot growth, which ceases sooner than normal and is followed by early autumnal leaf fall. Nitrogen-deficient trees with poor, weak growth do not initiate as many flowers as healthy trees; hence, the crop is usually light.

Nitrogen deficiency may occur in heavy soils with considerable water-holding capacity, even though nitrogenous fertilizers are applied regularly. Dense soil and waterlogging prevent aeration of roots. Roots are unable to absorb nitrate ions because soil organisms under anaerobic conditions reduce nitrate to nitrite ions and then to gaseous nitrogen. This reduction process, whereby soil nitrate is converted to elemental nitrogen and lost to the atmosphere, is known as denitrification.

Loss of nitrate is also a common occurrence in sandy soils because sand particles have little capacity to exchange cations and anions. That is, the particles adsorb only small amounts of positively or negatively charged ions to exchange for others of like charge. Hence, excess irrigation or a heavy rain will leach nitrates to depths below the root zone. Deficiency symptoms can appear very suddenly in these sandy areas, but they can be corrected readily by applying soluble potassium or calcium nitrate. In sandy soils nutrient deficiency can be avoided by applying before each irrigation small amounts of a complete fertilizer containing varying ratios of nitrogen, potassium, and phosphate.

The tendency to overfertilize tree crops, especially in years when nitrogenous fertilizers are relatively inexpensive, leads to excessive vegetative growth, shading, and a light set of fruits. The fruits that set are relatively large, soft, poorly colored, and flavorless. Cell walls in these large fruits are thin and flexible, making them more vulnerable to bruise than fruits from trees that were given moderate amounts of nitrogen. Overfertilized apricots ripen unevenly; the distal end of the fruit softens and becomes yellow, but the shoulders remain firm and green. The maturity of fruits with poorly developed color may be misjudged, resulting in excessive preharvest drop. Besides this loss, a grower faces additional pruning costs during the dormant season to remove the extensive vegetative growth.

If an excess of nitrogenous fertilizers is applied inadvertently, the addition of super-superphosphate or treble-superphosphate (rock phosphate treated with sulfuric and phophoric acids, respectively) fertilizers will correct the adverse effect of excess nitrogen. Phosphate ions antagonize the uptake of nitrate ions by roots.

Trees slightly deficient in nitrogen produce smaller but better-flavored and more highly colored fruits with firmer flesh than do trees with moderate to high nitrogen levels. Photosynthates, arriving from leaves, are metabolized into cell wall constituents and are accumulated in the vacuoles as sugars rather than as proteinaceous compounds. Because fruits on trees receiving small amounts of nitrogen tend to be smaller, the crop on these trees will need to be thinned more than that on trees receiving moderate nitrogen.

In sandy loam soils, ammoniacal fertilizers should be applied early in the fall so that they can be nitrified to nitrate and washed down to the root zone by the winter rains. The available nitrate can be absorbed whenever conditions for root growth are favorable.

Potassium. In some areas in California, the magnesium content of the soil is relatively high with respect to calcium. These soils tend to bind potassium very tightly near the soil surface. Therefore, surface applications of potassium fertilizers is ineffective for deep-rooted fruit trees. The fertilizer must be trenched into the soil to the depth of the roots.

French prunes growing in these areas suffer badly from dieback of shoots and scorching of leaves (see Plates 6 and 7). As fruits ripen and accumulate potassium and photosynthates, roots are deprived of sugars and are unable to grow and absorb sufficient minerals, especially potassium.

Potassium deficiency in prunes can be overcome by spraying trees three to four times with potassium nitrate at biweekly intervals during the growing season, beginning in mid-June. Early crop thinning coupled with applications of potassium may be the most effective and economical means of improving fruit size and yield. Adding potassium fertilizers to the drip irrigation system alleviates the problem because the potassium concentration in the soil solution is kept nearly constant.

Fruits on trees with incipient deficiency of potassium neither size nor color as well as fruits supplied with adequate amounts of this element (Fig. 8.2). Although leaves may not show symptoms of deficiency, when transport of photosynthates from leaves to the fruits is limited, the levels of sugars in the vacuoles remain relatively low. The low osmotic concentration restricts cell enlargement and anthocyanin synthesis.

Phosphorus. A deficiency of phosphorus stunts growth and turns leaves deep green early in the summer. By late summer the leaves have bronze to purplish hues. Young leaves on phosphorus-deficient trees tend to be narrow and straplike rather than ovate. Phosphorus deficiency in California occurs in marginal land in the foothills of the Sierra Nevada and the Coast Range but rarely in areas where soils are deep. This element is often deficient in soils derived from volcanic ash.

Phosphorus deficiency can be corrected by spreading crushed rock phosphate, which is a nearly insoluble form of phosphate, or the more soluble super-superphosphate or treble-superphosphate. Applying excessive phosphates can limit nitrate uptake by roots and precipitate iron phosphate. The unavailability of these elements may inhibit shoot growth and cause fruits to ripen early.

Phosphorus and zinc deficiencies often appear on seedlings growing in nurseries where the soil has been fumigated with a combination of methyl bromide and chloropicrin. The treatment eradicates mycorrhizal fungi, a group of symbiotic organisms which would normally become attached to roots. The organisms can absorb phosphorus and zinc from soils with very low concentrations of these elements and supply them to the trees. In nurseries where soil fumigation is a necessity, large quantities of ammonium phosphate and zinc sulfate are added to assure adequate growth of the seedlings. After the fumigants have vaporized, inoculating nursery soils with vesticular–arbuscular mycorrhizae, masses of fungus mycelia in symbiotic association with roots, has been shown to enhance greatly the growth of seedlings.

Calcium. A deficiency of calcium is rarely evident in foliage in good fruit-growing districts. Calcium deficiency induced experimentally in hydroponic trials has retarded shoot growth and caused puckering of leaves. Puckering seemingly occurs when the midrib does not elongate while the adjacent cells of the lamina continue to divide and expand.

Deficiency of this element is not apparent by soil or foliar analyses, but it causes bitter pit of apples (see Plate 8), which is the single largest source of loss to the apple industry.

Iron. Leaves on iron-deficient trees have a characteristic pattern of chlorosis (see Plate 9). The interveinal portion of the lamina appears blanched, and the region along the midrib and veins remains green. The first leaves to appear in the spring receive whatever amount of iron that is stored as reserve or is present in the transpiration stream. Once the iron is fixed as part of the cell components, such as an enzyme cofactor, it does not move. Hence, the basal leaves may be

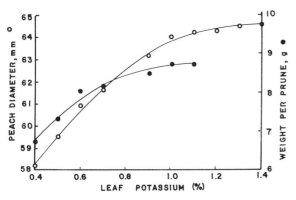

Figure 8.2
Relation between the size of peaches and the weight of prunes and the amount of potassium in leaves on a dry weight basis. (Courtesy of O. Lilleland)

slightly affected by iron deficiency, and the symptom worsens toward the shoot tip. Symptoms are commonly seen on *water sprouts* at bases of trunks and occasionally in tree tops. Iron deficiency in calcareous soils appears early in the season. In some parts of Greece, peach trees turn completely yellow by midsummer from lime-induced iron chlorosis. In these areas the soil reaction is alkaline, so that iron salts remain very insoluble. In acidic soils with a pH of 4.0 or less, aluminum becomes soluble and restricts iron absorption by roots.

If the soil is not calcareous but the pH of the soils is 7.0 or greater, the alkalinity may be gradually corrected by adding acidic fertilizers such as ammonium sulfate.

Acidic soils are often treated with crushed lime to neutralize the acid and increase the pH. Iron chlorosis is corrected by spraying plants with an iron chelate or an iron salt of ethylenediaminetetraacetic acid (EDTA), or by trenching the compound into the soil.

Zinc. Mature fruit trees require about 500 milligrams of zinc per year. Nevertheless, this element is still deficient in soils of many fruit-growing districts in North America. Lack of this element causes new growth to form rosettes with mottled and straplike leaves (see Plate 10). Thus, the disorder is known as "little leaf." Peach trees deficient in zinc produce flat and pointed fruits because the cheek diameter does not increase as much as the suture diameter (see Plate 11).

Driving zinc-coated nails about 3 centimeters in length or glaziers points (triangle pieces of zinc used by glaziers to hold window panes in place) into tree trunks at 3- to 5-centimeter intervals is an effective means of correcting the problem for several years. These zinc pieces should be placed in a helical pattern around the trunk so that the circumference is circled at least twice. They should be driven into the wood to assure that enough zinc will dissolve by the action of the sap and move up the trunk with the transpiration stream. Spacing them closely ensures that all scaffold limbs will receive some zinc.

Deficiency symptoms will not appear for one or two seasons if zinc sulfate or a chelate solution is applied just after budbreak, when some foliage is present to absorb the material. A heavy dosage prior to autumnal leaf fall also allows sufficient absorption of the material. Zinc applied in the fall is absorbed and translocated into the stems and spurs with the decomposition products of chlorophyll and other cell components and nutrients from the senescing leaves. This backflow mechanism enables plants to conserve nutrients in spurs for the following season. Care must be exercised not to cause excessive leaf injury because this will interrupt the backflow movement and defeat the purpose of the spray treatment.

Magnesium. Deficiency of magnesium manifests itself by marginal chlorosis. As the condition worsens, cells of the leaf tip and margins become necrotic (see Plate 12). The areas astride the midrib remain green, especially near the petiole. This element, like potassium, is translocated from older leaves to younger ones. As basal leaves age and chlorosis worsens, they may abscise prematurely.

The problem can be corrected by applying to the soil calcium magnesium carbonate (dolomitic limestone) or magnesium sulfate (epsom salts) or by spraying trees with magnesium sulfate after full bloom. In areas where the soil is sandy and potassium is already limited, applications of large amounts of dolomite may accentuate potassium deficiency.

Sulfur. The element sulfur exists as sulfate ions in soils and is rarely found to be deficient in tree crops because growers apply ammonium sulfate as fertilizer or calcium sulfate, gypsum, as a soil amendment to improve tilth.

Boron. Boron is required in quantities equivalent to zinc, about 500 milligrams per tree per year. When boron is deficient, shoots die back. Apple fruits on boron-deficient trees have zones that initially discolor and then become dry by midsummer (see Plate 13). These dry zones are called drought spots. Almond fruits grown in boron-deficient regions often exude gum, causing them to stick together and onto spurs. These dry, moldy fruits are called mummies or stick-tights because they do not abscise at harvest.

Boron deficiency can be overcome by spreading sodium borate (borax) or boric acid on the soil. Boron moves very slowly downward in soils; several irrigations are needed for the boron to reach the root zone. Growers can easily apply too much boron if they are not careful. A slight excess causes poor flower set and marginal burning of leaves, but an oversupply of this element can kill trees and make soil sterile.

Copper. A deficiency of copper seems to occur simultaneously with a deficiency of zinc, which makes diagnosis difficult. Copper may be deficient in soils where fruit trees are grown, but because trees are sprayed with antibiotics and fungicides that contain copper, a deficiency is rarely reported. Small chlorotic leaves, stunted shoot growth, and dieback, in severe cases, are symptoms of copper shortage (see Plate 14).

In alkaline soils copper is found in an insoluble form; soil applications in such sites may be ineffective. Injecting copper salts or placing gelatin capsules containing copper sulfate in holes drilled in the tree trunk is preferable; one treatment prevents the recurrence of deficiency symptoms for three to four years. Spraying trees with copper sulfate is effective for short periods; the treatment, although expensive, must be repeated.

Manganese. To an untrained eye, the foliar symptoms of manganese deficiency (see Plate 15) are very similar to iron chlorosis. Chlorophyll synthesis is restricted in the lamina; the interveinal parts therefore become chlorotic, but the zones about the midrib and lateral veins remain dark green. Shoot growth is stunted and some shoots die back. Trees manifesting these symptoms can be cured by spraying the foliage with a 1 percent mixture of manganese sulfate and lime or by injecting a dilute solution of manganese salt under pressure into holes bored in the trunk. Excess manganese in apple trees causes excessive roughening of the bark, along with swellings or pustules, a disease known as measles. Manganese toxicity has rarely been reported on fruit trees growing in commercial orchards.

Sodium and Chlorine. Sodium is not an essential element, but its presence in soils, especially in semiarid and arid regions, creates problems. Sodium and chlorine occur together as sodium chloride so that their excesses are difficult to distinguish. Both cause marginal necrosis. An excess of sodium in apricot causes upward curling of leaves.

Arsenic. About 40 years ago when lead arsenate was sprayed five to seven times during the growing season to control codling moths, arsenic accumulated in the soil to toxic levels in some areas. Although the use of this compound has long been discontinued, symptoms of arsenic excess still appear. Peach leaves with 2 ppm arsenic have marginal scorch and necrotic spots.

MINERAL ANALYSIS

Mineral deficiencies and excesses are very difficult to diagnose, especially by untrained eyes. The roots of plants having abnormal symptoms should be examined because nematodes and some scale insects that feed on roots often induce foliar disorders similar to those of nutritional imbalance. Symptoms of some viral infections mimic those of mineral deficiencies or excesses. If foliar analyses reveal any element to be deficient or in excess, in the absence of pathogens and pests, a corrective treatment should be initiated.

Because leaves contain what was absorbed by roots and translocated, foliar analysis is more reliable than soil analysis for determining the nutrient status of plants. The usual procedure for establishing the nutritional status of trees by foliar analyses is to collect separate 50-leaf samples from several healthy and unhealthy trees. Fully expanded basal leaves of spurs or shoots on well-

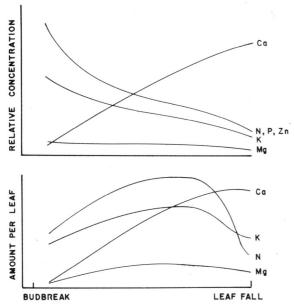

Figure 8.3
Seasonal changes in the amount of mineral elements in leaves based on concentration (upper) and on per leaf basis (lower).

exposed parts of trees are collected at two- to three-week intervals, starting at budbreak and ending at leaf fall.

Leaves with intact petioles are washed with soapy water and then rinsed with distilled water to remove any surface residue or contaminant that may give misleading results. Petioles of grape give better estimates of the nutritional status of vines than leaf blades do, especially for nitrogen. From these analyses, seasonal patterns (Fig. 8.3) and the limits for normality (Table 7) have been established. The values on the seasonal curves of samples taken from trees with moderate to adequate amounts and those with slightly deficient amounts of an element tend to converge as the season progresses. Hence, samples taken in early to midsummer are more reliable for diagnostic purposes than those taken late in the season when small initial differences disappear (Chapman, 1966). Because of the converging curves, postponing the initial sample in order to collect leaves from the midportion of shoots will result in lower values than those of basal leaves collected earlier. For those interested in more detail, the textbook *Fruit Nutrition* by Professor N. F. Childers (1966) is recommended.

Soil Amendments

Soil amendments are not necessarily fertilizers, but they are added to soils primarily to (1) improve their tilth, (2) increase water infiltration rates, (3) increase ion exchange capacity, or (4) alter the pH of the soil solution. In many semiarid and arid regions of the world where sodium salts from runoff water originating from surrounding hills and mountains have accumulated, soil tilth

Table 7. *Range of Concentrations of Macronutrients and Micronutrients in Healthy Fruit Tree Leaves Sampled in Late July, Four Months after full Bloom*

Nitrogen	Potassium	Percentage Phosphorus	Calcium	Magnesium
2.0–3.5	1.0–2.5	0.12–0.25	1.0–2.0	0.25–0.30
Boron	**Zinc**	Parts per million **Manganese**	**Iron**	**Copper**
20–80[a]	16–20	25–75[b]	50–75	4–10

[a] Walnut, 200.
[b] Pecan, 200.

and water penetration are generally poor. This is especially true where light, frequent irrigations have been applied year after year.

In areas where the sodium content of total exchangeable cations exceeds 15 percent, clay particles platelike in shape are dispersed or deflocculated. Soil structure is destroyed, and the surface tends to puddle or seal and becomes resistant to downward movement of water. Puddling leads to poor aeration of the root zone so that plants do not grow well. When the soil dries because of transpiration and surface evaporation, soil aggregates form hard clods. Frequent, light irrigations cause sodium to accumulate; occasional heavy irrigations are recommended to leach out the sodium from the root zone.

Calcium has an effect opposite to that of sodium: it flocculates the clay particles into tiny aggregates that improve tilth and water penetration. Calcium is added in the form of gypsum (calcium sulfate), lime (calcium carbonate), or dolomite (calcium magnesium carbonate). Lime and dolomite supply calcium, which displaces sodium on the clay particles so that the sodium can be leached. The carbonate ions also neutralize any acidity and buffer the soil solution. In the presence of calcium, root cell membranes are better able to discriminate between sodium and potassium. This preferential uptake of potassium leads to a healthier plant.

Solubilities of many of the chemicals found in soils are a function of the pH of the soil solution. Thus, pH plays an important role in plant nutrition. Iron, for example, forms at low pH an insoluble salt with phosphate so that neither iron nor phosphorus becomes available. At high pH, iron complexes and forms oxides. Thus, most temperate zone deciduous fruit trees do best when the range of soil pH is between 5.5 and 6.5; blueberries and cranberries will tolerate soil pH as low as 3.8.

Lime is often used to increase the soil pH, whereas sulfur and aluminum sulfate are applied to lower the soil pH. In areas where the soil reaction is alkaline, addition of sulfur and aluminum sulfate will slowly shift the soil pH toward the acid range. Sulfur is utilized by soil organisms to produce hydrogen and sulfate ions, thus acidifying the soil. Aluminum ion complexes with hydroxide ions to release hydrogen ions. Iron sulfate will do the same and concurrently supply soluble iron, but when large areas must be treated, the cost becomes prohibitive.

References Cited

CHAPMAN, H. E. 1966. Diagnostic criteria for plants and soils. University of California, Div. of Agric. Science, Berkeley, Calif.

CHILDERS, N. F. 1966. *Fruit Nutrition*, 2nd Ed. Horticult Publications. Gainsville, Fla.

LYON, T. L., AND H. O. BUCKMAN. 1949. *The Nature and Properties of Soils*. The MacMillan Co. New York.

SHEAR, C. B., H. L. CRANE, AND A. T. MYERS. 1953. Nutrient element balance: Response of tung interactions. U.S. Dept. Agric. Tech. Bull. 1085. Washington. D.C.

VAN SLYKE, L. L. 1932. *Fertilizers and Crop Production*. Orange Judd Publ. Co. New York.

CHAPTER 9

Pruning and Training of Fruit Trees

The intelligence of our prehistoric ancestors who fashioned a bow from a bough and a leather thong to propel an arrow, and who, using the same materials, invented a harp to accompany the songs of heroic accomplishments, cannot be underestimated. One might speculate that the practice of pruning trees was initiated by an equally intelligent and observant person who noticed that fruit trees damaged during a winter storm grew more vigorously and bore larger fruits the following season than did undamaged trees.

This speculation is more plausible than the Greek legend that attributes the practice of pruning or removing plant parts to the happenstance of a donkey whose owner forgot to feed and tether it one winter night. The animal, driven by hunger, fed on nearby grapevines, for which it was reprimanded the following day. But to the owner's amazement, the tattered vines on which the donkey had foraged produced grapes that were much better and were easier to harvest than did the vines left untouched by the beast. The legend may be a figment of someone's imagination, but the observation that pruning results in vines with greater vigor and better fruits has been developed into a routine cultural practice with clear-cut objectives. Hence, Gardner, Bradford, and Hooker remarked in their book *Orcharding* (1927), "Even though the legendary first pomological instructor was an ass, his followers in the orchards of the present need not be asanine in their pruning."

The fable bears out two basic horticultural axioms with respect to dormant pruning: (1) reducing the shoot : root ratio invigorates the tree, and (2) removing more flower buds than vegetative buds reduces the crop and improves the potential size of individual fruits, provided they set.

The ultimate aim of pruning is to obtain mature trees that have a fixed size and configuration and annually bear a crop. To attain this goal, a grower must know how, when, and why trees are pruned.

1. The manner in which pruning cuts are made may extend or shorten the longevity of a tree.
2. The invigorating effect of pruning depends on the time of the year it is done and how much is pruned.
3. The objectives of pruning change as young, nonbearing trees begin to initiate flowers and bear fruits. They change again as trees reach full size.

Theoretically, a well-manicured, aesthetically pleasing mature tree can be attained with a knife and a pair of hand shears, the tools a Japanese gardener uses on dwarfed bonsai trees. A dedicated amateur horticulturist with a few backyard trees may do such pruning, but an orchardist who has to borrow money to carry out the chore must often rely on workers using pneumatic shears or mechanical saws. Nevertheless, pruning should be considered as a disciplinary process—repressive to the tree at times but also invigorating at other times—a tedious and time-consuming operation, but necessary and rewarding in the end.

Principles of Pruning

A grower must decide on the species and cultivars to be planted during the planning stage and must adopt a suitable training system in order to control the ultimate tree size and shape. An orchardist, unlike a farmer who grows an annual crop, must live with decisions for the life of the orchard, which may be a period of 15 to 40 years. Any changes made after the basic framework of the trees is established will interrupt production. Hence, a great deal of study should go into the decision-making process.

The task of the orchardist is to shape the canopy in order to expose as many leaves to the sun as possible by training and pruning trees. To achieve this objective, the orchardist should be aware of leaf area and canopy area indices, foliar density (see Ch. 3), and the growth and bearing habits of different fruit tree species.

MECHANICS AND TYPES OF PRUNING CUTS

The types of cuts to make and selection of the branches to retain are dictated by the growth habit of the species and the anatomy of branching during the initial training period of three to four years. There are two kinds of pruning cuts, heading and thinning. The heading cut is made to shorten branches as a means of inducing additional branching near the point where the cut is made. A thinning cut is made at the base of an undesirable branch to reduce crowding. Branches in favorable positions will then fill in the developing canopy and start bearing a crop.

When a thinning cut is made, care should be taken not to leave short stubs. A stub either dies back to the base or gives rise to shoots from dormant buds located at the base of the stub. These regrowths will need to be pruned again at the end of the following season, adding to the cost of pruning. When branches are cut to the base, the cambium of the stock soon develops a circular ring of callus tissue which fills in the cut surface (Fig. 9.1). If a branch is too heavy to be held while being pruned, it should be removed in three steps to prevent damage to the bark (Fig. 9.2). The first step is to make a cut about one-third the diameter from the underside of the branch at a distance of about 50 centimeters from the trunk. (If the cut is made too deep, the weight of the branch will

Figure 9.1
A cut made flush with the trunk (right) develops callus tissue and heals readily, whereas a stub (left) dies back to its base and becomes a focal point for wood-rotting organisms.

cause it to bind the saw.) The second cut should be made on the upper side about 70 centimeters from the trunk. The weight of the branch will cause the branch to split from the second to the first cut and drop to the ground. The remaining stub should be sawed flush with the trunk. Any large wound should be covered with grafting wax to hasten its healing process.

ESSENTIALS OF A STRONG TREE FRAME

A good, strong tree framework has an upright vertical trunk and primary scaffold limbs which have a wide crotch angle, are evenly distributed about the trunk, and are separated vertically from each other, the uppermost scaffold being largest and in a dominant position. Conversely, the characteristics of a poorly trained and potentially weak tree are as follows. (1) The trunk of a weak tree leans (except in the bandera system in which trees are purposely planted in this fashion). (2) Three or more scaffold branches arise from the same junction. Such a crotch collects debris which harbors wood-rotting organisms (Fig. 9.3). (3) A weak tree has a narrow crotch angle which allows old bark tissue to accumulate between branches. The branches weaken and become readily subject to splitting. (4) Three limbs originate at a point but are aligned in the vertical plane (Fig. 9.4). In this conformation the middle branch does not receive adequate food and water and eventually becomes choked or overgrown. (5) One branch arises directly above another. The lower branch, even though it may be shaded temporarily, usually outgrows the upper branch, especially in yearling peach and plum trees. (6) A vigorous water sprout on the trunk will divert energy from the tree. If neglected, a water sprout may outgrow a trunk so that the tree will need to be reshaped later. Hence, a good pruning practice is to

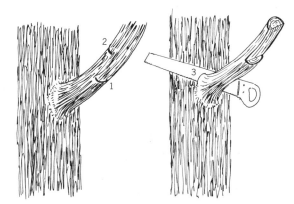

Figure 9.2
The trunk was damaged when the pruner attempted to remove a large limb with a single cut (upper). Such damage could be avoided by pruning in three steps (lower). The numbers indicate the sequence of cuts.

Principles of Pruning

Figure 9.3
Multiple branches arising from a single point forms a water pocket and collects debris (left). A narrow crotch often splits (right).

remove water sprouts early in the summer while they are succulent and growing rapidly.

Pruning During the Formative Years

The chief reason for pruning and training young trees is to develop a strong framework, one able to support profitable crops for the life of the tree. Fruit trees, usually purchased as grafted, bare-rooted seedlings, are planted during the dormant season. The first pruning cut is made on a newly planted seedling to reduce the size of the top in order (1) to compensate for roots that were lost during the digging operation in the nursery and (2) to establish the trunk height.

If trunk shakers are to be used for harvesting the crop, tree trunks should be at least a meter in height to accommodate the clamping device. If limb shakers are used, secondary and tertiary branches must be spaced far enough apart that the machine operator can position the shaking device without damaging limbs.

Since most trees delivered from nurseries do not have branches at desirable positions, the trunk is trained to a single stem or whip by heading back all lateral branches to one- or two-node stubs. Selection of primary scaffold branches is made during the first growing season when the trees are summer-pruned. If lateral branches are present on the trunk, those at optimum locations for a particular training system are selected to serve as primary scaffolds.

TIME OF PRUNING, AS AN INVIGORATING OR DEBILITATING PROCESS

Insofar as pruning in the formative years differs from one training system to another, the grower must clearly conceptualize the adopted system. The following questions raise other considerations. (1) Is the crop to be harvested mechanically with a trunk or limb shaker, or manually by workers on ladders or on mobile picking platforms? (2)

Figure 9.4
When three branches grow in the same plane, the center branch eventually becomes crowded or choked.

Do trees need to be trained so that ladders can be placed between limbs to facilitate the thinning and harvesting operations? (3) Are the primary scaffold branches high enough to assure good air drainage and clearance for the large equipment and safety of its operator? Before appraising the pros and cons of answers to these questions related to future orchard management, students of horticulture should be familiar with the kinds and techniques of pruning cuts that will yield a physically strong, healthy, and economical tree.

Dormant Pruning

Most dormant pruning that is done after autumnal leaf fall has an invigorating effect on the tree the following spring. The reserve food stored the previous summer in roots, shoots, and trunk is divided among fewer remaining growing points in the spring. Results from pruning trials have disclosed that compared to pruned trees, unpruned trees (1) make faster gains in girth, height, and dry weight because they have more leaf area (Chandler and Heinicke, 1925, 1926); (2) make more total shoot and root growth, (Head, 1967); and (3) come into bearing earlier because the first flowers to be initiated are not removed. These studies support the concept that pruning is dwarfing. They also point out two disadvantages. Unpruned trees branch randomly and freely so that crops become difficult to hand-thin and harvest. They also develop such thick, foliar canopies that lower, shaded branches soon become weak and eventually die. These branches are often invaded by insects and pathogens because thorough coverage with pesticides and fungicides is not possible. Although unpruned trees may bring an earlier income to growers, they soon become culturally unmanageable, unproductive, and short-lived.

Summer Pruning

Summer pruning is more debilitating than dormant pruning because it removes shoots and leaves when the reserve food level in the roots and stems is at its lowest ebb. Young leaves, which initially utilized the stored food in older, adjacent branches for their development, would normally begin to export photosynthates as they age. Elimination of leaves by summer pruning reduces the potential for shoot growth and delays the replenishing of reserve foods. Hence, summer pruning is said to be dwarfing.

The purposes of summer pruning are (1) to select, promote, and direct the growth of primary and secondary scaffold limbs; (2) to suppress or eliminate competing undesirable branches, for example, water sprouts; and (3) to minimize the amount of dormant pruning. Summer pruning yields the best results if primary branches are selected during the first growing season when they are about 15 centimeters long. Branches in undesirable positions are eliminated, or their growth is suppressed by pinching 2 to 4 centimeters of the succulent shoot terminal. Pinching induces the shortened stems to form many lateral sprouts, but in the meantime, the untouched primary branches continue to grow. Pinching should be repeated three to four weeks later to keep the new shoots suppressed. If summer and dormant prunings are well-coordinated for the first three to four years after the trees are planted, there will be a large gain in tree size because summer pruning reduces the severity of dormant pruning. When the total amount of wood removed is kept small, summer- and dormant-pruned trees come into bearing earlier than those that are just dormant-pruned.

In areas with a short growing season, summer pruning can prolong growth late into the season and may predispose shoots to injury at low temperatures. That is, the wood may not mature and become cold hardy by the time the first frost arrives.

Delayed Heading. A method of summer pruning that obtains scaffold branches with wide crotch angles and opens up the center of the tree is called *delayed heading*. A temporary head is developed after the tree is planted by leaving the trunk about 30 centimeters longer than the final desired height. Upright shoots arising near the cut are pinched back early in the spring, and lower shoots with wider crotch angles are encouraged to develop into scaffold limbs (Fig. 9.5). Pinching must be repeated two to three times during the first and second growing seasons so that during the summer a ball of foliage develops in the center of the tree. This bushy ball is removed by cutting back to the uppermost, strong, lateral branch when dormant pruning is done in the second or third growing season, leaving a well-developed, open-center tree.

Pruning of Mature Trees

DORMANT PRUNING

After trees reach maximum size, they are pruned to confine them within their allotted volume or space and to eliminate dead or interfering branches and water sprouts. How much and to what degree mature fruit trees are pruned depend on their bearing habits, the training system of the particular stock–scion combination, and the vigor of the cultivars. All these factors vary from species to species.

Some species bear most of the crop on spurs and the remainder on shoots; others bear their crops predominantly on shoots. Therefore, fruit trees are classified according to their predominant bearing habit as terminal or lateral, on spurs or shoots, or both.

Terminal:

a. On spurs, e.g., most apples, pear, and walnut.

b. On shoots, e.g., pecan, quince, and walnut.

Lateral:

a. On spurs, e.g., cherry and European plum.

b. On shoots, e.g., peach, fig, and persimmon.

Shoots and spurs:

Almond, some apples, apricot, and Japanese plum.

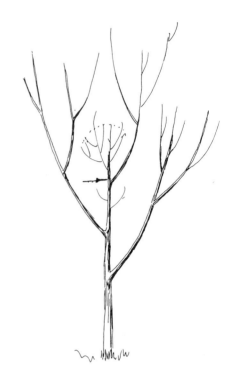

Figure 9.5
A three-year-old tree trained by delayed heading to encourage a wide-angled crotch. The temporary bushy head will be removed at the point indicated by the arrow.

Terminal Spur-Bearing Species

With species that bear terminally on long-lived spurs, for example, apple, apricot, and pear, most of the shoots that grew the previous season can be removed. In spite of moderate pruning, enough fruiting wood usually remains so that in most years the crop size will still need to be reduced at thinning time. Bartlett pear differentiates most flowers on terminals of spurs and some on long shoots. The long shoots should be headed back or removed entirely because if the flowers set, the weight of the developing fruits forces the shoots to bend downward. These fruits on pendulous shoots sway in the wind, often strike other bearing branches, and become bruised.

With new fruitful walnut cultivars, for example, Chico and Vina, annual pruning is necessary to maintain vigor, fruitfulness, and good nut quality and size. These cultivars set such large crops that once pruning is stopped, shoot growth is markedly reduced. Thus, when the trees attain desirable size, they should be pruned annually to be kept within bounds and yet maintain vigor.

In individual spur pruning multipronged, long-lived apple and pear spurs are shortened to prevent their breakage from excess weight of fruits. The heading back forces dormant lateral buds to grow and initiate flowers in due time. Spur pruning is so slow and tedious that it is rarely practiced except by an occasional expert who wishes to maintain professional competence. But as high-density plantings of apples, pears, and other spur-bearing species become more common, spur pruning should be incorporated as part of the routine dormant pruning operation.

Lateral Bearing Species

With species such as the peach and nectarine, which bear their crops laterally on last season's growth, about two-thirds of the new growth and all old weak wood should be removed, leaving only the longer vigorous shoots to bear the following season. Downward-growing branches or hangers on peach and nectarine trees should be thinned but not headed unless the limbs are interfering with cultivation. Fig and persimmon trees, which bear laterally on the current season's shoot, are occasionally pruned to remove dead limbs.

Wood Renewal System

After a tree has been fruiting for several years, older basal spurs on large scaffold branches either die from lack of light or are broken by careless workers when they are thinning or harvesting the crop. A wood renewal program can be initiated to replace some of these scaffold limbs which are relatively barren of spurs. This requires drastically heading back a large scaffold in the winter, thus forcing many latent buds into growth near the cut in the spring. All shoots are removed except a vigorous one in a desirable position; it is kept to develop a new spur system to replace that on the old barren scaffold. By following this routine annually, an orchardist can renovate an aging almond or apricot tree within five to seven years and return it to its former productivity.

Mechanically Hedging and Topping

The use of mechanical saws is gaining widespread acceptance because of the lack of skilled pruners and the speed at which trees can be pruned. Apple, peach, Japanese plum, sour cherry, and prune trees are topped horizontally with rapidly spinning, whirring circular saws or sickle-bar mowers. Since these saws make nonselective, heading-type cuts, vigorous regrowth from these pruned shoots produces numerous witches'-brooms, especially if the practice is continued for two to three years in succession. The resulting thick foliar canopy casts a dense shade over the lower part of the tree, so that lower spurs and hanger branches die from lack of light. Thus, canopies of mechanically pruned trees must be periodically hand-thinned to allow light into and between trees in order to maintain productivity throughout a tree.

As trees become crowded, they are often hedged to form a vertical fruiting wall or a tall trapezoidal shape in cross section. Mechanical saws work well with hedgerow plantings of apple, pear, and walnut. But besides making nonselective cuts, they have two other disadvantages. (1) If the soil is wet, which is usually the case during winter months, the saws are hazardous to the operator because they tend to topple over. (2) The large size and weight of the machinery tend to compact the soil.

SUMMER PRUNING

The practice of summer pruning was formerly confined mainly to (1) young trees for selecting and directing the growth of primary and secondary scaffold limbs and (2) older bearing trees with open centers for removing water sprouts. Currently, the practice includes stone fruits that ripen very early and high-density plantings. New cultivars of early-ripening peach, nectarine, and plum are thinned early, fertilized moderately, and irrigated frequently to obtain good marketable size by harvest, which may be in early June. After harvest, rank shoot growth may continue until Septemter.

In high-density plantings of apple and pear, the rootstock or native soil fertility may impart excessive vigor. The resulting crowded condition requires that branches be headed and thinned in summer to reduce vigor, improve fruit color, and keep spurs alive and bearing annually. A more permanent and economical solution may be to remove alternate trees and/or to reduce vigor by applying less nitrogenous fertilizers.

Summer topping or hedging of mature sweet cherry trees is not recommended because it is very debilitating, especially if it is done immediately after harvest when the reserve food supply is at its lowest level.

Summer pruning is then necessary: (1) to maintain a low tree profile, (2) to keep lower hanger wood alive, and (3) to promote flower initiation for the following season.

Training Systems

Vase-Shaped or Open-Center Tree

Ideally, in a vase-shaped or open-center tree, the lowest limb should be 1 meter above the orchard floor and the other two primary scaffold branches should be spaced 20 to 40 centimeters apart vertically, and as closely as possible to 120 degrees laterally around the circumference of the trunk. Two secondary scaffold limbs should arise about 1 meter above the junction of the primary branch and the trunk; two tertiary branches should fork another meter above the junction of the secondary scaffold branches, and should be oriented tangentially so that the tree will finally resemble an

Figure 9.6
Five-year-old sweet cherry trees trained to an open-center or goblet shape. Heavy annual pruning has delayed spur formation on branches. Compare with Figure 12.7.

inverted, hollow cone (Fig. 9.6). In a vase-shaped tree, the branches are developed to be nearly equal in size. This is achieved by suppressing the growth of the most vigorous scaffold branch through moderate pruning and by promoting that of the least vigorous through light pruning.

Two extreme variations of the vase shape for training peach trees were developed by two horticulturists, Sims and Dahlgren. The Sims system retains four to five scaffold limbs trained to a near upright position, with each branch having several pendulous hanger branches originating at different elevations. In the Dahlgren system six to ten wide-angled, primary scaffold branches are allowed to fork and fill in the assigned volume so that a mature tree is bowl-shaped.

Modified Central Leader

With species such as apple, apricot, and pear, the largest uppermost branch is selected as the leader and is headed at a height of 1.5 meters to a lateral. With the leader in a near upright position, two other primary scaffold limbs distributed vertically and laterally are selected (Fig. 9.7). The leader is kept dominant by pruning it the least and maintaining its height above other scaffold branches. As the tree approaches full size, the leader is gradually pruned more to suppress its growth while the others are pruned less to promote their growth so that their diameters eventually match that of the leader. The Winters system for apricots (Fig. 9.8), developed near Winters, California, initially starts the tree as a modified central leader with a strong frame. The tree is gradually trained to an exaggerated open-center shape so that the bearing zone is within a thin horizontal band.

With the vase and the modified central leader system, a heavy crop will put a stress on the crotch. Therefore, a ring of thick wire or nylon cord should be placed to encircle the tree at the tertiary branching level. This lessens the chance

Figure 9.7
An open-center tree started as a modified central leader. The uppermost branch is kept dominant until the tree approaches full size.

Figure 9.8
Open-center Winters system of training apricot trees.

of limb breakage or splitting at the crotch. Any long, lateral, horizontal branch arising from the secondary or tertiary scaffold branch should also be supported by cords tied to this ring. Thick metal or plastic guards should be placed between the wire and the branch to prevent the wire from girdling and weakening it.

With trees that eventually become large, for example, the Persian walnut, the leader is kept dominant and nearly upright. Ideally, large branches to be selected for scaffold limbs should arise from the trunk at different levels and face different directions so that no one branch is situated over another (Fig. 9.9). The lowest branch should be at a height of 2.5 meters to allow tractors and other equipment to pass underneath without endangering their operators or scarring the limb. When the tree reaches about half of its ultimate height, the leader is cut to a lateral and its dominance subdued so that other scaffold branches can eventually attain the same size as the leader. The basic framework of the young bearing tree is established by this time, and it is subsequently pruned as a mature tree.

Central Leader or Spindle-Shaped System

Certain species, such as the apple and sweet cherry, have a strong tendency to grow upright. Growers in the past have tried to take advantage of the strong apical dominance by shaping trees to a central leader. When this system resulted in

Figure 9.9
A walnut tree trained to a modified central leader. The leader was cut to a lateral to open the center for better light penetration into the canopy.

excessively large trees, it lost its popularity. Currently, standard and spur-type apple trees on size-controlling rootstocks, trained to the central leader system in the form of a spindle, are becoming popular. The slender spindle tree (Fig. 9.10) is trained to an upright taper form; the spindle bush (Fig. 9.10) has a low, spreading profile. Its leaf area index is ideal.

Closely planted trees trained to the spindle have gained wide popularity in Europe because they are easy to manage. There is the prospect of an early yield and, therefore, income to growers. In this training system, the central leader is headed back annually to 40- to 80-centimeter intervals to force several branches to grow below each cut. The uppermost, vigorous shoot is kept in the upright position to extend the central trunk. Superfluous branches are removed, and those in desirable positions are bent toward the horizontal position by using short pieces of laths (spreaders) with their ends notched or equipped with nails to keep branches in place. Orienting the lateral branches in the horizontal position promotes the early formation of spurs and the initiation of flowers while restricting terminal growth. The trunk is headed, and the lateral branches are bent to the horizontal until the tree attains its final height. Trees on M.9 are often trained on a three-wire vertical trellis, whereas those on more vigorous stocks are left freestanding or tied to a post for support. In northern Japan lower limbs are looped over to make them springy (Fig. 9.11), so that the weight of the snow will not break them.

The meadow orchard culture, a modification of the central leader system (Fig. 9.12), was conceived by Hudson and co-workers (1971, 1974) at the Long Ashton Research Station in England. In this yet experimental system, a highly fruitful apple cultivar is grafted onto a dwarfing rootstock. Trees are planted 30 centimeters apart within a row, and the rows are spaced 45 centimeters apart (71,700 trees per hectare). The central leader is trained to a whip the first season, during which it is sprayed with a growth retardant to enhance flower bud initiation. The same treatment is applied during the second growing season to divert assimilates to the developing crop. In the following dormant season, the leader

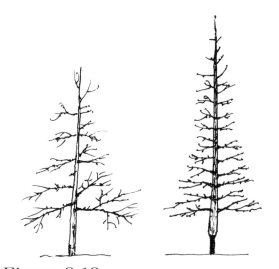

Figure 9.10
A tree trained to a spindle bush (left) and a slender spindle (right) grafted high on the trunk (high-worked).

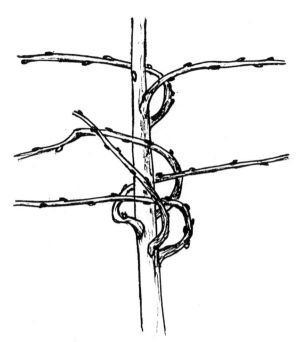

Figure 9.11
Low lateral branches on an apple trunk twisted into loops to avoid breakage by the weight of snow.

is cut back to a point just above the graft union. In the third growing season, a single shoot is allowed to grow from the scion to start a new cycle.

If the system proves to be economically feasible, it has two distinct advantages: tree size is controlled, and the harvesting and pruning operations can be entirely mechanized. The chief disadvantage is the cost of the trees and planting them. The cost might be reduced in the future if desirable cultivars can be propagated by cuttings or by meristem culture *in vitro*.

A similar training system coupled with a mechanical harvesting device is being tried with early-ripening peaches and nectarines in Israel and southern parts of the United States. Trees are started from cuttings, planted, and trained the first season, when flower buds are initiated. In the second season, after the fruits are harvested, the trees are cut down, leaving short stumps. A few, strong, upright-growing shoots from these stumps are trained upright to initiate flowers for the following season. Thus, unlike the apple grown under the meadow orchard system in which a crop is borne in alternate years, early-ripening peaches and nectarine can be treated as an annual crop with a perennial rootstock. Fruits can also be mechanically separated from the harvested trees and simultaneously sized on the harvester. But this system is usable only in zones where the growing season is long, so that sufficient regrowth can take place between harvest and leaf fall.

Espalier

The French word *espalier* originally meant to train a tree or vine to grow on a flat surface or trellis, but it is now often used to describe a tree pruned into a decorative architectonic sculpture. Thus, espaliered trees take on many forms

Figure 9.12
A single-stemmed apple tree trained for the meadow culture system.

Training Systems 213

through pruning, tying, and occasionally approach-grafting one to another (Fig. 9.13). Shaping and confining trees to a given configuration makes mechanization of the pruning, spraying, fruit thinning, and harvesting operations possible. Part of the success is attributed to the advent of concrete and chemically treated wooden posts and high-tensile wires.

Palmette System. In Italy the palmette system (Fig. 9.14) became popular after World War II when rural workers began moving to urban centers and farm laborers became scarce. The hedgerow planting system and picking platforms (see Fig. 6.37) not only hastened the harvesting, thinning, and pruning operations, but also made the work easier for women who constitute a large proportion of the labor force.

Trees, planted about 2.5 to 3.0 meters apart within hedgerows and with the rows spaced 4.5 to 5 meters apart, are trained to a three- or four-wire horizontal trellis and headed at 1-meter intervals annually. At each dormant pruning, a strong leader is maintained upright, and two lateral branches are tied to wires at an angle of about 55° from the horizontal. Lateral branches growing at right angles to the direction of the row are selected and bent to form the scaffold limbs, as they are less likely to split from the trunk.

Bending promotes budbreak on the upper side of branches and the formation of numerous water sprouts, which must be thinned out, an expensive chore. The number of sprouts can be minimized by bending the branch at a sharp angle rather than in a large looping arc. Budbreak is attributed to the evolution of ethylene as a consequence of bending.

Apple, pear, peach, and persimmon lend themselves to this training system because of their growth and bearing habits. Unless heavily

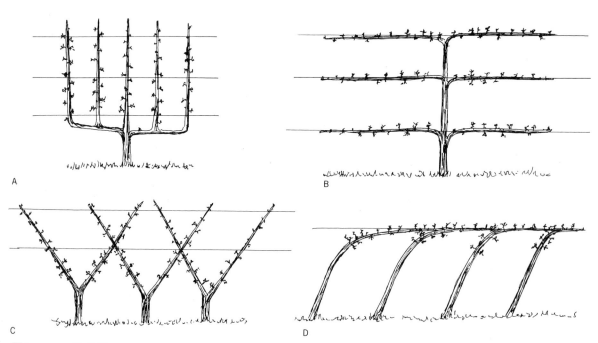

Figure 9.13
Espaliered trees trained to (A) a candelabra, (B) a horizontal palmette or Kniffin, (C) an ypsilon, and (D) a horizontal cordon.

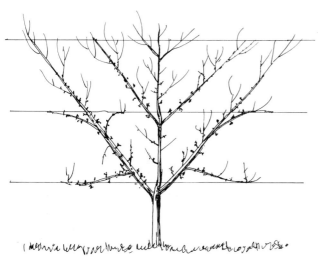 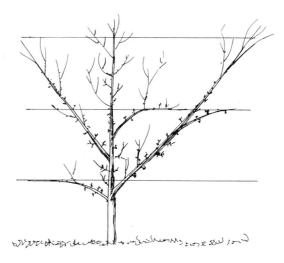

Figure 9.14
Trees of a spur-bearing species trained to a palmette (left) and a sprint palmette (right).

pruned, sweet cherry, Japanese plum, and apricot do not branch freely enough to fill in the allotted tree volume when lateral branches are spaced 1 meter apart. Hence, early heavy production is not achieved with these species.

The "sprint" palmette is a modification in which the leader is not headed annually but is allowed to grow until it reaches the desired height. All lateral branches that grow within the row are kept, but those that grow into the row and interfere with cultural operations are headed back or removed. With minimum pruning, these trees come into production earlier than those in the regular palmette system. Aesthetically, trees trained to the sprint palmette system are not as pleasing to the eye as those of the symmetrically pruned palmette system (Fig. 9.14). The height of hedgerow trees is maintained 1 meter higher than the row width.

Bandera (Flag) or Marchand System. Trees trained to the bandera system are planted at an angle of 55 to 60 degrees from the horizontal, and the trunks are tied to wires or stakes to support the trees. Maintaining the leader at this angle reduces apical dominance and slows the rate of shoot elongation. Lateral branches originating on the upper sides of these tilted trees are trained at about right angles to the trunk. When observed in the dormant season, the tree resembles a banner for which the system was named (Fig. 9.15). Tree spacing is equal to or closer than that used in the palmette system.

Tatura and Ypsilon Systems. In the Tatura system, which was developed in New Zealand, the trunk is headed at 30 centimeters; in the ypsilon system, the trunk is headed at 50 to 100 centimeters tall. One scaffold branch and its lateral branches are tied to one side of the trellis, and the same is done to a scaffold branch on the opposite side (Fig. 9.15). The tree has the configuration of a V- or Y-shaped trough. When the troughs or rows are set too close together, the system loses the advantage of an upright hedgerow because the lower side of the canopy receives little light. Prunings and brush fall into the trough; the arrangement of the horizontal wires makes their removal time-consuming.

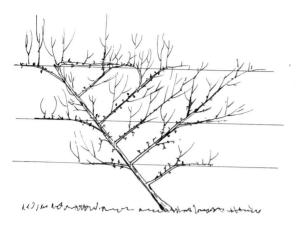
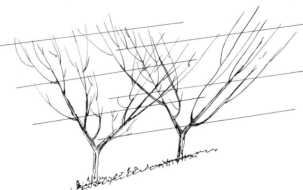

Figure 9.15
A tree trained to a bandera system on a three-wire trellis (left) and two trees trained to a Tatura system or a modified ypsilon (right).

Pruning of Bearing Vines, Brambles, and Bushes

GRAPEVINES

In the days of Leonardo da Vinci, grapevines were strung like cobwebs from tree to tree in Emilia-Romagna in northern Italy, a far cry from the way grapes are trained and mechanically pruned and harvested today. Grapevines are either spur- or cane-pruned according to the bearing habits of the cultivars. Within each group there are several training systems.

Spur-Pruning Systems

Cultivars that bear a cluster of berries opposite a leaf at the second, third, and fourth nodes on the current season's growth are spur-pruned. Dormant canes are headed back to two- or three-bud spurs, depending on the vigor of the cane. Those with larger basal diameters are headed back to three buds because they can support more clusters than canes with smaller diameters. Spur-pruned vines are trained to a head, bilateral cordon, Geneva double curtain, or pergola (see the next section, Kiwifruit Vines).

Head Pruning. A head-pruned vine is established by cutting the trunk at a height of 80 to 120 centimeters. Three to four lateral scaffold branches originating at different heights on the trunk are selected to bear the crop. After the vines begin to bear, canes on these branches are headed back to short spurs at each dormant pruning (Fig. 9.16). Most wine grape cultivars, such as Zinfandel, and some dessert cultivars, such as Flame Tokay, are head-pruned.

Bilateral Cordon System. A trunk on vines trained to the bilateral cordon system is headed back at a height of 1 meter to force lateral branches. Two vigorous shoots growing in the opposite direction from the trunk are selected and trained on the center wire to serve as the permanent horizontal cordons. Six to eight spur systems evenly spaced on each cordon are developed to bear the crop (Fig. 9.17). Shoots bearing the crop are allowed to drape over lateral wires of the trellis system. This system spreads the foliar canopy and the bearing positions of the clusters, making harvesting easier than head-pruned vines.

Geneva Double-Curtain System. The Geneva double-curtain (GDC) system (Shaulis and

Figure 9.16
A head-pruned grapevine before (right side) and after pruning (left side).

Cane-Pruning Systems

Certain dessert and wine grape cultivars have canes in which the basal seven to ten nodes are unfruitful. On cultivars such as the Thompson Seedless, syn. Sultanina, the fruitful nodes begin about nine nodes from the base and continue for another fifteen nodes. Thus, the pruning strategy is to (1) eliminate all weak and ill-positioned canes, (2) remove all two-year-old canes that bore a crop the previous summer, and (3) head back several long, vigorous one-year-old canes to 12 to 15 nodes and place them on the parallel trellis wires (Fig. 9.19). These will bear next season's crop. The last step is to head back three to four one-year-old canes to short, two- to three-bud spurs. Shoots arising from these short spurs will become replacement canes for next season's crop. Some growers tilt the crossarm on the trellis to facilitate fruit thinning, harvesting, and pruning of vines.

The canes of some vinifera wine grapes, for example, Semillon, are wound about two wires similar to trailing blackberry vines (see Fig. 9.22). Labrusca cultivars are commonly cane-pruned to a vertically oriented fan or candelabra system (see Fig. 9.13A), similar to espaliered trees or to a multistoried Kniffin system (see Fig. 9.13B). In all cane-pruned vines, all two-year old

Kimball, 1955) is a modification of the single-curtain cordon system. One double curtain starts from a single vine with a split trunk; each part of the trunk is divided to form one of the two cordons running bilaterally on parallel wires. In another method using several vines, the bilateral cordon of one vine is tied to one of the parallel wires, that of the next vine to the other wire (Fig. 9.18). The objective is to develop curtains of foliage from the cordons in order to increase the leaf area index and to place clusters at heights easily accessible to harvesters.

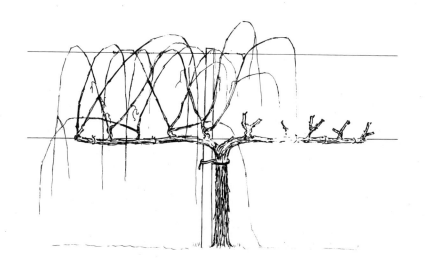

Figure 9.17
A grapevine trained to a bilateral cordon before (left side) and after pruning (right side).

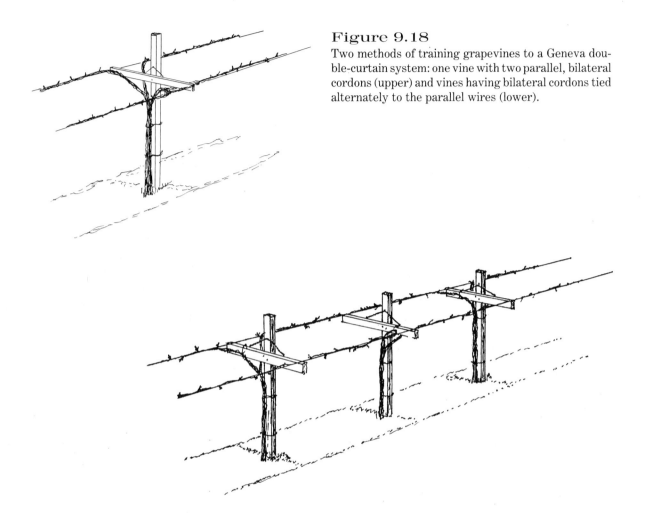

Figure 9.18
Two methods of training grapevines to a Geneva double-curtain system: one vine with two parallel, bilateral cordons (upper) and vines having bilateral cordons tied alternately to the parallel wires (lower).

canes that bore a crop the previous season are systematically removed; they are replaced with four to six long one-year-old canes. Two to four short two-bud spurs are left on the trunk, from which long renewal shoots will develop the following season.

Some spur- and cane-pruned cultivars are trained to the pergola or *tendone* system (Fig. 9.20), in which the entire vineyard is crisscrossed with high-tensile wires at a height of 2 meters. The cordons or arms bearing spurs or canes radiate from the central trunk axis to the wires.

KIWIFRUIT VINES

Kiwifruit vines have a bearing habit similar to that of grapevines, in that mixed buds on one-year-old cane give rise to long shoots bearing either pistillate or staminate influorescences from the third to the seventh node. The species differs from grapes in that kiwifruit inflorescences are borne in the axils of leaves rather than opposite a leaf; and all dormant buds on a kiwifruit cane are potentially fruitful unlike those of grapes. Kiwifruit vines, unlike grapevines, have no ten-

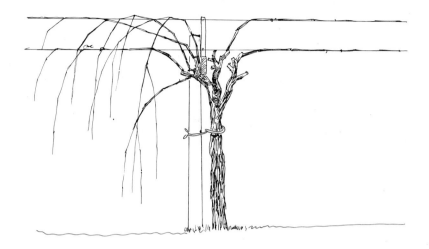

Figure 9.19
A cane-pruned grapevine before (left side) and after pruning (right side). Notice that few spurs have been left for cane renewal.

drils; instead, the shoot terminals become entwined about other branches and trellis wires to support the vine.

Vines are trained to the pergola (Fig. 9.20) or the T-bar system (Fig. 9.21). The Tatura or ypsilon trellis (Fig. 9.15) and horizontal espalier (Fig. 9.13B) systems are being tried experimentally to see whether fruit size might be improved by (1) better exposing flowers to pollen vectors, and (2) increasing leaf area index. Young kiwifruit vines that have not fully occupied their assigned volume are cane-pruned to extend lateral cordons. On mature bearing vines, long, vigorous canes are headed back leaving 12 to 15 nodes, and all short, spindly canes are thinned out. Canes should not be headed back to basal nodes that previously bore a crop because no axillary buds remain. Shoots that extend over the outside wires of a T-bar trellis are summer-pruned to promote better light conditions under the canopy and to facilitate harvest.

Male vines are pruned severely at the end of the pollination period to promote new growth and are repruned lightly during the dormant season to eliminate weak canes. Budbreak on dormant-pruned vines is earlier than that on unpruned vines.

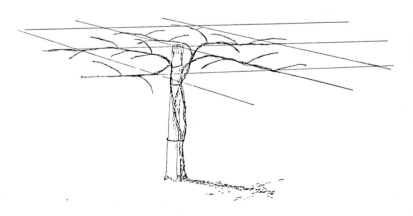

Figure 9.20
Pergola or *tendone* system of training grapevines and kiwifruit vines.

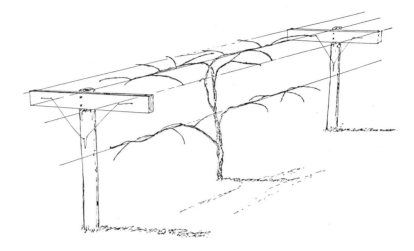

Figure 9.21
T-bar trellis system for kiwifruit vines.

BRAMBLES AND BUSH BERRIES

Blackberries and related species with trailing growth habits bear their crops on biennial shoots. Shoots that arise from the crown are called primocanes; the same canes bearing flowers and fruits the second season are known as floricanes. After bearing a crop, the floricanes die. These canes are then removed, and the new primocanes are placed on the trellis to bear next year's crop. Commonly, the canes are thinned, and the remaining canes are looped around two horizontal wires spaced vertically about 100 centimeters apart (Fig. 9.22). The lower wire is about 50 to 80 centimeters from the ground. An over-the-vine mechanical harvester has been designed to har-

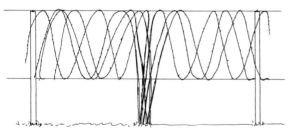

Figure 9.22
Trailing blackberry floricanes entwined on a two-wire trellis.

vest the crop. In some areas, canes are trained on a low horizontal T-bar system similar to a kiwifruit vine trellis in order to facilitate the hand-harvesting operation.

Dormant primocanes on upright growing blackberry and raspberry cultivars are thinned and headed back to about 1.6 meters. They are trained to be either upright or divided, and they are tied to two horizontal wires, resembling the Tatura system, but often they are left freestanding. Raspberry cultivars such as Hermitage and Amity form flowers and bear fruits on primocanes. During the dormant season, these plants are mown near the crown, from which several vigorous primocanes are selected in the spring to bear the next crop. Depending on the timing and dosages, spraying the primocanes with growth retardants promotes formation of flower buds and better fruit size.

Red currants and gooseberries bear on mixed and simple buds on one-year old wood. These mixed buds develop into short spurs that bear for two to three years. Older shoots lose their vigor and become barren while crowding younger shoots. Therefore, three- to four-year-old shoots are removed to avoid crowding and promote flowering.

Various species of blueberry and their interspecific hybrids should be pruned similarly to currants and gooseberries. Older shoots with low vigor are removed. Trailing cranberry shoots or runners are removed when the plants become bushy, leaving the upright canes to bear a crop the following season.

References Cited

CHANDLER, W. H., AND A. J. HEINICKE. 1925. Some effects of fruiting on the growth of grapevines. *Proc. Amer. Soc. Hort. Sci.* 22:74–80.

CHANDLER, W. H., AND A. J. HEINICKE. 1926. The effects of fruiting on the growth of Oldenburg apple trees. *Proc. Amer. Soc. Hort. Sci.* 23:36–46.

GARDNER, V. R., F. C. BRADFORD, AND H. D. HOOKER. 1927. *Orcharding*, 1st Ed. McGraw-Hill Book Co. New York.

HEAD, G. C. 1967. Effects of seasonal changes in shoot growth on the amount of unsuberized root on apple and plum trees. *J. Hort. Sci.* 42:169–180.

HUDSON, J. P. 1971. Meadow orchards. *Agriculture.* 78(4):157–160.

HUDSON, J. P., L. C. LUCKWILL, AND R. D. CHILD. 1974. Meadow orchards. Redesigning the apple tree with the aid of growth regulators. Long Ashton Res. Sta. Rpt.

SHAULIS, N., AND K. KIMBALL. 1955. Effect of plant spacing on growth and yield of Concord grapes. *Proc. Amer. Soc. Hort. Sci.* 66:192–200.

CHAPTER 10

Nursery Practices and Management

Propagation

To establish a stable civilization, our ancient progenitors had to change their life-style from that of nomads to that of plantsmen. They began to collect seeds not only for immediate consumption but also for sowing the following season. These early farmers settled along rivers because water was essential for their survival as well as for irrigating crops. Furthermore, the soils were more fertile. Among the various crops they gathered, they must have observed that certain plants yielded larger seeds that germinated better, tastier fruits, or stronger and softer fibers than others.

Ancient people also discovered that some plants were easy to propagate by cuttings, or by layering the tips of shoots, thereby obtaining daughter plants identical to the mother plant. Others, especially fruit trees, were found difficult to propagate by these methods. To preserve these trees which yielded superior fruits but were difficult to root by cutting, some unknown genius had to discover the art of grafting. These means of preserving plants of identical genetic makeup are called asexual, vegetative, or clonal propagation. All propagules derived from a single mother plant belong to the same clone.

Plant Propagation by H. T. Hartmann and D. E. Kester and *The Grafter's Handbook* by R. J.

Garner are recommended for their excellent treatment of propagation of fruit trees and vines.

SEXUAL PROPAGATION OF PLANTS

The two purposes of propagating trees and vines from seeds are to obtain rootstocks and to produce hybrid seedlings. Seeds of most fruit tree species may be air-dried after harvest, but they must be hydrated before being exposed to chilling temperatures to satisfy their rest requirement. The rest requirement of seeds cannot be satisfied when they are dry, presumably because the necessary enzymatic and nonenzymatic reactions to initiate the germination process require an aqueous medium (see Dormancy and Rest of Buds, Ch. 2; and Seed Development and Physiology, Ch. 5).

Stratification of Seeds

The term *stratification* originally meant placing alternate layers of sand and seeds in a screened cage. These boxlike containers were set in the soil in a shady spot where the seeds would get sufficient chilling but would not freeze. The screen kept out rodents but allowed rainwater to leach inhibitors in the endocarp, seed coats, and embryos (see Seed Germination and Storage, Ch. 5). The sand kept the seeds moist but not excessively wet. Like many other horticultural practices, it is unknown where and when this method of chilling seeds developed.

Sowing of Seeds and Managing Seedlings

After stratification, apple and pear seeds are sown thickly in a cold frame where they are grown for a season because the seedlings of these species do not grow to budding or grafting size the first year. The yearlings or liners are planted in a nursery row about 15 to 20 centimeters (6 to 8 inches) apart.

The endocarps of stone fruits are sown 2.5 to 4.0 centimeters apart in a row. Once the seedlings emerge, they are thinned to a desired spacing of 15 to 20 centimeters. This provides enough space for seedlings to grow and for the propagator to graft or bud them. Partially germinated walnut seeds are planted 25 centimeters apart because budded or grafted walnut seedlings grow rapidly and to a large size in the nursery. Nursery seedlings are irrigated and given a light dressing of nitrogenous fertilizers at nearly weekly intervals to stimulate their growth. If the soil surface is allowed to become dry and crusted, the elongating plumule may grow around a large clod and develop into a dogleg stem (see Fig. 4.2). If peach seed coats become dry, the plumule often curls within the seed coat and develops into a crooked stem. Such a seedling is usually culled by the budder or grafter.

ASEXUAL, VEGETATIVE, OR CLONAL PROPAGATION

Asexual means of propagating plant materials are practiced to obtain a uniform stand of genetically identical trees. The simplest means of achieving uniformity is to propagate plants by cuttage and layerage so that the scion and stock are identical. Most fruit trees do not root readily; hence, they must be grafted.

In the grafting process, a bud or a shoot or a scion is appressed against another shoot or root and forced to unite and grow together. Budding is a practice in which a bud is implanted into a stock; grafting is a practice in which a scion is inserted into a rootstock. It is customary to denote a scion and rootstock combination with a virgule, for example, Elberta/Nemaguard, Bartlett/Winter Nelis seedling, or Golden Delicious/M.126.

Some methods of budding and grafting trees and grapevines were described as early as 300 B.C.; others have evolved and have been handed down from one generation to the next. Except for the use of rubber bands and plastic tapes instead of raffia to tie the grafts in place, and substitution of cold synthetic wax emulsions in place of a hot beeswax–rosin mixture to prevent cut surfaces from drying, techniques have remained unchanged from the days of Aristotle. The biological basis of grafting has not changed. That is, the scion and stock must be compatible and must unite to form a strong graft union.

A functional graft union must be established between the scion and rootstock partners for the tree to survive. The lateral cambia of the graft partners must yield a mass of parenchymatous cells, a callus; these calli must then unite. Some cells in the interior of the callus differentiate into vascular elements with a common cambium so that a complete ring of lateral cambium is again formed at the union. Tapes or nails are necessary to bind the scion to the stock securely, for pressure seems to be essential for vascular elements to differentiate. If pressure is not applied, only the callus tissue proliferates. Without the vascular connections, the scions grows weakly or the weight of the top causes the scion to split from the stock.

The Need for Specific Rootstocks

THE FUNCTION OF ROOTSTOCKS

Early civilizations were founded along rivers where soils were usually deep and fertile. These settlements grew into large cities, forcing agriculturists onto poorer soils away from the rivers. This history repeated itself later as settlers immigrated to the New World, to Australia and New Zealand, and to South Africa. As horticulture expanded in all lands, various problems related to soil, climate, and pathogens confronted the orchardists. Fortunately, scion cultivars and rootstocks possessing the following desirable characteristics were available: resistance to pathogens, insects, nematodes; tolerance of poor soil conditions, for example, flooding, extremely heavy or sandy soils, salinity; cold hardiness; and the ability of rootstocks to limit or enhance the growth of the scion tops.

Plants possessing these desirable rootstock characteristics do not often bear high-quality fruits. Conversely, cultivars that bear good fruits rarely have good rootstock characteristics. Thus, such trees are often not satisfactory when grown on their own roots. Plant propagators have made rapid advances with micropropagation, a technique of producing own-rooted trees of scion and rootstock clones through meristem and tissue culture techniques. But the possibility of successfully combining various favorable rootstock traits with high-quality fruits in a single hybrid remains a challenge to future breeders. In the meantime, breeders might make gains much faster if they developed rootstocks and scion cultivars independently, assuming a priori that grafting is essential.

The distinct advantage of planting trees with clonally propagated rootstocks and own-rooted scion cultivars is that the trees in the orchard will be genetically uniform. Consequently, the tops of the trees will tend to attain the same stature at maturity and ripen their crop simultaneously while the rootstock wards off diseases and insects and/or tolerates adverse soil conditions. Any variability of tree behavior will be due to variations in soil depth and fertility, proximity to pollinizers, and other environmental conditions.

ECONOMICALLY IMPORTANT TRAITS OF ROOTSTOCKS

Tree Size Control. For centuries gardeners have maintained small fruit trees through horticultural manipulations such as top and root pruning. Horticulturists were aware that certain rootstocks dwarfed or invigorated the scion tops, but controlling the size of apple trees commercially was not widely practiced until clonal rootstocks were developed. The East Malling Research Station in England took the lead in introducing a series of apple rootstocks that control scion growth, producing extremely small to very large, vigorous trees (see Ch. 11).

Quince roots are used to reduce the size of pear

trees, but Bartlett and other pear cultivars are not graft-compatible so that an interstock of Beurré Hardy, Old Home, or Old Home × Farmingdale (OH × F) hybrid is used. Quince clones are excellent dwarfing stocks for pears, but in calcareous soils trees grafted on quince develop lime-induced chlorosis.

Prunus tormentosa and *P. insititia* seedlings have been tried for dwarfing peaches, but the low percentage of success in grafting and the weak trees produced have discouraged their use. A few size-controlling stocks have now been developed for stone fruits (see Ch. 12).

Rogers and Beakbane (1957) noted that dwarfing stocks often have willowy or pendulous growth habits. The bark to wood ratio in these stocks is relatively large compared to that of nondwarfing stocks. These observations hold true for two dwarfing sour cherry rootstock clones, the Stockton Morello and Vladimir. The dwarfing effect of the stock seems to depend on the height of the graft union. The higher the scion is grafted on the stock, the greater the overgrowth (see Fig. 10.13) and degree of dwarfing. The overgrowth impairs translocation by restricting passage of photosynthates to the roots and transport to the top of hormones and metabolites produced in roots. Such trees, however, display no signs of water stress or nutritional deficiencies.

Nematode Resistance. Infestation of roots by the root-knot nematodes *Meloidogyne incognita* and *M. javanica*, or by the lesion nematodes *Pratylenchus* spp., especially *P. vulnus*, is extremely debilitating to fruit trees. Upon hatching on the root surface, the larva penetrates roots in the region of elongation just behind the root cap and migrates to the zones where the vascular tissues are differentiating (see in Ch. 1 Fig. 1.8 and Morphology and Origin of Roots). The invasion causes cells to collapse and coalesce, producing a giant cell or syncytium. Root-knot nematodes invade both resistant and susceptible peach roots (Malo, 1967). In susceptible cultivars the syncytium continues to enlarge and allows the larva to grow and mature (Kochba and Samish, 1971a,b). The invasion by nematodes and the formation of the syncytia or knots limit water uptake and movement, so that heavily infested plants suffer from water stress. In a resistant host tissue, the giant cell stops developing after a few days and forms thick walls that effectively seal in the larva, thereby preventing it from completing its life cycle.

Peach and grape rootstock clones and tomato cultivars that are susceptible to nematodes contain more cytokinin than resistant ones (DiCollalto and Ryugo, 1982; Kochba and Samish, 1971a,b; Van Staaden and Dimalla, 1977). The hypothesis that nematode susceptibility is the result of an interaction between the high level of cytokinin in the host root and the infusion of auxin by the nematode is based on the observations of: (1) Kochba and Samish that injection or infusion of nematode resistant peach rootstocks with cytokinin caused them to become susceptible and (2) Sandstedt and Schuster (1966), that the radicle end of a carrot explant, infested with root-knot nematodes, underwent rapid cell division, forming callus tissue which resembled that induced by indoleacetic acid. Polyphenolic compounds have been implicated in the resistance mechanism, but since they react with auxin or interfere with indoleacetic acid oxidase, their roles may be indirect rather than direct.

Once nematodes become established in an orchard, they are difficult to eradicate. Soil fumigations prior to planting are effective in reducing the nematode population, provided the roots from a previous planting are completely removed. Soil fumigation of an established orchard is precarious because tree roots are also sensitive to these fumigants. Light dosages of fumigants applied periodically will help to restrict the nematode population.

Naturally resistant traits are found in *Prunus davidiana*, *P. armeniaca*, *Pyrus communis*, and *Cydonia oblonga*. The three cherry species *P. avium*, *P. cerasus*, and *P. mahaleb* possess some degree of resistance. The resistant factor is readily transmitted to hybrid progenies. Some interspecific hybrids, for example, Marianna 2624 and peach × almond, are resistant.

Tolerance for Flooding. Most rootstocks of deciduous fruit trees do not tolerate flooded conditions for extended periods. Rootstocks that tolerate heavy soils with relatively high moisture holding capacities are sweet and sour cherries, European pear, and Marianna 2624. Among walnut rootstocks, the Paradox walnut, a hybrid between the black and Persian walnuts, is the most tolerant of heavy soils. After the Christmas Eve flood of 1955 in Yuba City, California, orchards were under 2 to 4 meters (2 to 6 feet) of water for seven to ten days. Walnut and prune trees survived, whereas almond and peach trees, which do not tolerate "wet feet," died. Mahaleb seedlings and the nematode-resistant clonal peach rootstocks Okinawa and Nemaguard do not endure water-saturated soils.

Phytophthora Resistance. Waterlogging leads to anaerobiosis, but more often the secondary infection of the trunk at the soil level by such fungi as *Phytophthora* and *Pythium* causes the death of trees. These microorganisms have motile spores that require free water about the trunk in order to invade the bark. The chances of infection are reduced by planting trees on a berm and keeping water away from the trunk. French chestnut breeders have been successful in obtaining cultivars that are resistant to *Phytophthora*.

Resistance to Oak Root Fungus. The fungus *Armillaria mellea*, also known as shoestring fungus, is commonly found living on native plants growing along riverbanks. The large fungal mycelia grow from one tree to the next, invading the root and then spreading to the cambial area of the trunk. This fungus is very difficult to eradicate once it invades an orchard.

Most fruit tree species are susceptible to attack from this organism, especially stone fruits and quince. The European pear, black walnut, and Marianna 2624 are resistant; the pomegranate and fig are immune. Seedlings of Asian pears possess some resistance but less than the European. Removing the infected tree and as much of its roots as possible and then fumigating the soil help to stem the spread of the fungus.

Crown Gall Resistance. The bacterium *Agrobacterium tumefaciens* infects roots and crowns of trees, usually through wounds, to form galls, which are tumorous and knobby growths. Nearly all fruit tree species are susceptible. The Paradox walnut, quince, and sweet cherry are exceptionally susceptible; it is not unusual to find galls, 50 centimeters or more in diameter, on their roots. Mahaleb and the sour cherry rootstock clone Stockton Morello are moderately resistant.

Inoculating the roots of susceptible species prior to planting with a suspension of *Agrobacterium radiobacter*, Kerr 84, a native of Australia, offers some protection. Treatment of galls with bactericides is impractical because most galls are located below the soil surface.

Resistance to Fire Blight. Fire blight is caused by the bacterium *Erwinia amylovora*, which invades flowers and young developing shoots, especially of pomaceous species. The infection spreads rapidly in branches of susceptible cultivars, blackening the stems and leaves. Large diseased branches form cankers in which the organism thrives. Infected parts are removed, including 15 to 20 centimeters (6 to 8 inches) of healthy tissue surrounding them. The prunings and scrapings should be burned to eliminate the bacteria.

Old Home and Old Home × Farmingdale hybrids are used as rootstock and interstocks to avoid fire blight infection in pear trunk. With the recent outbreak of fire blight in Europe, the pace of breeding for fire-blight-resistant scion cultivars has accelerated. These breeding projects have produced some promising pear cultivars (see Ch. 11).

Methods of Vegetative Propagation

BUDDING

Clonal propagation by budding is usually done on seedlings, but cuttings can also be used for rootstocks. Except for the chip budding technique, all methods of budding require the lateral cambium of the rootstock to be active so that the bark separates easily from the wood. Cambial activity slows down if rootstocks are stressed for water or nitrogen. Therefore, it is essential that nursery trees be irrigated and fertilized lightly at nearly weekly intervals, especially if rootstocks are to be budded in late August and September when cambial activity normally slows down.

Before the budding operation is started, basal branches on seedlings are removed to provide 3 to 5 centimeters of clear space in which to insert buds. Commercially, budding is done by a team consisting of a budder and a tier. The budder removes a bud from the bud stick, 25 to 40 centimeters long, and inserts it into the stock; the tier ties the bud onto the stem (Figs. 10.1, 10.2). A third worker keeps the team supplied with budsticks, makes the labels, and maintains the record. (Buds do not need to be covered with grafting wax.) Buds are usually inserted on the shady side facing the adjacent seedling so that the

Figure 10.2
A sketch of T-budding, two methods of saddle grafting, and a short ladder used by propagators. (Taken from *Pomona or Fruit Garden Illustrated* by B. Langley, 1729)

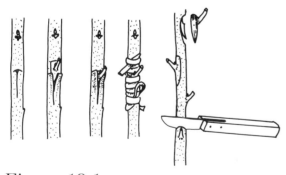

Figure 10.1
Sequence of removing bud shield and its placement and tying; on the right a shield bud is being removed, leaving the wood on the bud stick.

new shoot will grow upright in the row. This positioning protects the shoot from being injured by cultivating and spray equipment.

June Budding. The June budding method is used successfully with stone fruits in areas where the growing season is long. Seeds planted in December and January will attain heights of 40 centimeters or more by April when budding is started. The current season's growth is collected from the heavily pruned clonal mother tree and trimmed into bud sticks. A bud with a shield of bark is cut and lifted from a bud stick, leaving the wood behind to reduce the shield size (Fig. 10.1). The shield is inserted under the flap, keeping the top edge of the shield flush with the crosscut of the T. The bud is wrapped immediately with a rubber band or raffia to prevent it from drying. Four to seven days later the top is cut back to 5 to 8 centimeters above the inserted bud; this forces the bud to grow if it has united with the stock. When the new scion shoot is 15 to 20 centimeters tall, the remainder of the stock is cut back to the bud, leaving a short (5 to 8 millimeters) stub. Care should be taken not to injure the new shoot.

Without the stub, the new shoot will break or split easily at the graft union.

All growth below the graft union is removed or pinched (headed) to keep the water sprouts short and bushy. Leaves on these water sprouts provide the tree with photosynthates. The main advantage of June budding is that the trees are dug up after leaf fall, and the land is therefore occupied for only one season.

Dormant Budding. In dormant budding seedlings grown from seeds planted in the previous fall or winter are budded between July and September. A shield or chip bud is taken from a freshly collected bud stick, inserted or placed in a notch, and tied. The method is called dormant budding because the bud is not forced to grow the first season. The stock is cut back to the dormant bud in the winter; this forces the bud to grow the following spring. These standard-budded trees with one-year-old tops and two-year-old roots are dug after leaf fall. Standard-budded trees are larger than June-budded trees, but they require two years in the nursery.

Spring Budding. Bud sticks made from shoots that grew the previous season are collected in the winter and cold-stored for spring budding. The rootstock is budded as soon as its bark begins to separate or slip from the wood in the spring. A bud shield is cut with the wood attached because the cambium is not active. The rigidity of the shield makes budding easier but prevents close contact between shield and stock. But the stock soon produces sufficient callus to form a union. Two weeks later when the bud has united with the stock, the top is cut back to the bud, forcing it to grow.

June-budded trees will attain heights of 1.5 to 2 meters; dormant and spring-budded trees become 2 to 2.5 meters in height and 3 centimeters in diameter at the graft union. Larger trees are difficult to dig and cumbersome to handle and prepare for shipping. Besides, most of the top must be cut back at transplanting time to compensate for the roots that are lost at digging time. June-budded trees are preferable because they are large enough to market, yet they are small and easy to handle and transplant. All suckers and sprouts arising from below the bud union are removed when the trees are dug in the fall.

Chip Budding. Chip budding is commonly used with grapes; a small beveled piece of scion including a bud is cut and inserted into a notch made on the stock and tied (Fig. 10.3). The angle of the notch should be slightly smaller than the angle of the chip, so that the chip can be wedged in tightly. Since the bud is pressed against the stock instead of being placed inside of the bark flap, the cambium need not be active as in the methods already described. After a bud union is formed, the shoot is cut above the bud to force it to grow.

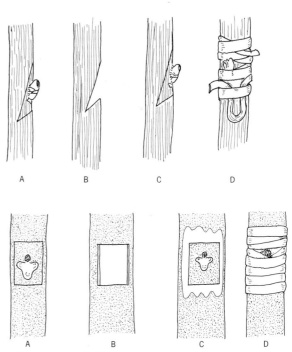

Figure 10.3
Sequence of making a chip (upper) and a patch (lower) bud graft. (A) The source bud sticks are on the left; (B) the stock is prepared to receive the bud; (C) the bud is in place, and (lower row) the bark on the stock is shaved; (D) the bud is wrapped.

Patch or Ring Budding. Two methods, patch and ring budding, which are modifications of June and dormant budding, are commonly used in walnut propagation. In patch budding, a double-bladed knife is used horizontally and vertically to obtain a square piece of bark with a bud. This patch is twisted laterally. If the bark of the walnut is lifted instead of twisted, the bud meristem often does not come free with the bark but remains behind on the bud stick. The patch is placed in the trunk of the rootstock where a square or rectangular piece of bark has been removed (Fig. 10.3) with the same double-bladed knife. The bark of the stock may need to be shaved so that the patch and the bark are equally thick.

In ring budding, parallel cuts are made with a double-bladed knife on the bud stick above and below a bud. This ring of bark is then twisted laterally off the bud stick. Using the same double-bladed knife, the budder pries a ring of bark from the seedling stock and replaces it with the ring taken from the bud stick. If the bark of the stock is thicker than that of the ring of bark taken from the bud stick, the bark of the stock is shaved down as in patch budding. The ring of bark is held tightly in place with grafting tape. The absence of the bud meristem on the patch or ring is the major source of failure of these two budding methods.

All the methods described can be used on established trees or vines if they are putting on new, vigorous growth. One or more cultivars or a pollinizer can be propagated to an existing tree by these methods. Buds should be placed only in vigorous current season's growth; they do not usually form unions if they are inserted into older branches.

Note: When the buds are checked after two weeks, the rubber bands should be cut if they have not deteriorated in the sun. Raffia, adhesive, and plastic tapes should also be cut after the bud begins to grow so that the trunk will not be girdled.

GRAFTING

A rootstock can be grafted or topworked with a desirable scion cultivar in several ways, depending on stock diameter, the time of year, and the species. The technique used when the tree is grafted in place in the orchard or in the nursery row may differ from that chosen when the stock and scion are to be bench-grafted indoors. Each method has its advantages and disadvantages.

Cleft Grafting. Cleft grafting (Fig. 10.4) is generally used to change the top variety on mature but relatively young bearing trees. The trunk or scaffold limb is carefully cut transversely, so as not to tear the bark away from the wood. The stock is then split down the middle with a special grafting tool driven with a mallet. The wedge on the tool is used to hold the two sides apart while the scion is put in place. A large meat cleaver or a chisel will substitute for the special splitting tool.

The scions are tilted slightly outward to assure that their cambia cross about 1 to 2 centimeters below the transverse cut on the stock. If the stock is small, only one scion is wedged in the cleft, held in place by a rubber band or tape, and waxed (Fig. 10.5). If the stock or scaffold limb is 5 to 8 centimeters in diameter, two scions are wedged in the cleft, one on each side. The scions are cut so that the beveled edge faces the center of the stock. Such a cut assures that the cambia of the scion and the stock touch and eventually unite. Trunks and branches of fruit trees usually have enough tension that they need not be drawn together to hold scions in place. Two advantages of this method are that it is easy and that it can be done in late winter when the bark is not yet slipping. The main disadvantage is that the cleft leaves a large wound which may allow infection by wood-rotting organisms should the grafting wax crack.

Bark Grafting. Scions are prepared for bark grafting by making a long beveled cut on one side

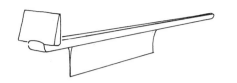

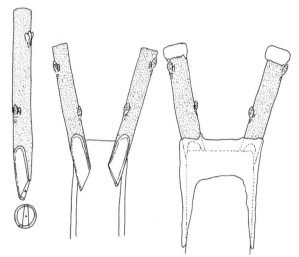

grafted in place about the cut surface. Wedge-shaped scions are cut (Fig. 10.7). A thin saw blade is used to make vertical slits on the edge of the stock. Then thin triangular wedges are removed from both sides of a slit with a leather worker's half-round knife. The kerf is gradually enlarged until the notch is slightly narrower than the bevel of the scion. The scion is then tapped into place. It is easier to shape the notch in the stock to fit the scion than to do the reverse.

This method produces a relatively small wound and allows more than two scions per stock, in contrast to cleft grafting. Another advantage is that it can also be used before and after the bark begins to slip in the spring. Half-round knives are not readily available, however, and matching the bevel of the scion to the stock is difficult for a novice. If the bark should lift and separate from the wood next to the scion, the bark should be nailed back in place.

Figure 10.4
Cleft grafting of a large stock. Beveled scion (left); in position (center) with stock cut away to illustrate where the two cambia cross; and when completely waxed (right). The splitting–wedging tool is shown above.

and a short one on the opposite side (Fig. 10.6). The scion is held against the stock, and a band of bark of equal width to the scion is cut; the uppermost 5- to 8-millimeter portion of the bark is then removed. The scion is slid behind the bark and nailed in place with three wire brads, 2 to 3 centimeters in length. The main advantage of this method is the ease and rapidity with which it can be carried out. Its disadvantage is that it cannot be used until the bark begins to slip in the spring.

Saw-Kerf Method. If the stock is 6 to 12 centimeters in diameter, three to four scions can be

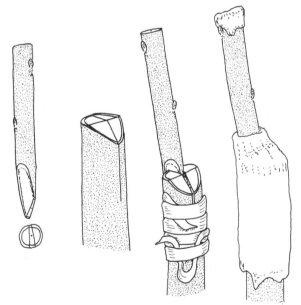

Figure 10.5
Cleft grafting onto a stock of unequal diameter. Beveled scion and the split stock (left); the scion in place and tied (center); and after being waxed (right).

Methods of Vegetative Propagation

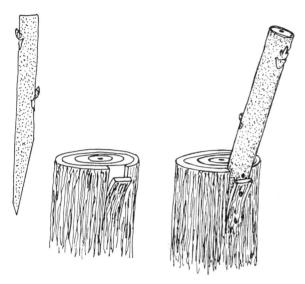

Figure 10.6
Bark graft with double-beveled scion (left); stock with bark flap peeled back (center); and scion in place prior to waxing (right).

Whip Grafting. The whip grafting method is also called the English or palm graft. It is called palm grafting because the graft resembles a handshake when the scion and stock are in place (Fig. 10.8). This method is utilized if the scion and stock are about equal in diameter. If the stock has a thicker bark, the scion diameter should be smaller so that the rings of cambia match. Since there is a little gap between the scion and stock, the close fit affords maximum opportunity for the two cambia to grow together and exposes only a minimum area to microorganisms to infect. Whip grafting can be done during the dormant season when the cambia are still inactive and can be continued late into spring after the bark begins to slip. Scions must be held in place with a rubber band, or plastic or adhesive tape, and covered with grafting wax. The binding material must be cut vertically about two weeks after budbreak so that the scion will not be girdled. Placed against the trunk is one long, vertical tape, which facilitates its removal, especially in grafted walnut seedlings. The graft is then wrapped with a second tape to prevent drying as well as to create pressure while the cambium forms a callus and the stem diameter increases. A third vertical tape is now superimposed over the wrapping and the first vertical tape. The horizontally wrapped portion is then covered with wax, leaving both ends of the vertically placed tapes free. When the time comes to cut the wrapping, the outer vertical tape is pulled off, leaving a strip of wrapping free of wax. One vertical cut with a knife will expose the inner vertical tape. When this inner tape is pulled free, the cut edges of the wrapping tape are raised. Thus, the tapes will not girdle the trunk, and the knife used for this operation remains clean and free of wax.

Bench grafting is a technique of whip grafting that was developed for areas where winters are so severe that propagation cannot be done outdoors. Seedling rootstocks or clonally propagated cuttings are dug up in the fall and held in a moist sawdust bin. The rootstocks are removed from the sawdust and whip-grafted with the desired scions; the grafted partners are tied and the union waxed. The trees are replaced in the bins and cold-stored to allow the unions to form a callus

Figure 10.7
Saw-kerf graft. Double-beveled scion (left); stock with a scion in place and a notch ready to receive the next one (right). The half-round knife is in position for the third notch.

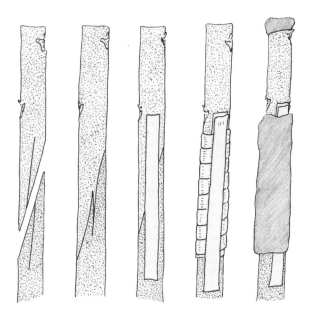

Figure 10.8
Whip or tongue and groove graft. Scion and stock (A) with long slanting cuts, (B) with tongue in place, (C) with first vertical tape in position, (D) laterally bound and second vertical tape in place, and (E) when waxed.

Figure 10.9
Matched scion and stock prepared for grafting with the "omega" knife (left), with a modified chip bud knife (center), and with circular dado saw blades (right). The dado saw blades are commonly used for grafting grapevines.

before planting. Grafted grape cuttings form the callus rapidly if they are stored at room temperature in moist sawdust or a comparable material for a few weeks. Mechanical knives and saws have been manufactured to expedite the bench grafting operation. These mechanical blades fashion precise cuts of equal diameters on the stock and scion (Fig. 10.9).

Nurse-Root Grafting. A nurse-root graft is another means of obtaining a desirable scion–stock combination, but it is also used to obtain own-rooted plants. If a cultivar is difficult to root by cutting or by layering, a scion is whip- or cleft-grafted onto a piece of root, a nurse root (Fig. 10.10). The graft combination is planted so that the union is below the soil surface. This allows the scion to initiate roots. For example, a piece of quince root is whip-grafted to a stem piece of Old Home pear rootstock. The graft combination is planted with the tip of the Old Home scion barely above the soil surface. As the Old Home shoot develops, the belowground portion initiates roots. The quince nurse root is eventually overgrown and dies.

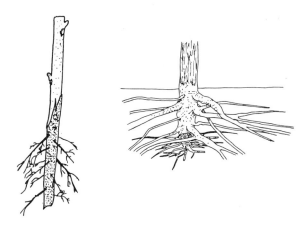

Figure 10.10
A nurse-root graft before wrapping with tape (left); own-rooted Old Home pear/quince combination after several years in the orchard (right). The quince roots nourish the tree until the Old Home root system develops.

Methods of Vegetative Propagation

Bridge Grafting. Bridge grafting is often used to repair rodent damage to roots or damage to trunks caused by cultivators and trunk shakers during harvest. The upper and lower edges of the damaged portion are trimmed smoothly. Flaps equal in width to scions are spaced 10 to 15 centimeters apart on the trimmed portion. Scions, slightly longer than the length of the damaged part and having a wedge cut on each end, are slipped under flaps as in bark grafting, nailed in place, and waxed.

Inarching. Inarching is a grafting method whereby a seedling is planted next to an established tree and headed back. The decapitated tip is shaped into a wedge and inserted under a bark incision on the established tree (see Fig. 6.52). The inarched shoot is nailed in place, and the grafted zone is covered with wax. The method is used to bypass portions of the damaged trunk when bridge grafting is impossible and sometimes to overcome partial graft incompatibility of the original union.

Special Precautions

Walnut, kiwifruit, and grape exude xylem sap in the late winter and early spring. Hence, they should be grafted after exudation ceases. In walnut, the seedling stock is headed back and allowed to bleed; after bleeding ceases, the stock is recut and the scion grafted. As a precautionary measure, slanting cuts are made through the bark into the wood so that should additional bleeding occur, it takes place at these cuts on the trunk rather than at the graft union. When bleeding occurs at the union, scions rarely unite.

Grafted trees should be whitewashed after planting to prevent sunburning of the scion and stock, for several weeks may pass before buds on scions grow enough to cast shade. Shoots arising from grafted scions are as fragile as those of budded seedlings. They should be tied to stakes to prevent breakage. On larger grafted trees, laths should be nailed on the trunk and developing shoots tied to them as they elongate.

Graft Incompatibility

Incompatibility between graft partners can be physiological, anatomical, or disease-induced. These categories are arbitrary because ultimately all incompatibilities are traceable to genes of the scion–stock combinations or of the host and its pathogen. Graft incompatibility causes loss of vigor or kills the tree, usually because a union (1) does not form between the scion and stock when they are taxonomically unrelated; (2) restricts transport; (3) is unable to metabolize or detoxify substances arriving from either the scion or the rootstock; (4) deteriorates as a result of an infection by a pathogen or its toxic metabolite; or (5) is anatomically weak and breaks.

Graft incompatibilities between a scion and stock that cause a slight loss of vigor or reduce stature may be beneficial to growers. They often facilitate tree size control and induce precocity.

Genetic Incompatibility. The farther the scion and rootstock diverge taxonomically, the greater will be the probability of graft failure. Just as an apricot pollen fails to germinate on a stigma of an apple flower, scions from a taxonomically unrelated tree do not make a functional union with the stock. Apparently, proteins of one species do not recognize those of another when the scion and stock are brought together. Thus, there is mutual rejection.

Among fruit tree species, the problem of graft incompatibility occurs even within the subgenus. Scions of apricot, almond, and peach do not unite with the common sweet cherry stocks. Peach scions grafted on seedlings of Nanking cherry, *Prunus tomentosa*, sand cherry, *P. besseyi*, and St. Juliens plums, *P. insititia* result in weak and short-lived trees.

On rare occasions, scions of distantly related species will unite with a rootstock to make a good, strong union. Pear cultivars have been grafted on dwarfing quince stock in Europe for centuries. Bartlett pear is incompatible with some quince stock so that an interstock of Beurré Hardy or Old Home is inserted between the two (Figs.

Figure 10.11
Transverse section of a bud graft of Bartlett pear (B) with a Beurré Hardy (B.H.) intermediate on a quince (Q) understock. Callus is being formed under the quince bark (arrow) and between the Beurré Hardy and quince tissues by the lateral cambium (C). (From Scaramuzzi, 1956).

10.11, 10.12). Early settlers in California, arriving with scions of chestnut, topworked them on native oak trees that belong to the same family, *Fagaceae*. They apparently knew that the combination was successful in Europe. There are some old chestnut trees grafted on oak still thriving in scattered areas.

Physiological Incompatibilities. If the cambia of the scion and stock do not divide at equal rates, the cambium of one graft partner will produce more xylem and phloem elements than the other and cause an overgrowth or undergrowth at the union (Fig. 10.13). Unequal growth retards transport of food elaborated in the top of the tree to the root. Unless the overgrowth is extreme and the tree is short-lived, it is not necessarily undesirable. Sweet cherry cultivars grafted on Vladimir, a sour cherry rootstock, are dwarfed and come into bearing earlier than those grafted on mahaleb or mazzard stocks, which form a smooth trunk at the union. One note of caution; when considerable overgrowth is expected, as in the case of Bing cherry grafted on Vladimir stock, only one scion per stock should be used. When two scions are positioned opposite each other, the overgrowth forces the scions apart or causes the rootstock to split and damage the tree. When Vladimir is used as a dwarfing interstock, the trunk becomes "cinched" (Fig. 10.13). When Bartlett scions were grafted on *Pyrus communis* seedlings that originated from seeds imported from France prior to World War I, the trunk acquired a "churn bottom" shape at the union (Fig. 10.13).

Unions of some graft combinations deteriorate because one graft partner produces metabolites that are toxic to the other partner or are catabolized to a toxic end product. Cyanogenetic glucosides are metabolites suspected of inducing graft incompatibility in certain pear and apple graft combinations; hydrolysis of these compounds yields hydrocyanic acid, a poisonous end product.

Anatomically Weak Graft Unions. Certain stock–scion combinations develop genetically compatible, physiologically sound, but anatomically weak unions. The lateral cambium apparently produces more parenchyma cells than fibers. The few fibers that are produced do not interlock to form a strong union. In such unions the scion frequently separates cleanly from the rootstock after the trees are several years old and have developed a full canopy. Breakages often occur during a strong wind when trees are bear-

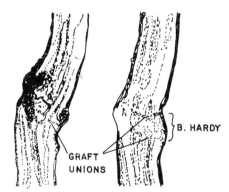

Figure 10.12
Bartlett pear/quince graft combination showing signs of incompatibility (left); that with the Beurré Hardy interstock (right) is healthy. (From Scaramuzzi, 1956)

Methods of Vegetative Propagation

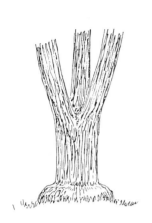  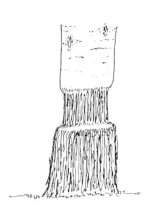

Figure 10.13
Undergrowth of scion forming a churn bottom trunk (left); overgrowth of scion on a dwarfing rootstock (center); and cinching of a trunk caused by a slow-growing interstock (right).

ing full crops. Almond, apricot, and peach are cross-graft-compatible and develop functional unions, but apricot trees on peach rootstock occasionally produce anatomically weak unions. Similarly, apricot, topworked on Myrobalan plum stock, is likely to break cleanly at the union.

Pathogen-Induced Incompatibility. Most pathogen-induced incompatibilities are caused by viruses that invade the scion top. The disorder does not become apparent until the pathogen arrives at the graft union and causes the cells of the rootstock to deteriorate. If the rootstock is tolerant of the pathogen, the cells at the graft union may turn brown, but the discoloration eventually disappears. Such a tree will be a symptomless carrier. Thus, trees with a tolerant scion–rootstock combination are not graft-incompatible.

Black line disease of Persian walnut, *Juglans regia*, is a pathogen-induced incompatibility caused by a pollen-transmitted virus. The virus infects the pistillate flowers, invades the spur, and moves downward to the graft union. If the tree is grafted on a California black walnut seedling, *Juglans hindsii*, the cells of the rootstock react spontaneously to the virus, yielding a black exudate as they succumb. Other rootstock species, resistant or immune to this virus, have the same black line reaction.

The formation of the black pigment is a hypersensitive response by the stock cells to the attempted viral invasion. The cells rapidly produce abnormal metabolites, known as phytoalexins. The production of phytoalexins kills the stock cells at the union and blocks the pathogen from invading the resistant tissue. The graft union continues to deteriorate as more cells die, so that the top eventually succumbs while the rootstock survives, sending forth many water sprouts. The appearance of water sprouts is an early symptom of the disease. That virus particles have not been isolated from water sprouts indicates that the virus has not invaded the stock.

Pear decline disease of French pear is caused by a mycoplasma that is transmitted by the insect pear psylla, *Psylla pericola*. The microorganism that resides in the salivary glands of infected psylla is transmitted to pear leaves. The mycoplasma multiplies and moves down the phloem to the graft union. Phloem cells at the graft union die, interrupting downward transport of photosynthates and thus causing the tree to die.

Nearly all Bartlett pear/Asian pear and Bartlett/Beurré Hardy/quince combinations are susceptible to the disorder, and 25 percent of Bartlett/Winter Nelis seedlings succumb. Own-rooted Bartlett pear trees may lose some vigor, but they do not die.

Bartlett pear trees expressing symptoms of pear decline have recovered if pear psylla was controlled with pesticides during the early stages of infestation or if the trees were injected with an antibiotic, for example, tetracycline. Sickly Bartlett trees have also recovered from the disease when trees were regrafted to an Asian pear cultivar leaving a Bartlett interstock. The recovery is attributed to the feeding preference of the psylla; it prefers to feed on European pear leaves rather than on Asian leaves.

Pesticide treatments and grafting trials indicate that if the trees are not reinoculated with new mycoplasma, (1) the microorganisms remaining from previous infections in the Bartlett top either do not survive the winter or lose their virulence after a period; and (2) the lateral cambial activity regenerates new xylem and phloem elements so that transport through the graft union can resume.

Buckskin disease of sweet cherry is caused by a mycoplasma that is transmitted by leaf hoppers and perhaps other vectors. Trees with this disease bear lusterless, leathery-skinned, pale fruits. The graft combination of sweet cherry on mahaleb seedlings, *Prunus mahaleb*, is highly susceptible. If the graft union is high on the primary scaffold branches, the rootstock of a diseased tree usually produces many water sprouts from the trunk and scaffold limbs. If only one scaffold branch is infected, it can be regrafted below the original union to redevelop the tree because the mycoplasma does not invade the mahaleb stock.

Injecting tetracycline, an antibiotic, under pressure into trees infected with mycoplasma has provided remarkable but not always complete recovery.

CUTTING

Plants can be multiplied from pieces of shoot, root, leaf, and even from a callus. Shoot and root pieces that can develop into entire plants are called cuttings. Most woody species that root easily are propagated by stem cuttings. Some species, for example, *Cydonia*, form sphaeroblasts, or preformed root initials in wartlike structures. Stem cuttings of these species form roots easily and can be taken any time of the year. Cuttings usually initiate roots more readily in early summer than in late summer or during the dormant season. Young buds and leaves produce and export more auxin than is essential for rooting compared to older mature buds and leaves, or they do not export as much rooting inhibitors as do the older organs (Kawase, 1964, 1965). Cuttings that are made from actively growing shoots are termed softwood or leafy cuttings.

Softwood Cuttings. Softwood cuttings, one to several nodes in length, are made from healthy shoots because sound leaves are essential for root initiation. To cut down on transpiration and prevent wilting, the propagator removes the basal leaves and reduces large terminal leaf blades to half or a third of their original size (Fig. 10.14).

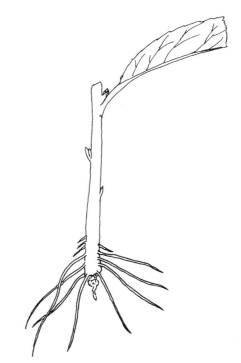

Figure 10.14
A softwood cutting of a peach ready to be transplanted. The basal part of the stem immersed in indolebutyric acid dissolved in 50 percent ethanol often becomes necrotic because ethanol is phytotoxic.

After the bases are treated with a solution or a dry preparation of indolebutyric acid (IBA), a synthetic rooting hormone, the cuttings are imbedded in flats at least two nodes deep because the first roots to emerge are usually located near these nodes. Leaves should not touch the soil, for they can become infected by some soil-borne organisms.

Flats, containing a well-drained rooting mixture, are placed in a high-humidity misting chamber where the cuttings are fogged for 5 to 10 seconds at 10- to 15-minute intervals. The beds of these misting chambers are often warmed with a heating cable to about 25° C. The addition of heat hastens callus and root formation. After several roots are formed, cuttings are transplanted to pots and replaced under the mist for a few more days; they are moved onto a greenhouse bench to harden.

Semihardwood Cuttings. Leafy semihardwood cuttings are made in middle to late summer after shoot elongation ceases and stems mature. They are prepared and handled in the same manner as softwood cuttings.

Hardwood Cuttings. Hardwood cuttings are made during the dormant season, usually treated with a rooting hormone, and stored temporarily in damp peat or sphagnum moss while the bases of the cuttings form callus. They are then transplanted into the nursery row and they develop roots and shoots the following growing season. Grapevines are commonly propagated in this manner.

The Biology of Root Initiation

Cuttings made from juvenile shoots root more readily than do those taken from older mature wood. An explant taken from a developing plumule, or a cutting made from a young seedling, or a water sprout arising from the trunk near the soil surface initiates roots more readily than does a cutting prepared from a bearing shoot collected from a mature tree.

Buds and leaves play important roles in the rooting process. If buds on a cutting are excised, few or no roots are formed, suggesting that a rooting hormone is produced in buds and translocated down to the base (Went, 1938). The term *rhizocaline* was given to the root-forming substance by Bouillenne (1933), a colleague of Professor F. C. Went. This substance, like the flowering hormone florigen, has not been isolated and identified. The presence of leaves hastens the rate of root formation, the number of roots formed being proportional to the leaf area. Petioles of some herbaceous species, for instance, African violet and common bean, form roots, indicating that the leaves of some species produce rhizocaline but to a lesser extent than buds.

When apical buds of Marianna 2624 cuttings were treated with indolebutyric acid, the cuttings rooted as well as those in which the bases were treated with IBA but better than those that did not receive the hormone treatment. When $^{14}CO_2$ was administered to leaves of Marianna 2624 cuttings in a concurrent experiment, radioactivity accumulated in the node to which the leaf was attached. Radioactivity was detected in bases of cuttings only after they became swollen and formed root initials (Breen and Muraoka, 1974). These experiments demonstrate that (1) root initiation is the result of an accumulation of basipetally transported rooting hormone in the base; and (2) the initiation may rely on reserve food but does not require current photosynthates.

Hess (1962), using the mung bean hypocotyl bioassay, demonstrated that easy-to-root cultivars contain phenolic cofactors that favor rooting, but difficult-to-root cultivars contain more polyphenolic substances which inhibit root initiation.

Kawase (1964, 1965) discovered that root initiation was enhanced by wrapping the basal parts of cuttings with black plastic tape or by centrifuging leafy cuttings. His analyses revealed that (1) the parts of stems from which light was excluded contained higher auxin levels than the exposed portions; and (2) the liquid remaining in the centrifuge cup contained substances that inhibited rooting of mung bean hypocotyl. His findings support the thesis that light causes photooxidation of auxins and that rooting inhibitors are diffusible.

Nemaguard peach cuttings treated with indolebutyric acid evolve ethylene gas instantaneously; the amount of ethylene gas evolved and the number of roots initiated are proportional to IBA concentration. Nemaguard cuttings fail to initiate roots if treated simultaneously with IBA and aminoethoxyvinylglycine (AVG), a compound which inhibits ethylene evolution. This suggests that ethylene plays an essential role in the rooting process, but that exposing basal ends of Nemaguard cuttings to ethylene or treating them with ethylene-generating compounds alone does not induce rooting. Thus, there are some unanswered questions. What is the role of auxin-induced ethylene in the rooting process? Does ethylene merely lower membrane permeability by allowing IBA to diffuse more rapidly to target sites? What are the bases for the difference in the responses to ethylene of Nemaguard cuttings and of mung bean hypocotyl since mung bean hypocotyl undergoes rhizogenesis when exposed to the gaseous hormone alone?

Hardwood cuttings of some species such as the Persian walnut form roots, but their buds often fail to grow. If the buds grow, the shoots wither soon afterward so that cuttings do not survive. The inability to survive is attributed to the depletion by the developing roots of reserve food stored in the basal portion of the stem piece. Fewer cuttings survive if bottom heat is used to induce the formation of callus on the cut ends, for heat hastens the depletion by increasing the respiration rate. Subsequent shoot development may utilize the remainder of the stored food before newly formed leaves are able to support the new plant. Forcing budbreak on cuttings with topical application of cytokinin on buds when roots are just emerging improves the survival rate of these cuttings, possibly by partitioning the reserve food to both root and shoot. After the new leaves begin to export photosynthates to the stem, the plants become self-supporting or autotrophic.

Easy-to-Root and Difficult-to-Root Species

Grape, quince, blackberry, raspberry, fig, pomegranate, gooseberry, currant, and kiwifruit initiate roots readily from softwood or hardwood cuttings. The rates of rooting and root development improve when the base of the cutting is treated with IBA or commercially prepared rooting hormones.

Most stone and pome fruit cultivars and nut species are difficult to establish on their own roots by cuttings. The exceptions are softwood cuttings of Marianna 2624, Stockton Morello, Nemaguard peach, and some peach–almond hybrids that readily initiate roots under mist.

LAYERAGE OR LAYERING

Many layerage or layering techniques were widely known and practiced before the Roman period. These are vegetative means of propagating daughter plants while they are still attached to the parent plant. Plants are manipulated to induce roots on their stems so that the resulting daughter plants are own-rooted. The daughter plants may be scion cultivars or clonal rootstocks.

Mound or Stool Layering. Clonal rootstocks that are difficult to obtain by cuttings, such as some East Malling apple rootstocks, are propagated by mound layering. A small tree of a desired cultivar or rootstock clone is planted and headed back to force numerous shoots the first season. At the end of the growing season, the bases of these shoots are slashed upward, and a wad of cotton soaked in a solution of IBA is inserted into each wound. A short length of copper wire may be looped twice about the stem below the treated zone so that the shoot will become girdled as growth begins the following season. The bases of the shoots are then covered with a mound of light soil or sawdust, which is kept in place with upright boards (Fig. 10.15). Roots are initiated during the following growing season. The propagules are severed from the parent plant during the subsequent dormant season.

Trench Layering. In a modification of mound layering, called trench layering, a tree or vine of a desirable cultivar is headed low and allowed to grow vigorously the first season. Long branches are girdled, 5 to 8 centimeters apart, with each girdled segment having one or two buds. The girdled portion may be treated with IBA before burying the branch in a shallow trench and covering the horizontal portion but leaving the shoot tip exposed. In the spring, the buried buds develop into shoots, and roots are initiated on the stem segments between girdles. The rooted pieces are separated from each other during the dormant season. Filberts, apples, and plants with long canes, for example, trailing blackberry, are multiplied in this manner.

Tip Layering. Brambles are commonly propagated by burying tips of canes during their growing season. If the soil is kept moist, the shoot continues to grow while roots are initiated below

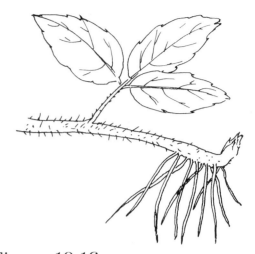

Figure 10.16
A tip-layered bramble shoot ready to be severed from the mother plant and transplanted.

the soil surface (Fig. 10.16). These newly rooted shoot tips are severed from the parent plant and transplanted. Grapevines may be propagated in this fashion to fill in an adjacent missing vine. The daughter plant derived by tip layering should not be severed from the mother plant for two or more years to assure that it has rooted and is growing.

Aerial Layering. In aerial layering, a branch is usually stripped of lateral branches and leaves in the spring, and a girdle is made at the base of the cleared region. This portion of the stem is covered with a moist medium in which roots are formed. Because it is an ancient technique, many ways are illustrated in old propagation handbooks and magazines. In one illustration a stem is pushed through a hole in the bottom of a terra cotta pot on a platform. The pot is then filled with a light-textured rooting medium. An old Japanese propagation manual shows the stem portion covered by a piece of large bamboo stem which had been split in two, joined, and filled with a moist rooting medium (Fig. 10.17). In a more recent modification, the girdled portion is treated with IBA and covered with a plastic sleeve or a sheet of aluminum foil, which is filled with moist sphagnum moss.

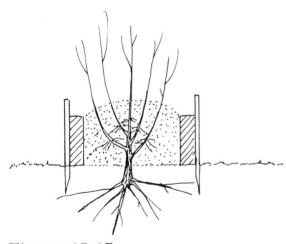

Figure 10.15
A mound-layered tree with rooted shoots at the end of the second growing season.

Figure 10.17
An aerial layerage using a large bamboo stem (From *Japanese Grafter's Manual*, Anon.)

When sufficient roots develop, the rooted stem is severed from the mother plant below the girdle and transplanted into soil. Some species require two growing seasons to form a good root system. This method is not commonly used for fruit trees because it is time-consuming, but it may be the only way to propagate some cultivars that do not readily form root initials by other techniques.

PROPAGATION BY CROWN DIVISION, SUCKERS, AND STOLONS

Crown Division. Species that form suckers or tillers at the crown, such as raspberry, dewberry, and strawberries, are propagated by separating the daughter plants from the mother plant. Usually dormant plants are dug up, and the crown is divided so that each subdivision has a bud and some roots. Although crown division is a common means for multiplying ornamental plants, it is not used to propagate brambles because these species can be rooted readily by softwood or hardwood cuttings.

Rooted Suckers. Suckers that arise at the base or at some distance from the trunk, as in Stockton Morello and Vladimir rootstocks, may be excavated and transplanted. Spraying or dipping the roots of suckers in an IBA solution produces a better stand of plants than if they are untreated.

Runner or Stolon Daughter Plants. Facultative short-day strawberry cultivars form runner plants when they are grown under long-day conditions or when they are exposed to excessively long chilling periods. Hence, strawberry plants are multiplied by maintaining them in a highly vegetative stage in areas where summers are long and warm. Some day-neutral or ever-bearing types that do not form stolons pose a problem to propagators.

IN VITRO CULTURE, OR MICROPROPAGATION

Meristem, callus, and single-cell cultures for propagation of plants have gained wide popularity in recent years (Street, 1973). With both callus and single-cell cultures, cell division occurs so rapidly that the chromosomes, rather than divide equally as in mitosis, often divide erratically, giving rise to somatic mutations. Hence, when plantlets are regenerated from these calli or cells, they frequently differ from the mother plant from which the callus or cell was originally obtained.

Meristem Culture. In meristem culture, a terminal or lateral bud is grown aseptically in an agar or liquid medium containing various salts, vitamins, nutrients, and hormones. The apical bud elongates, developing several lateral shoots which are excised and subcultured in a medium containing the rooting hormone IBA. After the roots form, the plantlets are removed from the aseptic environment, acclimated temporarily, and transplanted outdoors. Theoretically, meristem culture provides a means of clonally propagating plants because terminal and lateral buds should give rise to plantlets identical to the parental material.

Meristem culture has been used successfully to obtain virus-free strawberry plantlets. An apical meristem of a rapidly elongating stolon of a virus-infected plant is isolated and grown aseptically. The meristems are cultured under optimum growing conditions, forcing them to divide and elongate rapidly. The virus particles are unable to multiply as rapidly as cells in the apical meristem; the basal portion of the growing meristem therefore remains infected while the tip becomes virus-free. Plantlets excised and grown from these virus-free tips are then multiplied vegetatively by forcing stolons.

The inability to induce roots *in vitro* on lateral shoots derived from meristems is the biggest obstacle in regenerating complete plants by micropropagation of woody species. Species such as the chestnut and walnut with high levels of phenolic substances (the agar in which the shoot is imbedded turns gray after a day) are especially reluctant to form roots. Some success in avoiding the toxic effects of the diffusates has been obtained by daily transferring shoot meristems into a fresh culture medium, rich in cytokinin, for a period of several weeks. Or the meristems can be suspended above a liquid medium with a filter paper "bridge." When inhibitory phenols are allowed to diffuse out of the meristems and the tender lateral shoots are exposed to a fresh supply of cytokinin, the shoots behave as though they were juvenile organs. Vieitez and Vieitez (1980) excised chestnut meristems from plumules of germinating seeds and subcultured them in a dark chamber for several days. They dissected the elongating lateral shoots and treated them with IBA. Weeks later, the stem pieces initiated roots.

Callus Culture. If, under optimum culture conditions, small pieces of callus are dissected from the base of cut stems, bark tissue, or cortical cells and then cultured on an artificial medium, some cells of the callus will undergo various kinds of organogenesis. Harada (1975), using a modified culture medium (Murashige and Skoog, 1962), was able to regenerate entire kiwifruit plants from callus derived from cortical cells of a shoot (Fig. 10.18). He found that with kiwifruit callus cytokinins favor the formation of shoots. Certain combinations and dosages of auxin and gibberellins favor continued cell division and callus enlargement.

Single-Cell Culture. Single-cell cultures are obtained by treating callus with tissue-macerating enzymes which dissolve the middle lamella and parts of the cell wall. Steward (1958), who successfully regenerated a whole carrot plant from a derivative of a carrot callus, was the pioneer in this field of research. Various artificial media and their modifications have been developed for multiplying shoot meristems and regenerating plantlets from single-cell cultures.

Figure 10.18
Kiwifruit plant (right) regenerated from callus derived from cortical tissue (left) by manipulating the zeatin and indolebutyric acid concentrations in a modified Murashige–Skoog medium. (Courtesy of I. Okuse)

References Cited

BOUILLENNE, R. 1933. Néoformation des racines sur hypocotyls chez les plantules de *Impatiens balsamina*. *Ann. Jard. Bot.* 43:25–28.

BREEN, P. J., AND T. MURAOKA. 1974. Effect of leaves on carbohydrate content and movement of ^{14}C-assimilate in plum cuttings. *J. Amer. Soc. Hort. Sci.* 99:326–332.

DiCOLLALTO, GIOVANNI, AND K. RYUGO. 1982. Citochinine endogene e resistenza allo Xiphinema index in alcuni portinnesti e specie di vite. *Vignevini* VI:39–41.

GARNER, R. J. 1979. *The Grafter's Handbook.* 4th Ed. Oxford University Press. New York.

HARADA, H. 1975. *In vitro* organ culture of *Actinidia chinensis* Planch. as a technique for vegetative multiplication. *J. Hort. Sci.* 50:81–83.

HARTMANN, H. T., AND D. E. KESTER. 1983. *Plant Propagation.* 4th Ed. Prentice-Hall. Englewood Cliffs, N. J.

HESS, C. E. 1962. A physiological comparison of rooting in easy and difficult-to-root cuttings. *Twelfth Proc. Plant Prop. Soc.* 265–268.

KAWASE, M. 1964. Centrifugation, rhizocaline and rooting in *Salix alba* L. *Physiol. Plant.* 17:855–865.

KAWASE, M. 1965. Etiolation and rooting in cuttings. *Physiol. Plant.* 18:1066–1076.

KOCHBA, J., AND R. M. SAMISH. 1971a. Level of endogenous cytokinins and auxin in roots of nematode-resistant and susceptible peach rootstocks. *J. Amer. Soc. Hort. Sci.* 97:115–119.

KOCHBA, J., AND R. M. SAMISH. 1971b. Effect of growth inhibitors on root-knot nematodes in peach roots. *J. Amer. Soc. Hort. Sci.* 97:178–180.

LANGLEY, B. 1729. *Pomona or Fruit Garden Illustrated.* G. Strahan. London.

MALO, S. E. 1967. Nature of resistance of "Okinawa" and "Nemaguard" peach to the root-knot nematode *Meloidogyne javanica*. *Proc. Amer. Soc. Hort. Sci.* 90:39–46.

MURASHIGE, T., AND F. SKOOG. 1962. A revised medium for rapid growth and bioassay with tobacco tissue cultures. *Physiol. Plant.* 15:473–497.

ROGERS, W. S., AND A. B. BEAKBANE. 1957. Stock and scion relations. *Ann. Rev. Plant. Physiol.* 8:217–236.

SANDSTEDT, R., AND M. L. SCHUSTER. 1966. The role of auxins in root-knot nematode-induced growth on excised tobacco stem segments. *Physiol. Plant.* 19:960–967.

SCARAMUZZI, F. 1956. Ricerche sull'innesto "A doppio scudo" quale mezzo proposto per superare la disaffinita del pero sul cotogno. *Ann. Speri. Agraria.* n.s. Rome.

STEWARD, F. C. 1958. Growth and development of cultivated cells. III. Interpretations of the growth from free cell to carrot plant. *Amer. J. Bot.* 45:709–713.

STREET, H. E. 1973. *Plant Tissue and Cell Culture.* University of California Press. Berkeley, Calif.

VAN STAADEN, J., AND G. G. DIMALLA. 1977. A comparison of the endogenous cytokinins in the roots and xylem exudates of nematode-resistant and susceptible tomato cultivars. *J. Exp. Bot.* 28:1351–1356.

VIEITEZ, A. M., AND E. VIEITEZ. 1980. Plantlet formation from embryonic tissue of chestnut grown *in vitro*. *Physiol. Plant.* 50:127–130.

WENT, F. W. 1938. Specific factors other than auxin affecting growth and root formation. *Plant Physiol.* 13:55–80.

PART III

Species Characteristics

Various fruit species and cultivars grown in temperate zones of the world are described in this section. Some species such as figs, persimmons, and pomegranates are minor crops in the United States and Europe, but they are commercially important where they originated. These and other species such as the kiwifruit are now being planted more widely because their acceptance as dessert fruits is broadening. These exotic fruits not only add color and nutrition to our diet but also elevate our standard of living.

This section is divided into groups by horticultural fruit types. Hence, the almond, a drupe, is included with the stone fruits, but the pistachio, which is also a drupe, is included with the nut crops. The *Ribes* and *Vaccinium* are grouped with the berries.

The chromosome numbers for the species were taken from *Chromosome Numbers of Flowering Plants* (1969).

CHAPTER 11

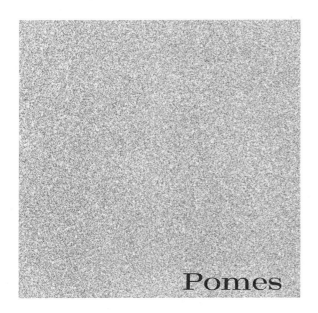

Pomes

Fruits classified as pomes belong in the subfamily or tribe Pomoideae in the family Rosaceae. Species that bear pomaceous fruits have an inferior ovary with a lignified endocarp. They include the apple, pear, quince, and other ornamental genera, for example, *Cotoneaster*, *Pyracantha*, and *Eriobotrya japonica*, the evergreen species popularly known as the loquat.

Apples, *Malus* spp.

The nomenclature of the common dessert apple has undergone several changes since Linnaeus named it *Pyrus malus*. Taxonomists have changed the name to *M. sylvestris*, *M. pumila*, and *M. domestica* in the past. Today, it is generally classified as *M. domestica*, although many current cultivars apparently originated as interspecific hybrids, involving *M. pumila*, native to southwestern Asia, *M. sylvestris*, native to Europe, and other species as well.

There are many species and interspecific hybrids of crab apples: *M. baccata*, Siberian crab apple, var. *M. mandshurica* and *M. cerasifera*, and *M. prunifolia*, ringo crab apple, var. *M. robusta*. But some are hybrids between *M. ioensis* and the cultivated apple. Some species of crab apples are utilized as pollinizers and dwarfing rootstocks or as ornamentals, for example, *M. floribunda*.

The cultivated apple is a hexaploid that behaves like a functional diploid ($2n = 34$). The presence of diploid, triploid, and tetraploid apple cultivars as well as pollen-sterile cultivars makes the selection of cultivars and pollinizers complicated. Yellow Transparent, Grimes, and Rome Beauty are usually self-fruitful. Golden Delicious and Jonathan are partially self-fruitful because not all of their pollen is viable. With these cultivars, a better set is obtained with cross-pollination. Partial self-fruitfulness is a desirable character because it reduces the need for pollinizers and often the cost of thinning. Others, such as McIntosh, Delicious, and Starking, are self-unfruitful. Gravenstein, Mutsu, Stayman Winesap, and Winesap are pollen-sterile; these cultivars require cross-pollination for setting.

The apple is reported to have been grown in Egypt as early as the twelfth century B.C. It is mentioned in Greek literature dating back to 600 B.C. and in the Bible. Over the centuries, countless numbers of chance seedlings were deemed worthy of cultivation and given varietal names. American nursery catalogs in the nineteenth century listed several hundred clones, more than 20 clones of which were of superior quality and historical interest and are still cultivated.

Fameuse or Snow is believed to have been brought into Quebec by French explorers in the early seventeenth century and distributed from there into New England. Thus, it is probably one of the earliest known apples to be introduced into North America. The tree is hardy and vigorous. The fruit is almost purplish black when mature, but the flesh is white, tender, and very aromatic. The fruit tends to be small and subject to apple scab infections.

McIntosh, a seedling of Fameuse, was discovered in 1870; like its hardy, vigorous parent, it is susceptible to apple scab. The cultivar comes into bearing early but has an alternate bearing tendency. The fruit is highly prized for its dessert quality. It is best suited for the northern latitudes of the United States.

Gravenstein was discovered in 1750 in a part of northern Germany that originally belonged to Denmark; hence, it still goes by its Danish name in Denmark. It is an early-maturing cultivar known for its excellent dessert and culinary qualities. Oldenburg or Duchess was introduced into the United States in about 1835 from Russia where it was prized for its texture, keeping qualities, and culinary attributes. The tree is reputed to be easy to grow and to bear early and heavily in regions bordering the Great Lakes and eastward to New York.

The Yellow Newton (Newton Pippin, Albemarle Pippin) is one of the older apple cultivars grown in the United States, having originated on Long Island in the early part of the nineteenth century. The cultivar, best known for its keeping quality, was a choice product for export. It has lost favor among growers because the trees take several years to come into bearing. The fruit attains a full, round shape when grown in Hood River, Oregon, but near Watsonville, California, the fruit is truncated. The fruit tends to russet if fog condenses on it during the growing season.

Baldwin, which originated in Massachusetts in about 1740, was once the most widely planted cultivar in the eastern parts of the United States. Mature trees are inclined to bear biennially. In light crop years, the fruits are likely to develop symptoms of Baldwin spot or bitter pit (*stippen* in German), the calcium deficiency disorder.

Northern Spy is a chance seedling discovered in East Bloomfield, New York, in about 1800 (Hedrick, 1925). It is grown widely throughout the apple-growing districts of Michigan, New York, and Canada. When it is properly grown, the Northern Spy develops into a high-quality fruit with good keeping qualities. The cultivar prefers a cool climate and tolerates relatively heavy soil. It is hardy and vigorous but is subject to infection by apple scab fungus and comes into bearing late.

Esopus Spitzenburg or Spitzenburg was a popular cultivar in the apple-growing districts of Washington and Oregon, but because of its susceptibility to fire blight bacteria, it has lost favor with growers. The shoot forms few lateral branches and bears its crop on long canelike branches. It has excellent cooking, dessert, and keeping qualities.

Jonathan, a seedling of Spitzenburg, was first described in 1826 in New York. Apple growers in the north-central and western United States favor it for its high quality and intense red pigmentation. Like its parent, Spitzenburg, it has a drooping growth habit and a susceptibility to fire blight.

The first plantation of Rome Beauty was made in 1817. The tree is vigorous, comes into bearing relatively early, and is not attacked by many diseases and insect pests. The fruit is not of the highest dessert quality, but it is especially desired for baking.

Cox's Orange Pippin is a popular English cultivar because of its excellent flavor. Granny Smith and Cox's Orange Pippin are grown in Australia and New Zealand primarily for the export market. Granny Smith is being widely planted in the Central Valley of California since the trees tolerate the excessively hot summers. It is a tart (acid) cultivar with a firm, crisp flesh, making it ideal for pie fillings. Its vigorous shoots must be headed back during dormant pruning to force lower basal buds to form spurs. Anna, another heat-tolerant cultivar, needs only a short chilling period and has good eating quality; it is widely grown in Israel.

Delicious, a chance seedling, was found in Iowa in 1881. The surface color of mature fruit developed attractively striped red markings. The weakness of the cultivar is its susceptibility to bitter pit, especially on lightly cropped trees. It was widely planted until some of its bud sports and their derivatives (Table 8) with highly red skin pigmentation and spur-type growth habits were discovered in recent years. These naturally occurring, semidwarfing cultivars and their progenies are preempting the apple market. Other spur-type clones have been obtained by ionizing radiation.

Other red cultivars such as Idared, Jonared, Fuji, Hokuto, and Senshu are making inroads into apple culture, especially in Japan. The last two are progenies of Fuji.

Golden Delicious, another chance seedling, was discovered in West Virginia and introduced in 1916. It is a widely grown cultivar because it seems to adapt to many different environmental conditions. Golden Delicious trees tend to set heavily and, therefore, require more thinning than other cultivars. It also tends to shrivel in storage because its thin cuticles form fissures. Currently, bud sports and hybrid progenies of Golden Delicious (Table 9) are becoming popular. These yellow-skinned types do well in warm areas where many red cultivars do not develop their best color.

Apple breeders at numerous national, state, provincial, and private experiment stations are striving to incorporate into new cultivars the compact growth habit with good dessert quality and scab and mildew resistance. Rootstocks possessing disease and insect resistance and dwarfing tendency, coupled with efficient mineral nutrient-absorbing capacity, are also sought by apple breeders.

Adaptable to a wide range of climatic conditions, the apple is widely planted in Europe, North and South America, New Zealand, Australia, and Asia. Most older cultivars require a relatively long chilling period to break their rest period, but newer cultivars that need very little chilling have been developed. These selections have extended the limits of apple production to the tropical areas, such as the highlands of Indonesia and northern South America, and Israel. Of all deciduous fruit trees, some apple cultivars are the most resistant to winter injury, withstanding temperatures as low as $-35°$ C $(-31°$ F), if the trees are well acclimated prior to the cold spell.

Some cultivars can be grown in areas where the growing season is short because their fruits mature within three months of full bloom. Others require about six months to mature so that their cultivation is limited to lower latitudes where the summers are long. Although such genetic diversity enables successful apple growing in a wide range of climates, apples are best adapted to

Table 8. *Bud Sports of Stark Delicious and Their Derivatives*

Parents			
Stark Delicious		Starking	
Standard	Spur Type	Standard	Spur Type
Gardner	Atwood Spur	Chelan Red	Ervin Red
Lalla	Okanoma	Evarts Red	Hardispur
Red Prince	Starkspur Red	Harrold Red	Miller Sturdyspur
Richared		Hi-Early	Red Spur
Shotwell		Hi-Red	Sali Spur
Starking		Houser Red	Starkrimson
Turner		Heubner	Wayne Spur
Vance		Imperia Del. Red	Wellspur
		Red King	
		Red Queen	
		Ryan Red	
		Skyline Supreme	
		Sterling	
		Super Starking	

Parent	Mutant
Richared	Royal Red
Wellspur	Earlistripe
Skyline Supreme	Sky Spur
Shotwell	Topred
Red King	Oregon Spur
Ryan Red	Earlibrite

Courtesy of S. Sansavini.

areas where winters are cold and summers moderately warm and relatively humid. Cool night temperatures and intense light during the maturation period foster the formation of anthocyanin pigment, but cloudy days accompanied by the precipitation of dew cause poor color formation and russeting.

Most apple trees were formerly grafted onto seedling rootstocks; the seeds were obtained from canneries or cider operations. Trees, grafted on seedlings, are usually vigorous but do not grow uniformly. For trees to be of desirable, uniform size at maturity, they should be propagated on a selected clonal rootstock. The Malling stocks are the best-known clonally propagated ones. In reality, they are apple cultivars that were collected in Europe and assembled at the East Malling Research Station, Kent, England, by the then director, R. G. Hatton. After years of trials as rootstocks for English cultivars, they were assigned Roman numerals, which were recently changed to the present Arabic numerals (Gourley and Howlett, 1947). For example, the dwarfing French Paradise was named EM VIII, and Jaune de Metz or Yellow Metz was labeled EM IX. These were later changed to M.8 and M.9, respectively. Semidwarfing rootstocks, M.2 and M.5, and the invigorating rootstocks, M.1

and M.16, originated in a similar fashion. The dwarfing stocks, M.27 and M.26, developed through breeding, were added later. The dwarfing, semidwarfing, and invigorating stocks, MM.106, MM.111, and MM.104, respectively, were developed cooperatively between Research Stations at East Malling and Merton. Stocks of the MM series not only control size but are also resistant to the woolly apple aphid. Some of the dwarfing stocks tend to sucker more than others.

Numerous size-controlling apple rootstocks have been bred and selected by research stations and private breeders. They are identified by their coded initials: East Malling–Long Ashton (EMLA), Michigan apple clone (MAC), Kentville stock clone (KSC), Cornell–Geneva (CG), and Oregon apple rootstock (OAR). Cuttings of easy-to-root strains of *M. prunifolia* are high-worked in Japan for dwarfing. Seedlings of the Russian cultivar Beautiful Arcade are being tested in Kentville, Nova Scotia, for cold hardiness.

Clonal rootstocks are propagated as softwood or hardwood cuttings or by layerage. Currently, there is much interest in propagating these and other stocks by meristem culture on artificial medium, primarily because many viruses can be eliminated.

The advent of dwarfing stocks has led to high-density plantings in which 620 to 2000 trees per hectare (250 to 810 per acre) are planted, whereas only 70 to 120 standard-sized trees per hectare may be planted. Trees in high-density plantings are trained on trellises of various designs or pruned to a slender spindle or a spindle bush, whereas standard trees are trained to a modified central leader. Because the crop is borne on long-lived spurs (Fig. 11.1), pruning consists of removing crowded and crossing branches, one-year-old shoots, and vigorous water sprouts.

Apples are ready for harvest when they can be separated from the spur by lifting, when they show full color, and when their flesh is crisp with some characteristic flavor of the cultivar. They should not taste starchy when fully ripe.

Most apple cultivars keep well, especially under controlled-atmosphere storage with efficient scrubbers to remove ethylene. The optimal composition of the atmosphere, which extends storage life and yet retains the eating quality, varies from one cultivar to another (see Ch. 6, Table 4).

To enumerate the advantages and disadvantages of the numerous scion cultivar–rootstock combinations, with and without an interstock or stem builder, is beyond the scope of this text. To learn which cultivars are best suited for a particular soil and climate of the area, prospective apple growers should consult with farm advisers from a

Table 9. *Golden Delicious and Its Bud Sports and Hybrid Seedlings*

Bud Sports		Hybrid Seedlings	
Standard	Spur Type	Standard	Spur Type
Bovey	Gold Spur	Cherdan	Mutsu
Delbard	Elliot Spur	Ed Gould	Ozark Gold
Dorsett	Frazier's Spur	Ein Shemer	Pacific Gold
Golden Russet	Morspur	Griffith Seedling	Skinlight
Lutz	Nugget	Honeygold	Stark Blushing
Simon's Russet	Perleberg	Hosei	Sungold
Smoothee	Stark Spur	Kinsei	Virginia Gold
Sundale	Tester's Spur	Magnolia Gold	Thew Gold
	Yellowspur		

Courtesy of S. Sansavini.

nearby Cooperative Extension Office and representatives of reliable nurseries.

Pears, *Pyrus* spp.

Over 20 species of pears are found in the Northern Hemisphere, ranging from southern and western Europe, northern Africa bordering the Mediterranean Sea, and eastward through Asia Minor, Iran, China, and Japan. All are diploids and have 34 chromosomes. The three species of economic importance for fruit production are *Pyrus communis*, European or French pear; *P. nivalis*, Snow pear; and *P. serotina* (syn. *pyrifolia*), Asian, Japanese, or Sand pear. *Pyrus calleryana* and *P. betulaefolia* are selected for ornamental shade trees. The Eurasian hybrid species, which inherited resistance to the fire blight bacteria, *Erwinia amylovora*, from its Asian parent, *P. serotina*, is grown in warm, humid climates. Either the fire blight bacteria has mutated and become more virulent, or the Asian pear cultivars are susceptible under certain climatic conditions, for some cultivars have been infected in recent years.

EUROPEAN OR FRENCH PEAR, *P. communis*

The cultivation of European pear predates the Christian era by several hundred years. The French and English colonists introduced it to the American colonies in the early seventeenth century. Most popular cultivars originated in France and Belgium where the fruit has been grown and relished for centuries. Like the apple, hundreds of pear cultivars were grown in the nineteenth century, but today fewer than ten are commercially grown. Of these, Bartlett makes up about 75 percent of the world's acreage and production, because (1) it can be grown under widely different climatic conditions and (2) it regularly produces high-quality fruits.

The Bartlett pear was discovered as a chance seedling by a Mr. Stair in Berkshire, England, from whom a nurseryman, Williams, acquired and propagated it as the Williams pear. It was

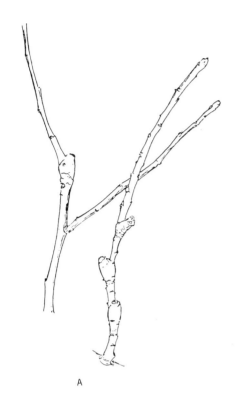
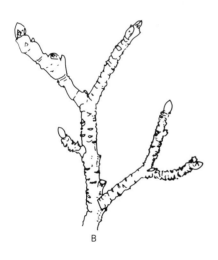

Figure 11.1
Apple spurs from (A) a young and (B) an older bearing tree.

brought to Boston in 1797 or 1799 and established under the name Williams, on the Brewer estate in Roxbury, Massachusetts. In 1817 this property passed into the hands of Enoch Bartlett, who released the cultivar under his name. In the meantime, it was introduced into France as Williams Bon Chrétien. Thus, Bartlett is grown and marketed under two other names, Williams and Bon Chrétien.

In most pear-growing districts of the world, Bartlett is self-sterile and requires a cross-pollinizer, for example, Winter Nelis or Beurré Hardy. Along the Sacramento River delta of California, the Bartlett is self-fruitful, setting crops of parthenocarpic fruits, presumably because the temperature during bloom is relatively warm.

Although Bartlett has a chilling requirement of about 1500 hours, it usually blooms before Winter Nelis. However, after a mild winter, not only is the bloom sequence reversed (Fig. 11.2), but Bartlett also produces many late blooms, which predispose the tree to fire blight infection. Under California conditions, the crop matures in early July in the Sacramento River district and in mid-August in the foothills.

Beurré Bosc, Beurré d'Anjou, Beurré Hardy, Doyenne du Comice, and Winter Nelis are excellent late fall and winter pears. Beurré Bosc is a large, attractive, russeted pear with a tapering neck. It is favored among gourmets for its excellent flavor. The cultivar was first described in 1807 in Belgium and introduced into the United States in about 1832. It is partially self-fruitful but sets a better crop with pollination. Although the tree is slow to come into bearing, once it begins to flower it bears regularly and abundantly.

Beurré d'Anjou is an old French cultivar that was introduced into North America in about 1842 by Marshall P. Wilder, one of the founders of the American Pomological Society and the person for whom the Wilder Medal is named. The fruit is large and has a distinct obovate–pyriform shape. Its flavor does not rank high but, according to Hedrick in *The Pears of New York* (1921), the fruit is never poor in quality. Trees of Beurré d'Anjou are spreading, vigorous, and relatively free of fire blight. They have a reputation for being uncertain croppers.

Beurré Hardy, a lightly russeted, attractive fall pear, is sufficiently self-fruitful and is planted as a pollinizer for Bartlett. Some pear connoisseurs enjoy the slightly astringent flavor of the ripe fruit. It is used as an interstock between Bartlett and the dwarfing quince rootstock.

Doyenne du Comice, which was first described in 1849 in Angers, France, is rated highly for its flavor and texture by pear fanciers. It is subject to fire blight infections and is less cold hardy than most pear cultivars. It is sufficiently self-fruitful to bear a good crop, but it yields better with cross-pollination. The cultivar thrives and produces a valuable commercial crop in the Pacific slope of the United States, but its trees require the best of soil, climate, and care.

Winter Nelis originated in Belgium in the early nineteenth century. It has as many good points as bad. It is self-unfruitful but cross-fruitful with Bartlett. The fruit is small, rough, and russeted, but it keeps and ships well. The juice is rich, sweet, and perfumed. A baked Winter Nelis fruit has the most delectable of all pear flavors. Because the cultivar is not graft-compatible with quince, it must be double-worked with Beurré

Figure 11.2
Winter Nelis trees (left) blooming ahead of Bartlett trees (right) after a mild winter. The sequence is usually reversed after a cold winter.

Hardy as an interstock to obtain a small tree. The tree branches freely and has a willowy growth habit.

Packham's Triumph is grown in Argentina, Chile, and South Africa for its good shipping and eating qualities. The tree regularly bears fruits that resemble Bartlett except for the slightly flattened calyx end.

Conference, an English cultivar, is commonly grown in France. The fruit is slightly long and gourd-shaped, its smooth skin dotted with prominent lenticels. It manifests hard and black end symptoms when grafted on some Asian pear seedlings (see Physiological Disorders, Ch. 6).

Passé Crassane, another French cultivar, produces well in France and Italy. The fruit is rough-skinned and has a pale green to yellow finish. The tree is highly fruitful and susceptible to fire blight.

Precoce Morettini, developed at the University of Florence in Italy, resembles Bartlett in appearance but does not quite match it in flavor. Its main economic advantage is that it matures about two weeks earlier than Bartlett.

Italy is currently the leading pear-producing country, followed by the United States; here over 90 percent of the production is located in the Pacific Coast states of California, Oregon, and Washington. Pears are important export crops in Argentina, Chile, Australia, and South Africa.

Pears are grown in areas where winter temperatures remain below 7° C (45° F); they require from 1200 to 1500 hours to overcome the rest influence in buds. If trees are fully dormant, they will withstand temperatures as low as $-26°$ C ($-15°$ F) without injury. Pear is less hardy than most apple cultivars but hardier than peach.

Currently, trees are propagated on Winter Nelis seedlings, or on Old Home × Farmingdale selections, or on rooted quince cuttings for dwarfing. Some cultivars are incompatible with quince. Thus, Beurré Hardy or Old Home interstock is used to overcome this problem (see Economically Important Traits of Rootstocks, Ch. 10). Bartlett pears on seedlings derived from imported French seeds or double-worked on quince stocks have not only a smooth finish but also an excellent eating quality. Own-rooted Bartlett trees propagated under mist will grow to a size nearly equal that of trees double-worked on quince stocks. These trees are not susceptible to the pear decline disorder because they do not have graft unions.

Prior to World War I, pear seeds for rootstocks were imported from France by American nurseries. When this source was curtailed by the armed conflict, Asian pear seeds were imported from Japan and China, partly because these species were touted as being blight-resistant. Trees propagated on Asian pear seedlings, however, developed black and hard end symptoms and also proved to be very susceptible to pear decline disease, which is caused by a mycoplasma (see Pathogen-Induced Incompatibility, Ch. 10). This organism is transmitted to the pear by the insect pear psylla, *Psylla pyricola*.

Pear crops are not thinned unless the set is exceptionally heavy. Cultivars may set as many as five fruits on a single spur. If so, the number should be reduced to one or two, for a heavy set may induce alternate bearing. With a moderate crop, growers can harvest selectively by picking the largest fruit first and leaving the smaller ones on the tree to gain in size.

Although the pear tree, like most fruit trees, grows best in deep, sandy loam soils, most rootstocks tolerate soils that are too heavy for peach and apple. Pear trees thrive behind the levees of the Sacramento River in California where the water table may rise within 2 feet of the surface.

Pear wood is very flexible, and most of the crop is borne on long-lived spurs (Fig. 11.3). To take advantage of these characteristics, orchardists train pear trees to diverse systems such as a modified central leader, Italian palmette, or the bandera (see Training systems, Ch. 9).

Preharvest drop causes considerable losses, particularly with parthenocarpic Bartlett fruits. The synthetic auxins 2,4-D and naphthaleneacetic acid have been used to reduce abscission. Pears develop their best quality if they are picked when mature, but still green and not yet ripe. Three criteria used to determine when to begin

Figure 11.3
A typical pear spur with an enlarged cluster base. Notice the terminal fruit scars.

harvest are flesh firmness, equatorial diameter, and surface color (see Maturation and Harvesting Indices, Ch. 6). The best eating quality is obtained by chilling harvested fruits at $-1°$ C ($30°$ F) for one to several weeks. The pears are then brought out into a ripening room, kept at $20°$ C ($68°$ F) and 85 percent relative humidity. The cold-storage treatment makes pears that were of different maturity at harvest ripen uniformly. To be prime for eating, the pulp should be buttery, as the French name for butter, *beurre*, implies. Pears are eaten fresh, stewed, or baked; they are preserved by canning or drying.

ASIAN PEAR, *P. serotina,* syn. *P. pyrifolia*

Members of the Asian pear species have been cultivated in the Orient for centuries. Cultivars were introduced into the eastern United States from Europe, whereas those on the West Coast were brought in by immigrants or agents of the Department of Agriculture.

Unlike the European pears, Asian pears ripen on the tree; they can be eaten immediately after harvest. They are watery, crisp, and somewhat gritty because of their stone cells, in contrast to the cloying character and buttery texture of French pears. When Asian pears were first marketed as Sand pears, they did not sell well. After new cultivars were introduced and their name was changed to Salad pears, they gained in popularity. Their popularity is also attributed to the increased number of tourists who visited Japan and had the opportunity to sample these fruits. Currently, they are sold as Asian pears in the American markets.

Ya-li and Tsu-li are old Chinese cultivars that bear pyriform greenish fruits. Many Chinese pears, sold under these names, may be seedlings because they do not appear true to type.

Several Japanese introductions are now being grown in California, Oregon, and Washington. One of the earliest cultivars to be introduced into California was Nijis-seiki (Twentieth Century), a yellowish-green cultivar with a tender skin that ripens at the same time as Bartlett. The cultivar must be heavily thinned early in the season because it tends to set a large crop.

Chojuro, a bronze-colored cultivar, is another pear introduced early; it ripens slightly later than Nijis-seiki. Its fruits tend to develop an off-flavor if they are left on the tree too long. Shin-seiki (New Century) and Kikusui, greenish-yellow cultivars, ripen in midsummer. Hosui and Kosui are the latest Japanese introductions. Their mature fruits have a yellow surface color with a slightly russeted finish. Their skin is not as tender as that of Twentieth Century; the flesh has a fine texture because it contains few stone cell aggregates.

Asian pears should be propagated on French pear seedlings to avoid the physiological disorder known as yuzuhada (pomelo disease), which is comparable to hard end disorder of Bartlett. Chojuro is highly prone to this rootstock-induced disease; on certain rootstock its flesh becomes so firm that it is inedible.

Nijis-seiki and Shin-seiki are partially self-fruitful and will set a crop without cross-pollination, provided the weather during pollination is favorable. Other cultivars can be cross-pollinated with Bartlett pollen or other Asian cultivars flowering at the same time.

The species grows well in sandy loam but also

tolerates heavy soils. The trees are trained to a modified central leader, but in Japan where typhoons are common in late summer, trees are lashed with bamboo poles to a shelf (pergola) construction to minimize windfall. They are easily trained to a palmette system.

Most cultivars are highly resistant to the fire blight bacteria, *Erwinia amylovora*. Thus, they may be grown where spring rains and high summer humidity limit the growing of Bartlett and other European pears. Most Asian pear cultivars keep well for one to two months at 0° C and 85 percent relative humidity without a noticeable loss of flavor and texture.

EURASIAN HYBRIDS

Owing to their resistance to fire blight, the Eurasian hybrids, Kieffer, Le Conte, and Garber, are grown in the middle and southern parts of United States. They are cold hardy and tolerant of hot, humid summers; their fruits are of poor quality. Kieffer and Le Conte should be propagated on French pear seedlings, for they also tend to produce fruits with black and hard end symptoms when grafted on Asian pear seedlings.

Quince, *Cydonia oblonga*

A species with 34 somatic chromosomes, quince has been cultivated from prehistoric periods in countries extending from Iran to India. The ancient Greeks and Romans grew quince for its attractive pinkish flowers and fragrant fruit. The time of floral induction is not known; flower differentiation is believed to occur simultaneously with the elongation of shoots. Since flowers are borne terminally on relatively long shoots (Fig. 11.4), they appear late in the spring. Therefore, losses from spring frosts occur less frequently than in most other fruit species. The large, pubescent, round or pear-shaped fruits with golden-yellow surface color mature during the months of August and September. Some cultivars are

Figure 11.4
A quince shoot with a terminally borne mature fruit.

Angers, Champion, Orange, Pineapple, and Smyrna.

Quince is adaptable to a wide range of climates and soils. It tolerates hot summers as well as cold winters. The species requires less winter chilling than pears and is nearly as hardy as peach. Its wood may be severely damaged by temperatures of $-26°$ to $-29°$ C ($-15°$ to $-20°$ F). It is susceptible to fire blight; therefore, the species is not grown in warm, humid areas of the United States. Quince tolerates soils that are somewhat too heavy for most other fruit trees species, but it develops iron chlorosis in calcareous soils.

Most cultivars are easily propagated as hardwood cuttings because stems possess preformed root initials. The few cultivars that root reluctantly are budded onto rooted cuttings of Angers. The shallow, fibrous roots are resistant to root-knot and lesion nematodes. The plant may be trained as a multiple-trunk bush or as a small tree which seldom grows taller than 4 meters (12 feet). Pruning consists of removing interfering branches, water sprouts on trunks, and suckers that arise from adventitious buds on roots.

The fruit is ready to harvest when the skin loses its greenish color and turns golden yellow. Its fragrance intensifies as maturity is reached. Although the fruit is quite hard, it bruises easily; it keeps well in cold storage. Too firm and mealy to eat uncooked, the fruit, rich in pectin, is made into an aromatic, honey-colored jelly and a stiff red jam. The jam is called *dulce de membrio* in Argentina and Chile.

Quince is used as a dwarfing rootstock for pears, but most pear cultivars are incompatible with quince stocks. Hence, Beurré Hardy and Old Home, which are compatible with quince, are used as intermediate stocks. Quince rootstock is subject to pear decline.

The Chinese quince, *Chaenomeles sinensis*, is a spiny bush which is cultivated primarily as an ornamental. Its bright red flowers bloom early in the spring before the leaves appear. The small, misshapen fruits are used for jams and jellies.

Loquat, *Eriobotrya japonica*

A member of the rose family and a native of subtropical Asia, the loquat is an evergreen species. It is commonly grown in gardens as an ornamental in warm regions of the United States. The thick, coarse leaves with serrated margins are sessile on shoots with short internodes. The flowers, which bloom in late winter, are borne in terminal panicles. In spite of inclement weather during bloom, a crop of fruits pyriform to round in shape usually sets. The loquat is the first tree crop to ripen in the spring, preceding the earliest sweet cherry cultivars.

The fruit, classified as a pome, possesses a membranous endocarp which encloses two to four brown, slippery seeds. The sweet but agreeably tart, juicy pulp is eaten fresh or made into preserves. The whitish pulp of the large-fruited cultivar Tanaka has a delicate aroma, and the skin has an attractive, deep yellow hue. The tree is propagated by grafting onto seedlings. The species is very susceptible to fire blight.

Pomegranate, *Punica granatum*

A species with 16 chromosomes in somatic cells, the pomegranate belongs to the family Punicaceae and is native to Asia Minor and Asia. It was introduced into Greece and the Roman Empire, where the storytellers put the fruit into their myths. The species include cultivars that bear attractive flowers with single- or double-whorled petals, ranging in color from bright orange-red to a variegated color with white petals and red splotches. The multistylar flowers (see Fig. 1.28) are pollinated by bees, but they are also visited by hummingbirds. These flowers are borne terminally on short shoots with shiny, lanceolate, oppositely arranged leaves. The fruits should be harvested before the thin, leathery epidermis splits. The size of the fruits depends on the number of seeds that form, a function of pollination. The outer integument of seeds or aril is the juicy, edible portion of the fruit. Some cultivars possess only seed traces, indicating that the embryos aborted. The principal commercially grown cultivar in California is Wonderful. When ripe its fruit has a deep red skin and similarly pigmented arils. Its juice is tart and rich in anthocyanin and is made into jelly or into a syrup called grenadine. There are other cultivars with white to pinkish rinds and sweet, insipid arils.

The plants are easily propagated by cuttings or seeds and can be trained as shrubs or trees that will grow to 4 to 5 meters. The species tolerates alkaline soils where most fruit tree species will not flourish; the trees thrive in neutral to acid soils as well. Pomegranate will withstand temperatures of $-8°$ to $-12°$ C (15° to 10° F) in the winter and 38° to 42° C (100° to 108° F) in the summer without showing stress. The species requires only a short chilling period for normal bud development. A dwarf form makes a good, compact ornamental house or garden plant.

The species is not attacked by codling moths and twig borers, but if the ripe fruit is not harvested on time, it will crack and attract fruit flies, ants, and earwigs. It is resistant to the oak root fungus, *Armillaria mellea*.

References Cited

Beach, S. A., N. O. Booth, and O. M. Taylor. 1905. *The Apples of New York*. Ann. Rept. New York Agric. Exp. Sta. J. B. Lyon Co. Albany, N.Y.

Bolkhovskikh, Z., V. Grif, T. Matvejeva, and O. Zakharyeva. 1969. *Chromosome Numbers of Flowering Plants*. Academy of Sciences of the USSR. Leningrad.

Gourley, J. H., and R. W. Howlett. 1947. *Modern Fruit Production*. The Macmillan Co. New York.

Hedrick, U. P. 1921. *Pears of New York*. Ann. Rept. New York Agric. Exp. Sta. J. B. Lyon Co. Albany, N.Y.

CHAPTER 12

Stone Fruits

Stone fruits belong to the genus *Prunus*, which has a haploid chromosome number of eight. The genus is subdivided into three subgenera: *amygdalus*, which includes almond, peach, and nectarine; *prunophora*, which includes the European (syn. French) and Japanese plums and apricot; and *cerasus*, the sweet and sour cherries. Although members of the genus originated in areas of diverse climates, they have been selected and bred for adaptability to specific localities.

Almond, Prunus dulcis

The species *Prunus dulcis*, formerly known as *P. amygdalus*, has been cultivated from prehistoric times and probably originated in the area between Asia Minor and India. The crop is widely grown in countries surrounding the Mediterranean Sea. The first trees in California were planted by Spanish padres at Mission Santa Barbara, but the planting apparently failed. A few hectares planted in 1843 expanded to over 120,000 hectares within 140 years.

All almond produced commercially in the United States is grown in California. Large quantities of these almonds are exported to Japan and northern Europe annually.

The species is the earliest to bloom among stone fruits. The Jordanolo begins in the last week of January if it has gone through the minimal period of chilling required. Thus, the species is very susceptible to spring frosts. All commercial cultivars in California are self-unfruitful, and some are cross-unfruitful, carrying the S factor (see Gametophytic Incompatibility, Ch. 5). Therefore, a selection of compatible cross-pollinizers is essential for high yields.

About 60 percent of the acreage in California is planted to Nonpareil, a thin-shelled cultivar. Thus, a common pollinizer arrangement consists of a relatively early-blooming cultivar, for example, Ne Plus Ultra, aligned with three or four rows of Nonpareil, the main cultivar, followed by a row of Drake, Eureka, or Mission (syn. Texas). This arrangement assures a good overlap of the bloom period of Nonpareil. Some self-fruitful cultivars are grown in the Mediterranean area.

Trees are more often propagated on peach seedlings because this scion–stock combination grows faster when fertilized and irrigated than does the almond–almond combination. Trees on peach stocks may weaken at the graft union at an earlier age than trees on almond stocks. A few nematode-resistant peach–almond hybrids are being tested for suitability as rootstocks. Almonds are compatible with Marianna 2624 (*P. cerasifera* × *P. munsoniana*) stock, which tolerates wet soils better than do peach or almond roots. Almond cultivars topworked on myrobalan plum (*P. cerasifera*) seedlings develop poor graft unions and become small, weak trees.

Trees are trained to a modified central leader with a 1-meter-high trunk to provide space for the clamp of the tree shaker. Young mature trees are given minimum pruning to promote and maintain high production. Dead limbs and those interfering with the placement of limb or trunk shaker are removed.

As trees become 15 to 25 years old, their production declines. Basal spurs on four- to six-year-old branches gradually become barren because some of their apical buds fail to elongate when the spurs initiate too many flowers that set. These spurs eventually die, but the peduncles persist (Fig. 12.1). About 15 percent of the branches with barren spurs should be dehorned annually or biennially, so that in seven to ten years, all secondary or tertiary scaffolds are renewed. This

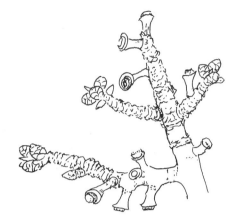

Figure 12.1
Almond spurs with persistent peduncles.

method of replacing barren branches and lowering the bearing area of a tree is known as the wood renewal system (see Pruning of Mature Trees, Ch. 9).

When the hulls begin to split, the crop is harvested by shaking the tree. A trunk shaker is used while the trees are small, but as the trees become large, the trunk shaker becomes less efficient in removing the crop. A limb shaker is then substituted for the trunk shaker because the limb shaker removes more of the crop and causes less damage to the tree. The fallen fruits are swept into a windrow, which is then collected and taken to a huller. The hull is sold for livestock feed, and the nut (embryo and shell) is dried down to a minimum moisture content and delivered to the sales outlet. The nuts are shelled, and the kernels consisting of 55 percent fat on the dry weight basis are sold as "meat" or prepared for the confectionery trade. A good proportion of the crop is flavored and sold in cans.

Problems associated with almond culture are (1) infection of flower and leaf buds by brown rot and shot hole fungi during a wet spring; (2) infection of roots by verticillium wilt, resulting in loss of leaves or even the death of the trees; and (3) invasion of the maturing nuts by larvae of the codling moth and the navel orange worm. The larvae overwinter in mummies (dried fruits) remaining on the tree; the moths that emerge in the spring attack the young developing almonds. A fourth problem is noninfectious bud failure, a serious physiological disorder caused by high summer temperatures (see Temperature-Related Disorders, Ch. 6).

Peach and Nectarine, *Prunus persica*

A member of the subgenus *amygdalus*, *Prunus persica* is native to China and was introduced to Persia, probably by silk traders, and then into Europe. Colonists brought the species to the New World. During their westward movement, pioneers spread it throughout North America.

Current cultivars are rather homozygous because peaches have been inbred for many generations. Hence, seedlings derived by self-pollination of popular cultivars bear fruits that are very similar to their parents. The size, time of ripening, and appearance may vary slightly. Through outbreeding with different strains, cultivars that thrive in the warm, humid climates of southeastern United States, Brazil, Japan, and Europe, as well as in the hot, dry regions of the world, have been developed.

Peach flowers buds withstand $-28°$ C ($-18°$ F) if the cold is preceded by several days of freezing weather ($-2°$ to $-6°$ C) and the bud tissues are thoroughly acclimated (Proebsting, 1970). In an average winter, however, the lethal temperature at which 50 percent of the flower buds are killed, LT_{50}, ranges from $-16°$ to $-24°$ C ($-4°$ to $-12°$ F). Within this temperature range Elberta flower buds are damaged, but living cells of stem tissues usually escape harm. The LT_{50} fluctuates with the mean daily temperature, but the degree of damage also depends on temperatures before the cold spell, the duration of the cold spell, nitrogen content of the tissue, the rootstock, and so on (see Winter Injury and Acquired Cold Hardiness, Ch. 7).

After pollination and until the calyx cup splits, the enlarging pistil tolerates temperatures as low as $-1°$ C, but below this critical level the ovule is easily killed and the flower abscises. Ovules of young developing fruitlets become slightly hardier and will tolerate $-2°$ C; many fruitlets will still abscise, but others will persist even if the ovules are killed.

Peach trees are propagated by spring, June, or dormant budding using the shield-bud technique (see Budding, Ch. 10) on seedlings of Lovell or nematode-resistant stocks, such as Nemaguard, Stribling-37, or Rancho Resistant. The last two are resistant to the root-knot nematode, *Meloidogyne incognita*, but not to *M. javanica*; Nemaguard is resistant to both. The resistance in these rootstocks is derived from *P. davidiana*, a close relative of the peach. Peach seedlings and especially Nemaguard are susceptible to waterlogging or "wet feet." Peach scions are somewhat graft-compatible with plum stocks which are tolerant to wet soils, but such trees are usually small and weak.

Trees are trained to a vase or goblet form, or to the palmette system. The ypsilon or the Tatura system has not been widely accepted for peaches and other species because of the cost of installing the trellis and the difficulties in pruning and harvesting the crop.

Healthy, vigorous trees produce lateral branches freely; each shoot has a potential of bearing one or two flowers at nearly every node (Fig. 12.2). Therefore, moderate dormant pruning is necessary to invigorate the tree, reduce the cropping potential (lower thinning costs), and keep the tree within bounds. Dormant pruning consists of removing weak shoots and at least two out of three upright, lateral branches. The terminal one-quarter to one-third of long, vigorous shoots should be headed back to a lateral branch. Hanger branches may be thinned but should not be headed back unless they are so long that they interfere with cultivation or other cultural operations. Heading back hanger branches tends to weaken rather than to invigorate them. Hanger branches are encouraged to grow because (1) they can support many fruits without breaking and (2) the fruits are easily accessible to pickers.

Nearly all peach cultivars are self-fruitful;

Figure 12.2
Peach shoot with two flower buds and a vegetative bud at each node.

abaxial side, have a creamy sheen which becomes yellow as the leaves senesce and the chlorophyll degrades. Leaves of white-fleshed seedlings have a light green hue that becomes blanched as the leaves senesce. Thus, flesh color can be predicted for a seedling population by examining the leaves before the progenies begin to flower.

Leaves of peach, like those of sweet cherry, have glands at the base of the lamina. These glands are inherited as a single Mendelian factor. The eglandular (glandless) character is recessive; the reniform (kidney shape) gland is dominant, and the globose gland is the heterozygous expression of the trait. These glands exude substances that alter the surface tension of water, so that raindrops do not adhere to leaves and fruits. This trait allows glandular cultivars to evade infection from mildew, provided the duration of the rain is short and high humidity does not persist. Leaves and fruits from eglandular cultivars, such as Paloro and Peak, retain water, predisposing them to mildew.

Most fruits for the fresh dessert trade are hand-harvested; those destined for the cannery are increasingly machine-harvested. Dessert peaches are "de-fuzzed" prior to packaging and shipping partly for cosmetic reasons but also because some people are allergic to the fuzz.

Uniformity of flesh color, texture, and flavor is an important criterion for canning clingstone peach cultivars. Fruits of a given cultivar ripen over a five- to seven-day period. Therefore, to deliver a steady flow of peaches to a cannery once it opens, growers plant cultivars that ripen at nearly weekly intervals. This coordination allows both the canners and growers to maintain a labor force of nearly constant size over the six-week canning season in California.

In California a legal minimum size limit of 2⅜ inches (60.3 millimeters) in diameter has been set to uphold the quality of canned peaches. Maximum saleable tonnage on full-bearing, mature trees is attained if the mean suture diameter is 68 millimeters at harvest. Because the size distribution on a tree follows a normal, bell-shaped curve,

some, such as J. H. Hale and Alamar, are pollen-sterile. Anthers of pollen-sterile clones lack the reddish-purple anthocyanin pigments and are not as plump at anthesis as those that bear viable pollen. Pollen-sterile clones bear many small, misshapen parthenocarpic fruits known as "buttons."

Several genetic characteristics are transmitted by simple Mendelian factors. Freestone, white-flesh, and pubescence traits are dominant over the recessive clingstone, yellow-flesh, and glabrous traits. The firm, nonmelting flesh texture is closely linked with the clingstone trait; the melting flesh trait is transmitted with the freestone trait. The factor for flesh color is exhibited by the inner cells of the calyx cup and leaves. Cells lining the inner epidermis of the calyx cup (floral tube) of yellow-fleshed seedlings develop an orange pigment. The same tissue in white-fleshed seedlings has a greenish-white coloration. Leaves of yellow-fleshed seedlings, especially the midrib on the about 15 percent of the fruits by count will be less than the minimum size. If there is a demand, these undersized fruit are diverted to pickles. Oversized fruits and those with minor defects that are not usable as halves are canned as sliced peaches or diced for fruit cocktail mix.

Apricot, *Prunus armeniaca*

In Linnaeus' time, the apricot was thought to have originated in Armenia, which is now a part of Turkey and southern Russia; hence, the name *Prunus armeniaca*. Today the apricot is believed to be a native of western China. Apricot cultivars grown in Russia are winter hardy and require moderate chilling for normal bud development. Commercial cultivars grown in California need more chilling than the average peach cultivar. After a cold winter that has satisfied the chilling requirement of both species, apricot flowers bloom earlier than those of the peach. Therefore, apricots are more likely to be damaged by frost than peaches. Following a relatively mild winter, more apricot flower buds abscise than those of peach. Apricot flower buds are equally as hardy as peach to cold, tolerating $-16°$ to $-24°$ C ($3°$ F to $-11°$ F) in midwinter, provided the tissues are thoroughly acclimated.

This inherent characteristic to bloom early limits the area where the species can be grown commercially to places that are relatively free of late spring frosts. For example, apricots are grown in New York and Pennsylvania, but the flowers are killed by frost in most years. Apricots are also not grown where spring rains are prevalent because their flowers are very susceptible to infection by *Sclerotinia laxa*. Most infection occurs through the pistil by means of spores that germinate in the free water within the calyx cup. The fungus grows down the spur, causing copious amounts of gum to exude and killing the shoots.

Apricot trees are propagated on Marianna 2624, peach, or apricot. Marianna 2624 rootstock is a hybrid plum stock that tolerates wet feet better than do peach and apricot seedlings. When the soil is deep and fertile, the trees will become relatively large and should be spaced 7.3 meters (24 feet) apart. Scaffold branches of apricot trees split easily, especially if the branch angle is narrow or if three or more branches originate from the same point. Splitting can be avoided by training trees to a modified central leader with adequate vertical distance between branches.

In early-shipping districts, trees are trained to the Winters system in which trees form a wide but shallow bowl shape (see Fig. 9.8; Training Systems, Ch. 9). This system of training exposes bearing spurs to more uniform light, minimizing the difference in ripeness of fruits between upper and lower branches, and reduces the number of times the trees must be picked. After a tree attains full size and spurs are well developed on branches (Fig. 12.3), dormant pruning consists of

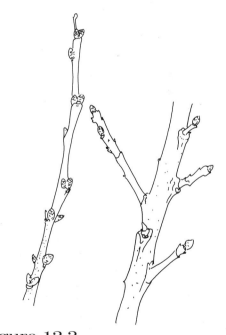

Figure 12.3
An apricot shoot terminal (left) and a section of a spurred branch (right).

removing nearly all new growth and interfering limbs.

Commercial cultivars are self- and cross-fruitful except for Riland and Perfection, which are self-incompatible and require cross-pollination. Spurs develop excess flower buds so that even when all long shoots have been removed, trees are usually thinned in the spring to obtain good fruit size.

Pitted halves of fully ripe apricots that are dried after exposure to sulfur dioxide fumes retain their flavor and color. Dried apricots are especially popular with backpackers because they are tasty, light in weight, and a source of quick energy. They can also be hydrated readily, cooked, and served hot.

Tree-ripened apricots are among the most delicious fruits available; unfortunately, they are too soft to be shipped any distance. In order to be shipped by train or truck, fruits are harvested at a firm-ripe stage before they develop their fine, rich flavor. Two new early-ripening cultivars, Castlebright and Pinkerton, are firm-fleshed and highly acid and develop a deep orange color, but they lack flavor. They are replacing Stuart and Derby for the fresh market trade. Blenheim (syn. Royal) and Tilton, which ripen in midseason or later, are marketed as fresh fruits or processed as frozen, canned, or dried fruits. Blenheim is the preferred cultivar for its dessert quality, but it is soft and does not ship well. Patterson and Modesto, because of their color and quality, are replacing Tilton. White-fleshed, clingstone cultivars are grown for the fresh market and processing in the Balkan states.

Firm-ripe fruits destined for the fresh market should be kept at 0° C (32° F) and 85 percent relative humidity. Under these conditions they will store for 14 days, after which time the quality diminishes.

More and more acreages of canning apricots are mechanically harvested with a trunk shaker and catching frame. A varying degree of ripeness is a problem of mechanically harvested fruits. Close scrutiny by graders is required to keep the unripe and overripe fruits from being processed and to maintain quality of the canned product.

Plums

The subgenus *prunophora* includes plums from every land area in the Northern Hemisphere where the climate is temperate. A large number of species are native to North America. *Prunus nigra* ranges from eastern Canada, southerly to the mountainous region of the eastern seaboard of the United States; it is very cold hardy. *Prunus americana*, which is less hardy than *P. nigra*, extends from the Atlantic seaboard to Colorado and southward to Florida and Texas. The species *P. hortulana* and *P. munsoniana* are natives of southern Mississippi Valley; *P. maritima* grows among the sand dunes from Virginia to Canada. The Pacific plum *P. subcordata* is native to the mountains of southern Oregon and northern California. Cold-hardy cultivars of these species have been developed, but their quality has not equaled that of the European and Japanese plums.

Red-leafed cultivars of *P. cerasifera* var. *psardii* have been developed as ornamentals. They produce a mass of pink blooms. The flowers are self-unfruitful, but if pollinizers are nearby, a tree will set many plums which can be made into a jelly. Cultivars with double flowers (flowers with petaloid anthers) produce little or no fruits and therefore are preferred as an ornamental tree.

EUROPEAN OR FRENCH PLUM, *Prunus domestica*

Prunus domestica is believed to have originated by the doubling of chromosomes of a hybrid between *P. cerasifera* ($2n = 16$) and *P. spinosa* ($2n = 32$) because it possesses 48 chromosomes. Cultivars of *P. domestica* have been grown in Europe for centuries. Spanish missionaries on the Pacific Coast and English colonists on the Atlantic Coast introduced these plums to North America. Spanish padres planted plums in the mission gardens as early as 1792. The Pellier brothers established a nursery in San Jose, California, and introduced the Petite d'Agen plum into California in 1856. It is the principal cultivar grown in California and is marketed under its synonym, French prune. The spur usually has short pointed buds (Fig. 12.4).

The term *prune* is applied to oval, purplish European plums that are dried. Fruits of other plum species are generally not suitable for drying because they (1) do not have the purplish-blue skin and amber flesh that people are accustomed to eating; (2) do not accumulate sufficient soluble solids; and (3) tend to ferment and, consequently, acquire an off-flavor.

This species tolerates as much cold as apple and pear and requires about the same amount of chilling. In areas where the winters are not sufficiently cold to break the rest completely, the trees bloom relatively late. Some cultivars manifest delayed foliation and their flower buds abscise after a relatively warm winter. The French prune thrives in hot, arid areas, provided the trees are irrigated and branches are not exposed to direct solar radiation that will cause sunburning.

Prunus domestica is grown for the fresh fruit market and the canning and dried fruit industries. Cultivars grown in the Pacific Northwest are either canned or shipped fresh; nearly all those grown in California are marketed as dried fruit. Whatever cannot be sold because of the small size or some defect after drying is made into juice.

Sun-dried prunes were first shipped in about 1860 from California to the eastern seaboard via the southern tip of South America. The industry flourished in California because of the region's hot, dry summers, but today, the crop is dried in forced-air tunnels heated by gas. Excellent cultivars for shipping and drying are Italian and Stanley, a hybrid between Agen and Grand Duke. Other old dessert cultivars are Reine Claude or Green Gage, Yellow Gage, Bavay, and Tragedy.

Some self-fruitful cultivars are Reine Claude, Petite d'Agen, and Sugar; partially self-fruitful cultivars are Imperial Epineuse and Grand Duke. The German Prune, President, and Golden Drop are completely or nearly self-unfruitful. Thus, the last group requires cross-pollination by a compatible pollinizer to set a crop. Unless the crop is light, cultivars that are grown for the fresh market or for drying should be thinned to attain marketable size.

Plum trees are propagated on myrobalan (*P. cerasifera*) seedlings or on cuttings of Marianna 2624, a hybrid between *P. cerasifera* and *P. munsoniana*. These stocks tolerate heavy soils, but some cultivars are graft-incompatible with some myrobalan seedlings. Domestic plums are graft-compatible with peach rootstocks and thrive if the soil is well aerated. Although the Marianna stocks produce undesirable suckers, they are somewhat resistant to nematodes.

The trees become rather large, and their limbs are subject to breaking. Hence, they are trained to a modified central leader. The trunks on prune trees should be at least a meter in height to ac-

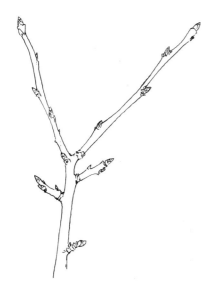

Figure 12.4
French prune with short spurs.

commodate the trunk shaker at harvest. After the basic framework is established, most of the crop is borne on short spurs, and very little pruning is done besides removing dead and broken branches. Trees bearing plums that are to be shipped are pruned moderately, but not as much as peach trees bearing fruit for shipping. Shoots need to be headed back to force branching. But the basal nodes of Imperial Epineuse and President shoots may have only flower buds, a bearing habit similar to that of sweet cherries. Pruning to short spurs is therefore not recommended for these cultivars because pruners may head back to a node bearing only flower buds. This results in short stubs with fruits but no leaves. The shoot then dies back to the base after harvest. Thus, to force branching, pruners should not head back long shoots more than half their lengths.

European plums for shipping do not have a long shelf life; they keep for two to four weeks at $-1°$ to $0°$ C ($30°$ to $32°$ F). Some cultivars may keep longer if they are harvested when slightly immature, but the flavor at this stage is not optimum.

Plums destined for drying are harvested when fully colored and the mesocarp turns amber. If the crop is harvested too early and the moisture content is still high (few soluble solids), the ratio between the fresh and dry weights (drying ratio) may be as large as $3.7:1$. A batch of fully mature fruit has a drying ratio of $3.0:1$. Fruits with a high drying ratio require a longer time in the dehydrator. This increases costs and gives smaller returns to the grower.

JAPANESE PLUM, *Prunus salicina*

A native of China, *Prunus salicina* was introduced into Japan no earlier than 1500. Many cultivars were introduced from Japan into California around 1870 and subsequently to Europe; therefore, they are commonly called the Japanese plum. The person most responsible for the popularity of this species is Luther Burbank, the noted horticulturist, who imported many cultivars and hybridized them in Santa Rosa, California. Among his cultivars are Santa Rosa, Burmosa, Formosa, and Wickson, an open-pollinated progeny of Kelsey. Some recently introduced plums to be eaten fresh are Casselman, Laroda, Nubiana, a truncated jet-black cultivar, and Queen Anne, a low-acid clone. They are interspecific hybrids that have become popular among growers and shippers, but Santa Rosa is still a favorite because it ripens early and the consumers like its characteristic plum flavor.

Japanese plums are propagated by budding onto seedlings of myrobalan plum or peach, or rooted cuttings of Myrobalan 29C, Marianna 2624, or Nemaguard peach. The last two are resistant to nematodes. Plums grafted on peach roots should be planted in areas where the soil drains well.

Most Japanese plums are self-unfruitful, so that a suitable pollinizer must be planted nearby. The exceptions are the partially or wholly self-fruitful Santa Rosa, Beauty, and Climax.

The trees are trained to an open-center vase shape, to a modified central leader, or to a palmette system. The shoots are spindly and at each node form short spurs bearing numerous flowers (Fig. 12.5). Since most of the crop is borne on spurs, nearly all long shoots may be removed without materially reducing the crop. A fully

Figure 12.5
Santa Rosa shoot just prior to bloom.

grown tree may produce about 100,000 flowers, of which only 1 percent must set to ensure an economical crop. Thus, thinning with a suitable chemical is necessary, preferably during the blossom period, to attain fruits of marketable size. One danger with blossom thinning is the high probability of losing a crop to frost, for the species blooms very early in the spring.

Cherries

The subgenus *cerasus* includes the sweet cherry, *Prunus avium;* sour cherry, *P. cerasus;* Duke cherries, which are hybrids between *P. avium* and *P. cerasus;* Nanking cherry, *P. tomentosa;* Early Chinese cherry, *P. pseudocerasus;* Western Sand cherry, *P. besseyi;* mahaleb cherry, *P. mahaleb;* and European dwarf cherry, *P. fruticosa*. The last two are European species used for rootstocks. Within this group is a host of ornamental flowering cherries from Asia, for example, *P. yedoensis* and *P. japonica*. The sweet, sour, and Duke cherries are grown commercially for their fruits.

The sweet cherry, a native of the Caucasus in southern Russia, has been cultivated in Europe from the pre-Christian period. Its diploid chromosome count is 16; that for the sour cherry is 32. The *P. cerasus* is believed to be a tetraploid interspecific hybrid between an unreduced gamete of *P. avium* and a haploid gamete of *P. fruticosa*, which also has 32 chromosomes. Some Duke cherries have 32 chromosomes.

SWEET CHERRY, *Prunus avium*

The sweet cherry was introduced into North America by the English and Dutch colonists and by the Spanish missionaries. In 1820 Russian fur traders planted seedlings at Fort Ross, California; some still thrive. Black Tartarian, Bigarreau Napoleon (syn. Napoleon, Royal Ann), and other cultivars were brought around the Horn by ship in 1850 and planted in the Napa Valley. About the same time, identical cultivars were brought overland by a covered wagon train to Oregon. From Oregon, Black Tartarian and Bigarreau Napoleon trees were transported and established in California. A nurseryman, Seth Lewelling, gave Bigarreau Napoleon its synonym, Royal Ann, causing a confusion that still exists. The term *Bigarreau*, meaning firm-fleshed in French, is rarely used by the trade.

The sweet cherry bears its crop on relatively long-lived spurs which have many simple buds (Fig. 12.6). Each bud contains an inflorescence with two to five preformed flowers. The species withstands cold better than most peaches, but it is not as cold hardy as the European pear or the European plum. The cultivars Windsor, Lyon, and Wood are more cold-tolerant than others. Where winters are not cold enough to satisfy the rest requirement completely, sweet cherry is the last of the stone fruits to bloom and is, therefore, able to avoid frost injury in most years. After a

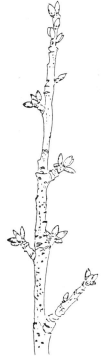

Figure 12.6
A well-spurred sweet cherry branch.

mild winter, a cultivar may bloom over a seven- to ten-day period, which allows for better cross-pollination. If the winter is exceptionally mild, a cultivar, such as Lambert, which needs a long chilling period, will have a straggly bloom that will not be adequately pollinated.

Sweet cherry trees are propagated on four rootstock species: *P. avium*, also known as mazzard; *P. mahaleb*; *P. cerasus*, commonly called the morello; and *P. fruticosa*, the European dwarf or ground cherry. The last species is used to a limited extent. Nursery trees are propagated by budding or whip grafting.

Mazzard stocks do well on deep, rich soil. A Black Tartarian tree on mazzard stock introduced from France to California in 1854 attained a height of over 65 feet (20 meters), a spread of 60 feet (18 m), and a girth of 10 feet (3 m) in 37 years. The species is resistant to root-knot nematodes, but it is susceptible to lesion nematodes and crown gall bacteria. Seedlings are budded or whip-grafted because cuttings taken from mature trees do not root. Colt, an interspecific hybrid between *P. avium* and *P. pseudocerasus*, reduces tree size under some conditions but does not do so in parts of the western United States and Italy.

The mahaleb stock is used widely in the western United States where the soils tend to be sandy and gravelly. It is deep-rooted and withstands drought better than mazzard or sour cherry stocks (see Root Distribution, Ch. 4). Conversely, it is very vulnerable to excess water in the root zone. Continuous late spring rains or ill-timed irrigation has caused trees to die. Sweet cherry trunks overgrow semidwarfing mahaleb rootstocks, but the graft union is mechanically strong. The mahaleb stock is tolerant and/or resistant to nematodes and to crown gall bacteria. The crown gall bacteria form small galls on roots but rarely kill the tree. In areas where buckskin disease, which is caused by a mycoplasma, is prevalent, the mahaleb stock is budded high in scaffold branches. If the sweet cherry top becomes infected, the scaffold is removed and the resistant rootstock is reworked with disease-free scions (see Grafting Methods, Pathogen-Induced Incompatibility, Ch. 10).

Sour cherry seedlings vary in their capacity as dwarfing rootstocks for sweet cherry. Stockton Morello, a clonal sour cherry selection, dwarfs the sweet cherry top, but not all cultivars are graft-compatible. Chapman, an early-ripening pollinizer for Bing, is not compatible with Stockton Morello, so that an interstock must be used. Vladimir, another sour cherry selection, imparts greater dwarfing tendency than the Stockton Morello and promotes early flower formation (Fig. 12.7). Both Vladimir and Stockton Morello tolerate heavy clay soils better than do the mazzard stocks; they are also more resistant to disorders induced by wet soils, such as crown rot

Figure 12.7
A Bing sweet cherry/Vladimir (*P. cerasus*) scion and stock combination at the beginning of the third season.

and infections by *Phytophthora* species. They produce suckers profusely from their shallow root systems, so that they are propagated from these suckers or by cuttings. Vladimir has been successfully propagated by meristem culture *in vitro*.

All commercial sweet cherry cultivars exhibit gametophytic incompatibility and are, therefore, self-unfruitful and cross-unfruitful, with cultivars possessing the same S factors (see Gametophytic Incompatibility, Ch. 5). Cross-pollination with a compatible pollinizer is, therefore, necessary to set a crop. Stella is a self-fruitful cultivar, but it is not planted widely.

Bing is the mainstay of the industry in California, Oregon, and Washington, as well as British Columbia, Idaho, and Utah. Early Burlat is next in importance because it ripens ten days earlier than Bing and serves as a pollinizer for it. Black Tartarian is the earliest commercial cultivar and an excellent pollinizer for Bing, but it is not as firm-fleshed as Early Burlat. Van is another excellent pollinizer for Bing, but it tends to set too many small fruits that ripen concurrently with Bing. After several heavy crops, Van trees become debilitated and short-lived unless they are pruned moderately during the dormant season. Van competes well with numerous European cultivars because it sizes well in France and northern Italy. King and Tulare are two new cultivars that form few or no doubles (see Physiological Disorders, Ch. 6). They can be cross-pollinated with Bing.

White-fleshed cultivars with an attractive blush, for example, Napoleon and Emperor Francis, were never as popular as the highly pigmented ones, but with the introduction of Rainier and others with a light-colored flesh of high quality and good flavor, the trend has been reversed. Rainier, which was introduced by the United States Department of Agriculture, has excellent size, quality, and bearing capacity. The cultivar is not used by the processing industry because it has a relatively soft endocarp, which tends to split or crack with slight pressure.

The canning industry processes both red- and white-fleshed cultivars. Fruits made into maraschino cherries are initially preserved in brine, decolorized with sulfur dioxide, and then steeped in marasca, a liqueur distilled from the fermented juice of a bitter wild cherry.

Sweet cherry trees are most commonly trained to an open-center vase shape (see Fig. 9.6). A common planting scheme for sites with deep, fertile soils is to plant trees on the diagonal, 7 meters apart. When the trees begin to crowd each other, alternate rows are removed, forming a square planting in which trees are 10 meters apart (see Planting Systems, Ch. 7).

Primary and secondary branches on upright growing cultivars, such as Bing and Rainier, are headed back to force branching. Some cultivars, such as Early Burlat, branch freely each year, but if the annual extension is more than 0.8 meter, shoots should be headed back to obtain branching at desirable heights. After the trees occupy their allotted volume, they should be dormant-pruned regularly to maintain vigor and provide light to lower limbs.

Trees propagated on dwarfing stocks can be spaced closer and trained to a hedgerow system. Since sweet cherry shoots develop long-lived spurs at close intervals along the shoots (Fig. 12.7), trees can be espaliered to resemble the upright candelabra system (see Training Systems, Ch. 9), or they can be trained to a single central leader or to the French *axe* (spindle) system (Lespinasse, 1983).

Sweet cherry cultivars destined to be sold on the fresh market are selected for quality and firm attachment of the fruit to the pedicel. Since fruits are customarily sold with the pedicels attached, they must be hand-harvested, a slow and tedious process. When mechanical harvesters were tried experimentally, excessive numbers of fruits were damaged because fruits, being attached firmly to the pedicel, were whipped about and struck spurs, limbs, and one another. Therefore, only fruits destined for freezing, canning, or brining are currently harvested with a trunk shaker and catching frames (see Harvesting Techniques, Ch. 6). Vittoria, a red-fleshed cultivar, and Bianca di

Verona, a white-fleshed one, both of which abscise readily from their pedicels, were developed in Italy to withstand mechanical harvesting and to be used in the processing industry.

SOUR CHERRY, *Prunus cerasus*

Prunus cerasus originated in what is now southern Russia near the Black Sea and was introduced into the Mediterranean basin before the Christian era. In Germany, small groves of sour cherry are found in areas that were in ancient times Roman encampments, suggesting that cherries were part of the legionnaires' diet. Currently, better selections from these wild types are being tried as dwarfing rootstocks for the sweet cherry. Because the trees thrive without care, they are being planted in forests as a source of food for forest birds.

Three important self-fruitful cultivars that were introduced into the United States from England and France are Early Richmond, English Morello, and Montmorency. Meteor and North Star are newer introductions. Sour cherries tolerate heavier soils, colder winters, and higher summer humidities than the sweet cherry. In the eastern United States where the weather is humid during the summer, fruits attain 20 millimeters in diameter, but under arid conditions of the western states, they reach only half this size. Although the sour cherry industry is centered in Michigan and New York, the species is also grown in Utah and other south-central and eastern states. Most of the crop is canned, and the product is used for pie fillings; some fruit is made into maraschino cherries. In Denmark a well-known liqueur is made from sour cherries.

Trees, propagated on mazzard and mahaleb rootstocks, reach heights of 7 to 8 meters. They are trained to a vase shape or a central leader, leaving a tall trunk to accommodate the clamps of trunk shaker; most of the crop is mechanically harvested onto a catching frame.

References Cited

LESPINASSE, J. M. 1983. Apple tree management in flat, vertical-axis, and palmette forms, by cultivar fruiting type. Experiments with other species: plum, peach, pear, and cherry. *Colloque Scientifiques* No. 15, pp. 103–130. Ed. L. D. Tukey. Montreal, Canada.

PROEBSTING, E. L., JR. 1970. Relation of fall and winter temperatures to flower bud behavior and wood hardiness of deciduous trees. *HortScience* 5:422–424.

CHAPTER 13

Nut Crops

Walnuts, *Juglans* spp.

Members of the family Juglandaceae have a haploid chromosome number of 16. They are found in China, Japan, southcentral Asia from Iran to India to Turkestan, and in North, Central, and South America. Among the nut crops, the Persian walnut, *J. regia*, is the most widely grown species. The nut of the eastern American black walnut, *J. nigra*, is more flavorful than that of the Persian walnut, but the black walnut kernel is difficult to extract from its thick, hard shell. Although the nut of the northern California black walnut, *J. hindsii*, is prized for its unique flavor by the confectionery trade, it is not grown commercially for this purpose. The nuts are collected from drought-resistant trees planted along roadsides to provide shade.

Paradox seedlings are the preferred rootstock for the Persian walnut. The Paradox trees are open-pollinated hybrids between the northern California black walnut and the Persian walnut. They are themselves nearly sterile, producing very few nuts.

Lumber cut from walnut wood is excellent for making furniture. European furniture makers prefer the light-colored wood of the Persian walnut and Paradox, whereas American woodworkers favor the darker wood of the black walnut.

PERSIAN WALNUT, *J. regia*

Persian walnut, a native to Iran and north to the Carpathian Mountains, was introduced to Greece and then to the Italian peninsula during the pre-Christian era. Later, its culture spread into northern Europe, growing as far north as Denmark. The English settlers introduced walnut culture to the American colonies, and the Franciscan monks brought it to what is now California.

The Romans called the Persian walnut *Jovis Glans*, meaning Jove's acorn, from which the genus derived its name, *Juglans*. The Carpathian strains tolerate cold as well as pears and apples do. Other strains are slow in acquiring cold hardiness, especially if limbs in the tops of trees have grown late into fall. These branches are killed when the temperature drops to $-6°$ C ($22°$ F) in November (see Winter Injury and Acquired Cold Hardiness, Ch. 7).

The species is heterodichogamous, that is, cultivars either shed pollen before the pistillate flowers are receptive (protandrous) or their flowering sequence is reversed (protogynous). Most newly introduced cultivars are protandrous (Fig. 13.1). Terminal-bearing cultivars such as Scharsch–Franquette and Hartley are still planted, but older ones, such as Franquette and Eureka, which become very large and produce about 1 metric ton per hectare, are no longer planted except as pollinizers in occasional orchards. Newer cultivars possessing lateral buds that are 80 to 90 percent fruitful and yielding over 2 metric tons per hectare are gradually replacing them.

In lateral-bearing cultivars, the petaloid structures, or tepals, on pistillate flowers differentiate in the axillary buds even while the leaves are expanding (see Flower Differentiation, Ch. 5). Flower bud differentiation occurs and pistillate flowers develop even when the light is not intense, but in overcrowded orchards a large percentage of immature flowers abscise (see Physiological Disorders, Ch. 6). As shoot extension slows down, pistillate flowers emerge and the bi-

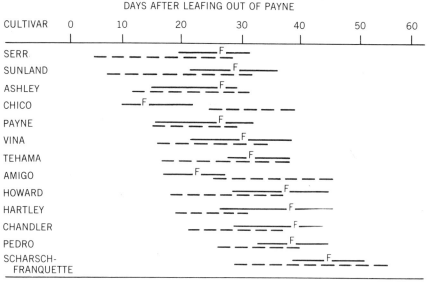

Figure 13.1 Sequential blooming periods of protandrous and protogynous Persian walnut cultivars grown in California.

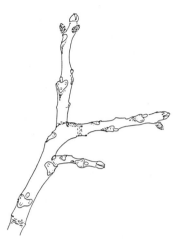

Figure 13.2
Dormant spur of a Persian walnut.

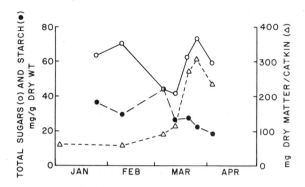

Figure 13.3
Seasonal changes in the total amounts of dry matter, sugars and starch in developing Serr walnut catkins. (From Ryugo et al., 1985)

furcate stigma becomes feathery and receptive to pollen. Although many grains of pollen may be trapped on the stigmatic surface, only one pollen tube grows through the transmitting tissue to the embryo sac. There is some evidence that an excessive number of pollen grains on the stigma may cause flower abscission.

Catkins are differentiated on the current season's growth (Fig. 13.2), but they expand and shed pollen the following spring. Considerable stored reserve food and mineral nutrients are utilized during catkin expansion (Fig. 13.3; Table 10).

Trees of fruitful cultivars are spaced 10 meters apart with appropriate pollinizer rows interspersed at certain intervals (see Pollinizers, Their Selection and Placement, Ch. 7). The trees are trained to a modified central leader with the lowest branch at least 2 meters above the orchard floor. This space allows for movement of harvesting and spraying equipment about the tree without endangering the operator.

The lateral bearing cultivars lend themselves to a hedgerow planting. The hedgerow is maintained by mechanically pruning the sides and top, annually or biennially, depending on the amount of vegetative growth. If these highly fruitful cultivars are not pruned, they stop growing and produce many small nuts. Hence, tree size can be controlled by pruning, and size-controlling rootstocks are not necessary for these cultivars.

In some areas of the world where grafting is found to be difficult, Persian walnut seedlings are grown. Therefore, neither nut quality nor size is uniform. In walnut production centers such as California, France, and Italy, trees are propagated on seedlings of Persian walnut, northern California black walnut, or Paradox. Propagation is by (1) whip-grafting dormant scions or (2) ring or patch budding on an actively growing seedling

Table 10. *Mineral Composition of Fully Developed Serr Walnut Catkins*

Percent of Dry Matter					Parts per Million	
Nitrogen	Potassium	Calcium	Magnesium	Phosphorus	Zinc	Manganese
3.61	2.67	0.45	0.34	0.39	41	28

From Ryugo et al., 1985.

stock after exudation of the xylem sap ceases (see Grafting Methods, Special Precautions, Ch. 10). If the exudate appears at the graft union, no callus is formed between the scion and stock; the union then fails to knit.

Not only are Persian walnuts difficult to root by cutting, but their bare-rooted cuttings rarely survive transplanting. Success in rooting by meristem culture *in vitro* has been limited.

All cultivars are attacked by bacterial blight, but those needing only a short chilling period are highly susceptible because they leaf out and bloom early in the spring when the humidity is still relatively high. Hence, walnut breeders strive to obtain late-leafing clones to avoid this disease.

Walnuts are susceptible to two other bacterial diseases known as shallow- and deep-bark cankers, which are caused by *Erwinia rubrifaciens* and *E. nigrifluens*, respectively. Both diseases, recognized by copious oozing of a black exudate from branches and trunks, are associated with water stress and poor soil conditions. Black walnut and Paradox rootstocks are resistant to the oak root fungus, *Armillaria mellea*, but they are highly susceptible to crown gall bacteria.

Seedlings of black walnut do not tolerate excessive soil moisture, for it leads to secondary infection by crown rot fungi. The Paradox rootstock imparts vigor to the scion cultivar and seems to be adaptable to shallow soils.

Trees on black walnut, Paradox rootstock, and other related species are highly susceptible to the pollen-transmitted virus that induces the graft incompatibility known as black line disease (see Grafting Methods, Pathogen-Induced Incompatibility, Ch. 10). Persian walnut topworked on a seedling of its own kind does not show any black line symptom but becomes a symptomless carrier of the virus.

The crop is harvested when the husk begins to split and the seed coat or pellicle turns tan-colored. The market prefers a light-colored pellicle, but as it turns brown, the kernel becomes sweeter and less astringent. Ethrel, a synthetic ethylene-generating compound, has been used to hasten husk splitting and to advance maturity. An overdose of ethrel causes premature leaf abscission and sometimes budbreak and an untimely flowering in October. Harvest is completely mechanized in larger holdings, starting with the trunk shaker, followed by a sweeper which concentrates the nuts in windrows between trees. The pickup machine scoops up the nuts and delivers them to a waiting gondola. The gondola is emptied into a huller which separates the nut from the husk. Nuts are dried in a forced-air drier to a moisture content of 8 percent, which is equivalent to 4 percent moisture on a kernel basis. At this moisture content, oil is slow to become rancid and the shelf life may be prolonged. With the increased cost of energy, driers are being redesigned to use ambient air and solar heaters. The nuts are bleached and sold in-shell, or the shells are removed and the kernels packaged and marketed as "meats."

Pecan, *Carya illinoensis*

The genus *Carya* belongs to the same family as the walnut, Juglandaceae. The scientific name was changed from *Hicoria pecan* to *Carya pecan* and then to its present one, *Carya illinoensis*. The nut of this species is called pecan; the edible nuts of the other two species, *C. ovata* and *C. laciniosa*, are called hickory nuts. The pecan, native to North America, was named for its northernmost native habitat, the state of Illinois. It is indigenous to an area bounded by Illinois, Tennessee, Nebraska, and the Mexican border. Pecans were a part of the diet of native Indian tribes in these areas.

Most pecan cultivars were discovered as chance seedlings growing in the wild or in orchards planted with seedlings. New superior cultivars that were bred by controlled hybridization at the Pecan Field Station at Brownwood, Texas, are Apache, Barton, Mohawk, Shawnee, and Wichita. Trees of precocious cultivars, Cheyenne, Shoshone, and Wichita, are initially planted at 11 × 11 meters. Later when trees begin to crowd one another, orchards are thinned. Cultivars require 140 to 210 days from budbreak to mature a crop in the southern states. Some winter chilling is necessary for normal growth of buds in the spring, but the length of the chilling period depends on the cultivar.

Pecans have dormant shoots with prominent lenticels and lateral buds with bent tips. The species is monoecious. Numerous catkins bearing staminate flowers are produced basally on a lateral compound bud, the distal portion of which abscises after a short growth (Wood and Payne, 1983). Pistillate flowers are borne terminally on the current season's growth arising from the shoot apex. The stimulus to flower is imparted to the terminal bud during the summer previous to blooming, but the differentiation of floral parts does not occur until budbreak (Wetzstein and Sparks, 1983). Although the pecan is self-fruitful, a dichogamous blooming habit warrants the placement of suitable cross-pollinizers to improve fruit set.

Pecan trees are propagated by grafting or budding on seedling rootstocks obtained from selected trees; the performance of a cultivar is influenced by the behavior of the stock. The first known cultivar was propagated vegetatively in 1846 by a slave in St. James Parish, Louisiana. In 1876 the cultivar was appropriately named Centennial at an exposition held in Philadelphia.

Alternate bearing is one of the major problems in pecan culture. A large nut crop in one growing season is followed by a light crop the next year. The maturing nuts create a large demand for current photosynthates and potassium. Potassium is accumulated rapidly by the maturing shuck, to the extent that the terminal leaves of some cultivars become chlorotic and abscise, symptoms typical of potassium deficiency. The depletion of reserve food and potassium in late fall and early winter prevents a complete restoration of nutrients before leaf fall. This seriously limits pistillate flower production and vegetative growth the following spring and causes the crop to be light (see Alternate Bearing, Ch. 5). Premature defoliation of trees by early fall frost and insect infestations will also hinder formation of flower buds and may hasten the onset of alternate bearing.

Pecan rosette, a foliar symptom attributed to zinc deficiency, is common in the southern states. Foliar sprays with zinc salts are effective in areas with alkaline soils. Where the soil reaction is acidic, digging trenches in the soil and applying zinc salts ameliorate the problem, presumably because the zinc remains soluble. Either treatment must be repeated periodically to prevent foliar symptoms from appearing.

Harvest is started when the shuck begins to split (Fig. 13.4), usually in October for early-maturing cultivars and is continued into December for many late-maturing cultivars. Maturation may be advanced several days with a foliar spray of ethrel. Harvesting is becoming increasingly more mechanized, utilizing trunk shakers, sweepers, pickers, and cleaners to separate the nuts from loose hulls, leaves, twigs, and other debris.

Figure 13.4
Cluster of mature pecan nuts ready for harvest.

Pecan, *Carya illinoensis*

Chestnuts, *Castanea* spp.

Chestnuts, with a diploid chromosome number of 24, belong to the family Fagaceae. Four important chestnut species cultivated for their nuts are *C. sativa*, the Eurasian or Spanish chestnut; *C. dentata*, the North American chestnut; and *C. crenata* and *C. mollissima*, the Japanese and Chinese chestnut, respectively. According to Bailey (1930), the Eurasian species grows wild from the Alps, westward into Spain, southward down the Apennines, a mountain range in central Italy, and eastward through Yugoslavia and into the dry hills of Greece. The species even ranges into North Africa and western Asia. The American species ranges from New England to Alabama and westward to the Mississippi River. This species, valued for its nuts and timber, is nearly extinct because of its susceptibility to the chestnut blight, a disease caused by a fungal organism, *Endothia parasitica*. The organism, native to Asia, was introduced into the eastern United States in the nineteenth century. It recently spread to Europe where it is devastating chestnuts in orchards and in forests. Chestnut blight in the western United States has been controlled by eradicating diseased trees whenever they are discovered.

Interspecific hybrids incorporate the blight-resistant character of the Japanese and Chinese species with the good nut quality of the American and European species. Seedlings of the American chinquapin, *C. pumila*, and two other Chinese species, *C. sequinii* and *C. henryi*, are being tested as a potential source of genes for blight resistance.

English settlers introduced European cultivars into the New World. Thomas Jefferson imported scions of European chestnuts from France and had them grafted onto native seedlings as early as 1773. Early pioneers migrating west with chestnut scions grafted them onto certain oak species with which they were compatible. Chinese railroad workers and gold miners planted chestnuts during the same era in the Mother Lode country of California. Some of these trees still survive.

Some Chinese cultivars developed in the United States are Abundance, Crane, and Nanking. Colossal, a large-seeded cultivar, suspected of being a hybrid between Chinese and European parents, was probably developed in about 1880 by Felix Gillet, a nurseryman in Grass Valley, California. It is the main cultivar currently planted in California with the pollinizers Letora, a Spanish cultivar, and Silverleaf. Marigoule and Maraval, two French hybrids between *C. crenata* × *C. sativa*, are grown in central and western France. They are good cross-pollinizers and resistant to *Phytophthora cinnamomi*. Bergougnoux and his colleagues (1978) found that inoculating roots with a particular mycorrhizal flora imparted additional resistance to *P. cinnamomi*.

All chestnut species are monoecious; some cultivars are self-unfruitful, whereas others are only partially so. Without adequate pollination, many burs set parthenocarpically. Hence, a compatible pollen source growing nearby is required to set a good crop. The large mixed terminal bud gives rise to a branch with catkins in axils of leaves. The basal catkins bear staminate flowers, whereas those near the shoot terminal are bisexual (see Fig. 1.21). Pistillate flowers are located near the base of these bisexual catkins. Three and, rarely, six free carpels are imbedded in the spiny involucre; all can develop into nuts if pollinated (Fig. 13.5). Evidence is strong that pollen from a large-seeded cultivar is manifested by large-sized nuts, an example of xenia.

Some interspecific graft incompatibility exists, but seedlings of Chinese chestnuts are least likely to be graft-incompatible with cultivars of other species. Trees are propagated by whip- or cleft-grafting scions or by patch, ring, or shield budding onto seedling stocks. Softwood and hardwood cuttings of all species are difficult to root, even when the rooting hormone, indolebutyric acid, is used. Own-rooted cuttings have been obtained from French cultivars by inducing adventitious buds on root cuttings (Bergougnoux et al.,

Figure 13.5
A Chinese chestnut shoot with two burs and a persistent catkin.

1978). Plants have been multiplied through meristem culture of embryonic axes obtained from Spanish seedlings (Vieitez and Vieitez, 1980).

Three important criteria in breeding and selecting new cultivars are (1) ease of removing the pellicle from the embryo; (2) the splitting of burs at harvest to facilitate the separation from the nut; and (3) resistance to chestnut blight, to *Phytophthora* spp., and to the Oriental chestnut gall wasp. The bitter pellicle consisting of seed coats and inner pericarp is deeply invaginated in the cotyledons of the Japanese species, so that the embryo is difficult to extract. Burs that do not split at harvest require pressure to extract the nuts. If cotyledons are injured as burs are removed, they are attacked by several kinds of molds during storage.

Chestnut blight, a soil-borne organism, has nearly decimated the American chestnut; most Chinese strains are resistant to this disease. The fungus infects the trunk and causes large cankers, which girdle and kill the tree. Suckers arising from roots develop into saplings until they are attacked and killed by the fungus. If small cankers are plastered with moist soil taken from the base of the tree, the infection does not spread and the cankers heal over. This outcome indicates that other soil organisms attack the blight fungus directly or produce an antibiotic, limiting fungal growth.

Recently, a nonvirulent strain of the blight fungi was isolated and found to be harboring a virus. Inoculation of bark cankers with this infected strain transmits the virus to the virulent form and causes the fungi to become nonvirulent. If the canker is small, the healthy bark tissue gradually overgrows the canker.

Infestation of chestnut groves by the Oriental chestnut gall wasp has caused considerable damage in the southeastern United States. The female wasp lays its eggs in the terminal bud. The developing larva produces zeatin, which promotes gall formation and stunts shoot growth. Fruit breeders are assessing selections of wasp-resistant Japanese chestnuts.

Chestnuts are harvested by allowing the crop to fall naturally and then picking them up. The nuts are separated from the burs and momentarily dipped into hot water to remove surface contamination and to eliminate most fungal spores. Chestnuts maintained at 40 percent moisture content will keep for several weeks at 5° C (41° F).

Chestnuts are eaten raw, boiled, roasted, or made into glazed candy. They are among the ingredients for stuffing roast turkeys for Thanksgiving and Christmas dinners. In Italy, chestnuts are made into a fine flour for confections.

The wood of chestnut is resistant to rot organisms. Many New England barns constructed during and after the Revolutionary War were made from chestnut boards. Similarly, the base boards of homes in Japan used to be made of chestnut lumber.

Filbert, *Corylus avellana*

The filbert, also known as hazelnut and cobnut, is one of nine species belonging to the birch family, Betulaceae, with a diploid chromosome number,

$2n = 28$. Wild species of *Corylus* are distributed from Japan westward through Manchuria, China, Tibet, Turkey, and into Europe. Two species indigenous to North America are the American filbert, *C. americana*, and the beaked filbert, *C. cornuta*. Their nuts are small and thick-shelled. The European filbert, *C. avellana*, is cultivated commercially for its nuts. This species constituted the dominant vegetation of northern Europe several millennia before Christ and was cultivated by ancient Greeks. The American filbert has been crossed with the European species, but the progenies have not been widely accepted by the nut industry.

The species is monoecious but self-unfruitful, requiring cross-pollination for setting. Wind pollination occurs in midwinter when the catkins elongate and pollen are shed. During pollination the style and stigma protrude from a yet unexpanded bud. The pollen tube grows down to the base of the style and stops. In the spring the pistillate flower matures after the elongation of the mixed bud has taken place. The pollen tube recommences to grow into the ovary and finally into the ovule where double fertilization occurs, several weeks after pollination.

Filberts are grown where summers are cool and winters cold but not severe. They need a period of chilling about equal to that of Delicious apple. Most filbert trees are not irrigated but grown in deep, rich soils. Trees do not tolerate poorly drained soils or a high water table.

Turkey is the world's largest producer of filberts, followed by Italy, Spain, and the United States. The Willamette Valley of Oregon is the production center in the United States.

The natural growth habit of the plant is to form a bush with numerous suckers arising annually. These suckers readily initiate roots when mounded with moist soil, a technique by which they are propagated. A nonsuckering rootstock is desirable because removing suckers is an annual problem. The Turkish species, *C. colurna*, does not sucker, but the rootstock overgrows all but the most vigorous cultivars.

Filbert trees in Europe are allowed to assume a bush form. In the Pacific Northwest, trees are trained to a single trunk to facilitate mechanized cultural practices. Four to five scaffold branches are developed. When trees begin to decline in vigor, every fifth row is severely pruned every fifth year so that as much as 75 percent of the terminal growth is removed. There is a drastic reduction in yield in the first year after pruning, but yields and nut sizes in the second and third years increase; they more than compensate for the initial loss.

Since filbert plants do not become large, they are planted 5 meters apart with compatible cross-pollinizers. Every sixth tree in every third row should be a pollinizer, making the ratio of the main cultivar to pollinizer 17:1.

Barcelona, a self-sterile selection, is the principal cultivar in Oregon; its pollinizer is Daviana. The pollen-shedding period of Daviana coincides with the full bloom of Barcelona, which produces high-quality nuts. Barcelona is susceptible to bacterial blight and a physiological disorder known as brown stain. Its production of excessive amounts of blanks (parthenocarpic fruits) has caused its popularity to wane. Ennis, a more productive cultivar with larger nuts and fewer blanks, is gradually replacing Barcelona. Butler, which is pollen-compatible with Barcelona and Ennis, is replacing Daviana.

Filberts are harvested following their natural drop in the fall. The nuts are swept into windrows between rows of trees, picked up mechanically, separated from the debris, and then delivered to the processing plant where they are dried down to 8 to 10 percent moisture content. Fully grown trees in a well-managed orchard yield 1 to 1.5 metric tons of dry nuts per hectare.

Pistachio, *Pistacia vera*

Pistachio is a member of the family Anacardiaceae, which also includes the cashew, mango, poison ivy, poison oak, and sumac. The genus *Pistacia* has 11 other species, but only the fruits of *P. vera* are large enough to be edible; those of other species are about the size of a barley grain. The species, native to central Asia, has somatic cells with 30 chromosomes. The crop has been grown throughout recorded history in Iran, Turkey, and Afghanistan, and it is an important export commodity of these countries. In the United States the crop is grown from northern California southward into Arizona, New Mexico, and Texas.

The species tolerates long, hot, dry summers and moderately cold winters. Trees tolerate temperatures as low as $-10°$ C ($14°$ F), but their flowers and new growth are susceptible to spring frost. The species requires approximately 1000 hours at or below $7°$ C to satisfy its rest period. The slow-growing deciduous trees reach a height and spread of 7 to 8 meters.

The species is dioecious, the plants bearing apetalous pistillate and staminate flowers in large panicles. The female flowers, which do not have nectaries, are wind-pollinated. The ratio of male to female trees should be 1:8 or 1:12 (see Pollinizers, Their Selection and Placement, Ch. 7).

Kerman, the principal female cultivar (Fig. 13.6) in the western United States, has a strong tendency toward biennial bearing (see Investigations on Flower Initiation, Ch. 5) and formation of parthenocarpic nuts. Red Aleppo, another female cultivar, is grown in limited quantity. Peters is the main pollinizer for both.

Trees are propagated by budding seedling stocks of *P. atlantica*, *P. terebinthis*, and *P. integerrima*. Propagation of rootstock species by softwood and hardwood cuttings has not been successful. These stocks are adapted to many kinds of soils, but they grow best on relatively deep, light, sandy loams with a high lime content. They seem to be more tolerant of alkaline and saline soils than most other deciduous fruit tree species but not of prolonged wet conditions. In poorly drained soils, pistachio trees readily succumb to *Phytophthora* spp.

These stocks are more resistant to nematodes and other soil-borne organisms than seedlings of *P. vera*. With the exception of *P. integerrima*, the others are susceptible to *Verticillium* wilt

Figure 13.6
Pistachio branch laden with fruits. Notice that the number of leaflets per leaf is not uniform. (Photograph courtesy of J. C. Crane)

disease, so that sites selected for a pistachio orchard should be pretreated with a soil fumigant.

Cultivars of the Chinese pistachio, *P. chinensis*, are admired for their bright fall coloration and planted as ornamental shade trees. Chinese pistachio seedlings are occasionally used as rootstock.

The crop is mechanically harvested with trunk shakers and catching frames. Mechanical removal of the hull, and washing and dehydration of the nuts soon after harvesting, ensure unblemished, ivorylike shells.

The disease known as epicarp lesion is caused by the leaf-footed plant bug, *Leptoglossus clype-*

alis, which feeds on the drupelet. Fruits that are attacked early in the season by the insect shrivel and abscise, but those fed upon late in their development persist with a brownish-black necrotic spot. These injured fruits have poorly developed kernels and discolored shells. To camouflage the surface blemishes, foreign exporters dye the shells with a red pigment.

References Cited

BAILEY, L. H. 1930. *Hortus, A Concise Dictionary of Gardening and General Horticulture*. The Macmillan Company. New York.

BERGOUGNOUX, F., A. VERLHAC, H. BRESCH, AND J. CLAPA. 1978. *Le Chataigner Production et Culture*. Institut National de Vulgarization pour les Fruits, Legumes et Champignons, Paris, and Comite National Interprofessional de la Chataigne et du Marron, Nimes, France.

RYUGO, K., G. BARTOLINI, R. M. CARLSON, AND D. E. RAMOS. 1985. Relationship between catkin development and cropping in the Persian walnut, "Serr." *HortScience* 20:1094–1096.

VIEITEZ, A. M., AND E. VIEITEZ. 1980. Plantlet formation from embryonic tissue of chestnut grown *in vitro*. *Physiol. Plant.* 50:127–130.

WETZSTEIN, H. Y., AND D. SPARKS. 1983. The morphology of pistillate flower differentiation in Pecan. *J. Amer. Soc. Hort. Sci.* 108:997–1003.

WOOD, B. W., AND J. A. PAYNE. 1983. Flowering potential of pecan. *HortScience* 18:326–328.

CHAPTER 14

Vine Crops and Bushberries

Grapes, *Vitis* spp.

The grape species *Vitis vinifera* is believed to have originated in the area adjacent to Asia Minor and extending into southern Russia as far east as Turkestan. The diploid species with 38 chromosomes has been grown from prehistoric times, and for these millennia the berries of its vines have been used for desserts and for making wine, juice, and raisins. The New World species, *V. labrusca* ($2n = 38$) and *V. rotundifolia* ($2n = 40$), natives of eastern and southern United States, are now being developed for these purposes. The species *V. solonis* and the interspecific hybrid *V. solonis* × *V. othello* (1613) are nematode-resistant rootstocks.

OLD WORLD GRAPE, *V. vinifera*

The Franciscan monks who grew grapes in the mission gardens introduced viticulture, the cultivation of grapes, and enology, the study of wine making, into North America. The important cultivar was named Mission. The Russians probably introduced the Madeira when they colonized northern California. The importation of Zinfandel shortly thereafter was an important step in expanding the wine-making industry. By 1850 there was a brisk demand for wine, and large acreages were being planted to grapes. Colonel Agoston Haraszthy, reputed to be the father of the grape industry in California, purchased 1400 different varieties in Europe in 1861 to plant in the state. Additional vines were purchased in Persia, Asia Minor, and Egypt to meet local demand. In those days, some raisins were shipped to the eastern seaboard by sailing vessels, but with the completion of the transcontinental railroad and improved shipping facilities, grapes and other fruits grown in California were transported to eastern states by freight cars.

European cultivars were brought over and planted by the early American colonists, but the plantings failed because they were attacked by mildew and insects.

All important commercial varieties are self-fruitful; the flowers are perfect and wind-pollinated. At bloom, the calyptra dehisces from the circular receptacle, exposing the pistil and stamens (see Flower Morphology, Ch. 1). Some popular table grape cultivars grown in California are Cardinal, Ruby Seedless, Thompson Seedless (syn. Sultanina), Italia, Ribier, Muscat of Alexandria, and Flame Tokay. A myriad of wine grape cultivars are grown throughout the world. Among the important ones are Zinfandel, Cabernet Sauvignon, White and Gray Riesling, Chablis, Merlot, Pinot Noir, and Chardonnay. (Growers interested in trying to grow the grapes from which the varietal wines are made should confer with the representatives of the Cooperative Extension Service to learn which have been tried and which have proved successful.)

By noting the geographic localities where grapes are grown, one might assume that this species is adaptable to various soil types and climates. The soil, its physical characteristics, and the amounts of clay and organic matter present affect the growth of vines, but their vigor can be altered by the choice of rootstock, by the amount and kinds of fertilizers, by applying irrigation water, and by controlling the intensity of pruning and the amount of cropping. Vines may appear to grow equally well under different climatic conditions, but subtle differences in fruit quality are detectable in fresh berries and wines. Hence, some cultivars are grown in regions with cool summers, such as the Napa Valley of California, the Rhone Valley of Switzerland, and along the Rhine River of Germany. Other cultivars are better suited for growing in the San Joaquin Valley of California, the foothills of the Andes in Argentina, and around the Mediterranean Sea where the summer temperatures often exceed 35° C.

Choosing the best-suited rootstock for the particular site and judicious pruning are probably the most economical, cultural means of regulating production of high-quality berries and yet maintaining vine vigor. Excessive pruning enhances vine vigor but encourages strong vegetative growth and poor flower set so that total yield will decrease. The resulting large but soft berries may possess poor color and contain insufficient soluble solids. With underpruning too many shoots having clusters of small berries develop. Such overcropping delays ripening, depletes carbohydrate reserves, and debilitates the vines.

The secret of successful production lies in balancing yield with vigor. This is accomplished by pruning together with moderate applications of fertilizers and water. Vines should be pruned soon after leaf fall and before xylem sap begins to flow in late winter. Vines, pruned after the flow becomes copious, lose some vigor because the exudate contains nutrients. Cultivars with different growth and bearing habits are pruned and trained to various trellising systems (see Ch. 9).

In the tropics where temperatures are relatively warm throughout the year, grapes tend to remain evergreen. Pruning and defoliating these vines about two months after harvest force budbreak and flowering so that three crops can be harvested in two years. Such manipulations educe results under tropical conditions because (1) flower initiation occurs in the compound buds in the axils of newly expanded leaves; and (2) florets differentiate on the rachis at budbreak (see Ch. 5).

Seedless table grapes such as Thompson Seedless, Perlette, and the seeded cultivar Kyoho, a *V. vinifera* × *V. labrusca* hybrid introduced from

Japan, are routinely sprayed with gibberellic acid. The treatment thins the cluster, elongates the rachis, and promotes berry size (see Blossom and Fruit Thinning, Ch. 6).

Grapes are harvested after the berries are completely colored and/or have acquired sufficient soluble solids to make them palatable. The optimum sugar:acid ratio for wine grapes at harvest depends on the cultivar and the quality of the final product that the enologist wishes to achieve. For most varietals, the sugar content should range between 20.5 and 24 degrees Brix, and the acid content may vary from 6 to 10 gram-equivalents of tartaric acid per liter of juice.

Although more wine grapes are being harvested mechanically, those intended for the fresh dessert market are still hand-harvested and often packed in the field to reduce the handling and bruising of berries. The packed units are then placed in cold storage until they are ready to be shipped.

FOX OR EASTERN GRAPE, *V. labrusca*

Vita labrusca is highly resistant to powdery mildew and can be grown in areas of hot, humid summers where the disease is prevalent. Many American cultivars, such as Concord, Early Campbell, Delaware, and Niagara, believed to be interspecific hybrids of *V. vinifera*, carry mildew resistance. Thus, they are grown in the eastern United States and Japan where warm summer rains are common. Their berries, unlike those of *V. vinifera*, have a characteristic foxy flavor, and their epidermal and subepidermal layers slip freely from the inner pulp. Hence, they are known as slip-skin grapes. Concord, with its delightful flavor, has been a favorite for making juice and jelly. The juices of eastern grapes have also been made into sweet wines.

In northern latitudes of United States, the species lends itself well to the Geneva double curtain and to the Kniffen (fan or palmette) system of cane pruning. These training systems provide good foliar exposure to light, which is essential for forming flower buds and for uniform ripening of the berries. Under these systems the vines acquire the necessary degree of cold hardiness to survive the winter.

Grapes are propagated as a single-node leafy cutting under mist, by hardwood cutting, by layering, or by grafting scions onto desirable rootstocks. Propagation by hardwood cutting is a popular and inexpensive means of clonally multiplying grapevines because most cultivars initiate roots easily (see Hardwood Cuttings, Ch. 10). If the soil is infested with phylloxera or nematodes, it should be treated with appropriate soil fumigants prior to planting. For additional protection, scion cultivars should be topworked onto appropriate nematode-resistant rootstock cultivars such as Dog Ridge, Salt Creek, *V. solonis*, and *V. solonis* × *V. othello* (1613).

Kiwifruit or Chinese Gooseberry, *Actinidia deliciosa*

The kiwifruit is a member of the genus *Actinidia*, which consists of about 50 species and 100 taxa. The genus belongs to the family Actinidiaceae, formerly called Dilleniaceae (Ferguson, 1984). Members of the genus range from the mountains of south-central China eastward to northern Japan and Taiwan, westward to Russia, and southward to India and Indochina. In this expansive area, which includes some of the world's most rugged terrain, plant explorers are still discovering new species and forms.

Not only are the chromosomes of the genus *Actinidia* very small and tedious to count, but polyploidy and interspecific hybrids are common so that speciation is difficult. The cold-hardy species *A. arguta* ($2n = 116$) and *A. kolomikta* ($2n = 112$) produce small, smooth, green-skinned edible fruits which ripen from midsummer into fall. Two botanists, Chou-fen Liang of China and A. R. Ferguson of New Zealand, recently divided the species *A. chinenesis* into two groups. The original name was assigned to the soft-haired to nearly fuzzless type possessing 58 chromosomes and a yellow or green mesocarp tissue at maturity. The group that bears fruits with stiff epidermal hairs and has 170 chromosomes was classified as *A. deliciosa* var. *hispida*. The commercial cultivars such as Hayward and Bruno belong to this species.

Actinidia purpurea bears small ovate to oblong purple fruits; *A. melenandra* bears reddish-brown fruits. David Fairchild, an American plant explorer, developed the interspecific hybrid *A. fairchildii* between *A. arguta* and *A. deliciosa*. The hybrid species *A. ananasnaya*, developed by the Russian horticulturist Ivan V. Michurin, is a cross between *A. arguta* and *A. kolomikta*. The rich germ plasm pool of the genus *Actinidia* awaits exploration by cytologists, geneticists, and fruit breeders.

At the turn of the century, seedlings of *A. deliciosa* were introduced into New Zealand where current cultivars and cultural practices were developed. The principal cultivar grown is the Hayward, although Bruno, with its long, slender, highly flavorful fruit, is more productive. The female cultivars, Abbott, Allison, Monty, and Vincent, are grown to a lesser extent. The pollinizers are California Male, Matua, and Tomuri.

Kiwifruit vines are trained to a pergola system (see Fig. 9.20) or the T-bar trellis (see Fig. 9.21). Dormant pruning is rather simple, for nearly all buds on vigorous one-year-old canes are potentially fruitful mixed buds. Currently, the ypsilon or Tatura and palmette trellis systems are being tried experimentally to maximize interception of solar radiation by leaves and to minimize shoot breakage in the spring to prevent fruit loss. Shoot breakage not only reduces yield but also induces adjacent dormant buds to break and develop into vegetative shoots (see Flower Differentiation, Ch. 5). Some vigorous vegetative shoots are needed for cane replacement, but a large number of them will create excessive shade. Hence, tying flowering shoots in the spring and erecting permanent wind breaks around a kiwifruit plantation are essential for producing bruise-free fruits.

The species is now grown commercially wherever the climate is temperate. Mature vines toler-

Figure 14.1
Pistillate flowers of Hayward kiwifruit (left) and staminate flowers of California Male (right). Notice the vestigial parts in each.

ate midwinter temperatures of −12° C (10° F), but early fall or late spring frosts of −3° to −6.7° C (20° to 26° F) will damage the fruit and kill the bark tissue, usually just above the soil surface.

Kiwifruit vines are monoecious, although a few hermaphroditic plants exist. Inflorescences of pistillate or staminate flowers (Fig. 14.1) are borne from the second to the eighth basal nodes. The first flowers to open on a vine are the largest; those that bloom later are progressively smaller. Flower size is proportional to the number of fused carpels. Hayward produces "fans," which are formed by the fusion of the terminal flower with one or two lateral ones on the inflorescence (Fig. 14.2). Lateral flowers on an inflorescence usually bloom later and develop into smaller fruits than terminal flowers.

The fine pollen grains are carried by wind and insects. Unlike apple or peach bloom, kiwifruit flowers do not attract bees, but they seem to attract syrphid flies. For pollination to be adequate, 10 to 15 beehives per hectare should be set out during the bloom period.

The ratio of male to female vines should be 1:8 or 1:11 to assure maximum pollination. If the ratio is 1:8, a male vine is planted in every third position in the row and in every third row (see Planting Systems, Ch. 7). In areas where the chilling requirement is not completely satisfied, the bloom periods of the female and male vines do not overlap. When dichogamy occurs, fruit size is inevitably poor.

Kiwifruit vines should be grown in deep, well-drained soils. Although vines in full leaf utilize as much water as do large peach trees, their roots do not tolerate extended flooding. Excessive moisture or poor drainage causes vines to become chlorotic and susceptible to infection by crown rot organisms.

Young vines are irrigated twice a week during the first season. As vines develop full canopies, they are irrigated daily (see Irrigation Systems, Ch. 7). The use of overhead sprinklers with water containing salts causes marginal burning of leaves; dark streaks form on the surface of fruits, marring their appearance. The discoloration fades if har-

Figure 14.2
Abnormal "fan"-shaped Hayward kiwifruits.

vested fruits are dipped in a dilute citric acid solution. This problem of discoloration can be overcome by acidifying the water or converting the irrigation system to a low head sprinkler or a drip system.

Vines are propagated by softwood and hardwood cuttings, by whip grafting, or by dormant budding of a yearling rootstock (see Budding, Ch. 10). Since the rootstock bleeds freely in the spring, grafting or budding is done after xylem exudation has ceased and the stock has leafed out.

Kiwifruits are harvested when their soluble solids reach 6.25 to 7 percent. Fruits are brushed to remove the pubescence before they are sized and packed. When damaged fruits are removed and ethylene is eliminated in the storage facility, Hayward fruits can be stored from three to five months at 0° C and 85 to 90 percent relative humidity.

A well-grown Hayward fruit contains about 120 milligrams of ascorbic acid per 100 grams of fresh weight. The aromatic component of flavor becomes detectable after the fruit is soft and sweet. Although the fruit is best eaten fresh, it does make an excellent jam.

Cranberry and Blueberries, *Vaccinium* spp.

The genus *Vaccinium* includes numerous native North American species of blueberries and one of cranberry in the family Ericaceae. The basic diploid chromosome number is 24, but polyploidy is common. Some related but uncultivated species are found in the Alpine regions of Europe. The native American Indians introduced the trailing swamp cranberry and blueberries to early American colonists who quickly included them in their diet. Commercial culture of cranberry began in Massachusetts in about 1815 and spread westward to Wisconsin, to the Pacific Northwest, and northward into Canada.

CRANBERRY, *V. macrocarpon*

The species *V. macrocarpon* ($2n = 24$) has some strict cultural requirements that limit the sites where it can be grown. The region must have cool, clear summer days for the fruits to attain a large amount of anthocyanins. The soil must be acidic and rich in peat. The land should be flooded in the summer with fresh running water to provide weed and insect control and also during the winter months to protect the plants from freeze damage.

The evergreen shoots initiate roots readily from cuttings. Cuttings are transplanted into bogs overlain with sand. The commercial cultivars Howe and McFarlin, as well as less important cultivars, originated as open-pollinated chance seedlings. Flower buds are initiated and differentiated during the summer before they bloom. Flowers are insect-pollinated, and the fruits, whose sizes are proportional to seed count, mature in late fall to early winter. The crop is harvested by flooding the bogs and mechanically gathering the buoyant fruits or raking the vines by hand with a combed bucket. An important factor in cranberry quality is its large amount of anthocyanins. The fruit is processed into sauce, which has customarily been served with the Thanksgiving turkey; the juice is frozen, bottled, or blended with other fruit juices.

BLUEBERRIES

More than 30 blueberry and related species grow wild in the United States and Canada. The cultivated highbush cultivars possessing 48 chromosomes belong to *V. corymbosum*; the lowbush blueberry, also with 48 chromosomes, includes three species: *V. myrtilloides* ($2n = 24$), *V. angustifolium* ($2n = 48$), and *V. brittonii* ($2n = 48$). The rabbiteye blueberry, *V. ashei*, has 72 chromosomes.

Although newly bred cultivars produce fruits with excellent quality, berries growing in the wild are also harvested to supply local markets. Commercially, highbush plants are propagated by hardwood cuttings, whereas the rabbiteye plants are started as softwood cuttings under mist.

Highbush blueberries are cultivated in the northern latitudes of the United States, extending from the Atlantic Coast states to the Mississippi Valley and northward into Canada. Some plantings are located in Washington and Oregon. This species accounts for the bulk of the commercial plantings along the coastal plains extending from North Carolina northward to New Jersey. Blueberries are also grown in other parts of the world.

The species tolerates cold temperatures and requires as much chilling as the peach for normal development of flower buds. In areas where the winters are relatively mild and the summers hot and humid, the rabbiteye blueberry grows well. Interspecific hybrids between highbush and rabbiteye blueberries, have been selected for their adaptability to various ecological niches. Other desirable traits sought by blueberry breeders are resistance both to stem canker caused by the fungus *Botryosphaeria corticis* and to bud mite, increased aromatic constituents, improved color and better size, and ease of mechanical harvesting.

The species requires well-drained, sandy soils that are rich in organic matter. Plants do best in highly acidic soils ranging in pH from 3.8 to 4.5. Since the plants are shallow-rooted, they require frequent irrigation. The berms are mulched with 10 to 15 centimeters of sawdust to conserve water and reduce competition from weeds.

Urn-shaped flowers are borne on inflorescences arising from one-year-old shoots. Cross-pollination is essential to obtain good yields. Hence, rooted cuttings of different cross-fruitful cultivars are planted in double rows on hills or berms.

Harvest of highbush blueberries begins in late May along the southeastern seaboard of the United States, and it progresses northward. The earliest rabbiteye clones start to ripen about a month later. Fruits are harvested when they are still firm but fully colored. They are eaten fresh, canned, or frozen and utilized for pie fillings and other baked goods.

Currants and Gooseberries, *Ribes* spp.

Diploid species in the genus *Ribes*, belonging to the family Saxifragaceae, have 16 chromosomes. The ancient Greek and Roman herbalists and botanists made no mention of currants, probably because the species is limited to the northern latitudes of Europe where agriculture developed later. These ancient people neither relished nor sought gooseberries, for the thorny bushes bear tart, slightly bitter fruits, and they grow in the mountainous regions of the Mediterranean basin.

CURRANTS

The European red currants, *R. sativum* and *R. rubrum*, were introduced into the American colonies in 1629, shortly after the Pilgrims landed in Massachusetts. Plants of *R. nigrum* were probably introduced about the same time, but being susceptible to mildew, they did not flourish. Many states prohibit the cultivation of currants by law because several species are an alternate host to white pine blister rust, *Cronartium ribicola*.

Red currants thrive in the cool coastal climate of California, but because the berries are tender and difficult to harvest, they have essentially disappeared from commerce. Plants are easily propagated by hardwood cuttings. The rooted cuttings should be planted on hills or berms to keep free water away from the crown. Since they are shallow-rooted, they should be mulched with straw or bark material to prevent the soil from drying and to suppress weed growth.

Flowers are borne on long racemes on which 10 to 20 fruits may set (Fig. 14.3). The crop ripens in early summer and is either hand- or machine-harvested in northern Europe. The berries are made into jellies and jam. Bushes are kept vigorous by pruning out the older three- to four-year-old canes that become weak and barren.

GOOSEBERRIES

The European gooseberry, *R. grossularia*, like the European currants, was introduced to the New World by early American colonists, but it did not thrive. The species is very susceptible to the gooseberry mildew and serves as an alternate host to white pine blister rust. Therefore, the native American gooseberry, *R. hirtellum*, was domesticated and hybridized with the introduced species.

Figure 14.3
Racemes of red currants arising from a mixed bud. (Adapted from *The Small Fruits of New York*. Ed Hedrick, 1925)

One or two flowers arise from simple or mixed buds on short lateral branchlets (Fig. 14.4). Most cultivars are self-fruitful and insect-pollinated. The crop may be mechanically harvested when the fruit acquires a yellowish tinge. Gloves are essential for protection from spines during hand-harvesting. Some spineless hybrids with mildew resistance have been introduced, but they are not planted as widely in North America as they are in northern Europe. Like currant bushes, gooseberry bushes should be planted on mounds and mulched to protect the crown from soil pathogens and to minimize cultivation and irrigation.

Raspberries, Blackberries, and Dewberries, *Rubus* spp.

Members of the genus *Rubus* have diploid somatic cells with 14 chromosomes, but according to Darrow (1937), numerous polyploids ranging up to nonaploids exist. The genus is distributed throughout the Northern Hemisphere, southward into Mexico, and in the Southern Hemisphere along the Andean Mountain Ranges. Members of the genus make up a large group within the rose family, having adapted to various soils and climates from the Arctic to the tropics.

The genus is grouped into two subgenera, one of raspberry and the other of blackberry and dewberry. Linnaeus named the European raspberry *Rubus idaeus* because the species was growing on Mount Ida in Greece. This species is believed to have originated in Asia Minor. Early colonists brought it to North America where the native American raspberry, *R. occidentalis*, was already growing.

Different species have overlapped and crossed naturally. Fruit breeders added to the taxonomic confusion by hybridizing species from widely different climates and areas. Luther Burbank imported blackberry seeds from areas near the Hi-

Figure 14.4
Bearing habit of American gooseberry.

malaya Mountains. He grew the seedlings and bred them with a native California dewberry, naming one Himalaya. Loganberry is a hexaploid selection of a California dewberry, originated by Judge Logan in Santa Cruz, California. Mammoth, Cory, and Cory Thornless are classified in *R. titanus*; Laxton, a heptoploid, is a hybrid between an American and a European blackberry species.

Two characteristics distinguish raspberry from blackberry and dewberry. (1) Ripe raspberry drupelets come free from the receptacle, but those of blackberry and dewberry adhere to it. (2) Raspberry drupelets are hairy and adhere to one another, even without the receptacle; drupelets of blackberry are glabrous. The dewberry has a trailing prostrate growth habit, whereas the blackberry canes grow in an upright position. Raspberries are called red raspberry or black raspberry, depending on whether the drupelets accumulate red or purple pigments.

Plants are easily propagated by tip layerage, root and hardwood cuttings, or crown division. The roots are perennial but the tops are biennial. In dewberry culture, primocanes, the vegetative shoots that arise from the crown in the spring, are allowed to trail on the ground. The more vigorous primocanes are brought up and wound about the trellis wires (see Fig. 9.22) to bear the next year's crop. The same canes are called floricanes when they bear flowers. Immediately after harvest, the floricanes are removed and burned to get rid of any pests and diseases.

In the cultivation of blackberry and raspberry vines, the upright floricanes are removed and burned after harvest. The new primocanes that arise from below ground are thinned and tied to parallel wires to keep them upright and within the row.

The crop is usually harvested by hand for the fresh dessert market. Fruits intended for processing, for jams, jellies, and frozen products, are mechanically harvested by over-the-vine harvesters.

References Cited

DARROW, G. M. 1937. Blackberry and raspberry improvement. In *United States Department of Agriculture Yearbook for 1937*, pp. 496–533. The Macmillan Co. New York.

FERGUSON, A. R. 1984. Kiwifruit: A botanical review. *Hort. Rev.* 6:1–64.

HEDRICK, U. P., ED. 1925. *The Small Fruits of New York. Thirty-third Annual Report. J. B. Lyon Co. Albany, New York.*

CHAPTER 15

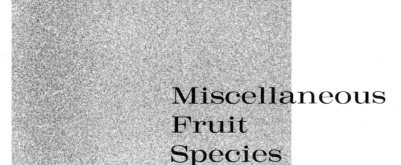

Miscellaneous Fruit Species

Persimmons, *Diospyros* spp.

Persimmons belong to the genus *Diospyros* within the family Ebenaceae. Members of the genus range from the temperate zones of China, Japan, Australia, and the United States to the tropical areas of Central America and Southeast Asia. *Diospyros kaki* and *D. lotus* are natives of China, although the term *kaki* is Japanese. Seedlings of *D. lotus* are used only for rootstock. The native habitat of *D. virginiana* ranges from New Jersey westward to Nebraska, growing wild in the woods of Missouri and Arkansas and extending southward to Texas and Florida.

Persimmon wood, especially the jet black heartwood from tropical species, yields ebony, a valuable timber for furniture. The hard sapwood from the American persimmon is still the main source of wood for golf club heads.

ORIENTAL PERSIMMON, *D. kaki*

The oriental persimmon has a haploid chromosome number of 15, but cultivars are tetraploid or hexaploid, possessing 60 or 90 chromosomes. The species has been grown in China and Japan for centuries; it was introduced into the United States, Israel, and Italy near the turn of this century. Japanese cultivars and their hybrids are commercially grown in California and Florida, in Israel and Italy, and more recently in New Zealand. In other temperate zone countries, they are grown as ornamentals for their bright red fall foliage and deep orange-colored fruits.

Kaki trees are propagated by budding or grafting seedlings of all three species. Kaki seeds are difficult to obtain because many cultivars tend to set parthenocarpically. But seeds are readily obtained from fruits of other species. Trees on *D. kaki* rootstocks are usually smaller than those on rootstocks of *D. lotus* or *D. virginiana*. Lotus seedlings are preferred for their vigor and graft compatibility with most kaki scions. Fuyu is graft-incompatible with some strains of lotus seedlings. Trees are trained to a modified central leader or to the palmette system. Vigorous trees, especially of the cultivar Hachiya grafted on lotus seedlings, tend to drop immature fruits. Vigor and losses from preharvest drop are reduced by withholding nitrogen and irrigating trees sparingly. Girdling of limbs and trunk in the spring lessens the problem of fruit abscission.

Oriental cultivars are monoecious or bear only

Figure 15.2
Staminate (upper) and pistillate (lower) flowers of the oriental persimmon, *Diospyros kaki*.

pistillate flowers. Monoecious cultivars, such as Zenji Maru, bear small male flowers on small cymes whose peduncles are persistent (Fig. 15.1); pistillate flowers with large fleshy sepals are borne singly near the bases of shoots (Fig. 15.2). Although fruits will set parthenocarpically, more are likely to set if flowers are pollinated and the resulting fruits develop seeds.

Cultivars are classified as pollination-constant (PC) if the flesh color of a seeded mature fruit turns orange-yellow; and as pollination-variant (PV) if the flesh surrounding the seeded locule becomes brown through polymerization and oxidation of tannins (see Pollination-Related Phenomena, Xenia and Metaxenia, Ch. 5). Cultivars are also classified as nonastringent (NA), for example, Fuyu, Gosho, and Jiro; or astringent (A), for instance, Hachiya, Fuji, and Saijo. Thus, Fuyu is classified as pollination-constant and nonastringent (PCNA); Zenji Maru as pollination-variant, nonastringent (PVNA); and Hachiya and Fuji are pollination-constant, astringent (PCA).

These classifications are arbitrary; environment–genotype interactions affect the expression of the constant–variant trait. Cultivars, although classified as being pollination-constant and nonastringent, for example, Fuyu and Jiro, remain astringent when they are grown in areas

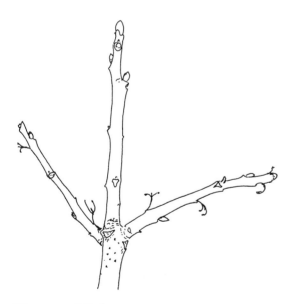

Figure 15.1
Dormant shoot of a monoecious Japanese persimmon tree with persistent peduncles of male flowers.

with short, cool summers. The tannins in the harvested fruits do not polymerize, even when they are treated with ethanol fumes.

Harvested fruits store well at 0° C and 85 percent relative humidity. Astringency may be removed by exposing the fruits of partially seeded and parthenocarpic PCNA and PCA cultivars to carbon dioxide, ethylene, acetaldehyde, or fumes of ethanol. Placing fruits in airtight containers with a small amount of whiskey on the calyx lobes or with a few ripe bananas or pears will cause the tannin to polymerize, rendering the fruit nonastringent. Ethylene- and alcohol-treated Hachiya fruits will become sweet and soft within a week. Fruits treated with 80 percent carbon dioxide for several hours become nonastringent, but they require more time to ripen compared to those given other treatments. The high concentration of carbon dioxide, which suppresses ethylene synthesis, delays ripening.

Persimmons may be eaten fresh after they lose their astringency. The melting-flesh PCA types such as Hachiya, Saijo, and Tanenashi are dried whole. The firm, fully mature fruits are peeled and strung up to dry. During the drying process the fruits are kneaded to keep them from becoming hard and woody. Peeling or slicing the fruits apparently promotes the evolution of ethylene, which in turn induces tannins to polymerize, for the prepared fruits lose their astringency rapidly. If the dried product is stored in a cool, dark place, a fine coating of powdery sugar appears on the surface. Firm-fleshed, nonastringent fruits can be thinly sliced and placed in a warm oven until the pieces become chewy. Ripe Hachiya fruit may be frozen and eaten as a dessert upon thawing. The soft pulp is also used for making bread, cookies, and puddings.

AMERICAN PERSIMMON, *D. virginiana*

Cultivars of the native American persimmon, *D. virginiana* ($2n = 90$), such as Early Golden, Ruby, and Miller, are chance seedlings grown for local consumption in the eastern and southern United States. Trees are susceptible to attack by the fungus *Cephalosporium diospyri*.

Diospyros virginiana flowers are apparently not cross-pollinated by pollen from *D. kaki* or *D. lotus*. The fruits are about 30 millimeters in diameter when mature and very astringent until they become soft, sweet, and gelatinous. The seeds germinate readily, and the seedlings make good, vigorous rootstocks for oriental cultivars.

Fig, *Ficus carica*

The fig, a member of the mulberry family, Moraceae, has 26 chromosomes. It is the only species in the genus *Ficus* cultivated for its fruit. The banyan tree, *F. benghalensis*, is known for its aerial roots which develop into multiple trunks to support a huge tree covering several acres. The strangler fig, *F. aurea*, grows on and around another tree for support and eventually kills it. Cultivars of *F. benjamina* are common house plants.

As early as 2700 B.C., the Egyptians cultivated the fig. Its culture spread throughout the Mediterranean basin, and it became a staple of the Greek and Roman diet. The Spaniards brought the fig to Hispaniola in 1520 and to Mexico in 1560. In 1769 the Spanish padres planted trees of the Spanish cultivar, Brebal, at a mission in San Diego, California. The cultivar is marketed today as Black Mission, Mission, or Franciscan Mission (Condit, 1947).

Commercial fig production in California started in 1885. Although the Smyrna fig was introduced in 1880, trees did not bear a crop. Researchers, learning that the tiny fig wasp, *Blastophaga psenes*, was required to transport pollen from the male or caprifig to the Smyrna fig to obtain fruit set, introduced the insect in 1900 (Tufts, 1946). Aristotle had noted in 340 B.C. that a wasp performed an unknown function in the setting of fig fruits.

The species is propagated by hardwood cut-

Figure 15.3
Kadota fig tree trained to a low, sprawling configuration.

tings of two- to three-year-old wood, 20 to 30 centimeters (8 to 12 inches) in length, and stored at 4° C and 85 percent relative humidity to induce callus formation. Cuttings are lined in the nursery row after the danger of spring frost is over. Rooted cuttings are dug the following winter and planted in the orchard.

Most cultivars are trained to a modified central leader to form a strong framework. After the basic structure is established, mature trees are pruned only to remove weak and dead limbs. Kadota (syn. Dottato) figs, grown for the fresh market or for canning, are trained and pruned to a low, wide-spreading head to facilitate harvesting without ladders (Fig. 15.3).

Fig trees grow in sandy loam to clay loam soils. They tolerate soil pH of 6 to 7.8, but they are sensitive to high levels of boron and sodium. Their water requirement is less than that of most deciduous fruit trees. They thrive best in areas with low humidity, intense sunshine, and hot summer temperatures. Young developing shoots in the spring and immature wood in the fall are injured by subfreezing temperatures of only −1° C (30° F), but after mature trees are fully acclimated, they will withstand winter temperature of −12° C (10° F).

Fruits develop and ripen at temperatures between 32° and 37° C (90° and 100° F), but they become leathery if the temperature rises to 43° C (110° F) during the ripening period. Nomads in Middle Eastern countries knew even in the pre-Christian era that if the ostioles of immature figs were injured and a drop of olive oil was dabbed on the wound, the fruits would enlarge and ripen early. The mechanism for the success of this treatment is better understood today because (1) any injury to fruits will cause ethylene evolution, and (2) olive oil is metabolized to ethylene by some fruit tissues.

Fig trees whose crops are to be dried are trained to a modified central leader tree with an open center. Fruits ripen and partially dry on the tree; they fall to the ground where they are mechanically swept into windrows. The fruits are then picked up mechanically and placed in bins. Since the crop does not mature uniformly, harvesting continues at weekly intervals for four to six weeks. At present, growers spray Mission and Calimyrna trees with ethephon, an ethylene-generating chemical, to hasten and promote uniform ripening.

Figs, grown in humid climates, are subject to several fungal diseases. Should it rain when fruits are ripening, the water is absorbed and causes the skin to split, allowing fruits to spoil. A rust organism attacks young leaves, and smut and molds, which are spread by the dried-fruit beetle, invade the fruit and cause spoilage. Endosepsis is a particularly serious problem of Calimyrna figs because it is spread by the wasp during caprification. The morphology of the fruit makes these diseases and insects difficult to control with fungicides and insecticides.

The trees are highly susceptible to root-knot nematodes. If the presence of nematodes is suspected, the soil should be pretreated with an appropriate fumigant.

The fig fruit is a syconium, consisting predominantly of peduncular tissue and the floral organs (see Fig. 1.32). As the fruit ripens, it enlarges, accumulating as much as 65 percent sugars on a dry weight basis. Four horticultural types of fig based on the kind and structure of flowers and the need for pollination and fertilization to produce fruit are caprifig, Smyrna, common, and San Pedro (Chandler, 1957).

CAPRIFIG AND CAPRIFICATION

The caprifig is the only type in which the winter fruit (Fig. 15.4) possesses both staminate and pistillate flowers. It is generally termed a "male," and the fruit is inedible. The staminate flowers line the inner side of the ostiole. The pistillate flower has a short style that is adapted to oviposition by the fig wasp. The larvae of the wasps overwinter in the mamme crop. The larvae undergo metamorphosis within the syconium, enter the pupal stage in early spring, and develop into adult wasps in April. The male wasp impregnates the female wasp while they are still in the syconium. As the female wasp leaves the mamme fig through the ostiole, the pollen adheres to her body. She locates the profichi or summer fig and lays her eggs in the stylar canal and then dies. Growers pick these profichi figs in June before the wasps emerge and place them in baskets which are hung in Smyrna- and San Pedro-type trees so that their flowers can be pollinated. This process of pollination is known as *caprification*.

The female wasp, which emerges from the profichi figs left on the caprifig tree, locates a mammoni or autumn fig and lays her eggs. The adult wasp emerging from the mammoni crop seeks the mamme fig, the overwintering crop in which to lay eggs and to start another cycle.

SMYRNA FIG

The Calimyrna is the only Smyrna type that is grown in California. The pistillate flowers have long styles and are not adapted to oviposition by the wasp. Hence, the wasp crawls about the flowers, effecting pollination, and then leaves because it is unable to lay eggs. The cultivar does not set without pollination and seed formation.

COMMON FIG

The common-type fig has long-styled pistillate flowers which set parthenocarpically. Mission, Kadota, Magnolia, Adriatic, and Brown Turkey belong to this group. The first crop is borne on one-year-old wood, but the second crop ripens on the current season's wood (Fig. 15.5).

SAN PEDRO FIG

Figs of the San Pedro type combine the traits of the Smyrna and the common types. The fruit of the first crop may set without pollination, yielding parthenocarpic fruit. The second crop requires pollination by the fig wasp. This type is therefore not recommended because a caprifig tree must be grown nearby.

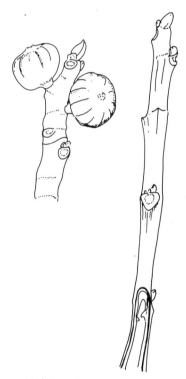

Figure 15.4
Overwintering crop of caprifigs (left) and dormant shoot of Mission fig (right).

Fig, *Ficus carica*

Figure 15.5
First crop of fig maturing on one-year-old wood and second crop maturing on the current season's growth. The leaves were removed to illustrate the bearing habit. (Photograph courtesy of J. C. Crane)

Mulberries, *Morus* spp.

Mulberries, belonging to the family Moraceae, consist of many species found in the temperate zone of the Northern Hemisphere, ranging from Europe to China. The Black mulberry, *M. nigra*, may have originated in China; it bears many sweet, tender, pleasantly flavored fruits in axils of leaves. Early-ripening cultivars have been planted around cherry orchards to detract birds and to serve as windbreaks. The species tolerates temperatures as low as $-8°$ C during the dormant season. The Red mulberry, *M. rubra*, a species native to the eastern United States, bears fruits similar to that of *M. nigra*. Fruits of *M. alba*, the White or Russian mulberry, are rather insipid.

M. multicaulis and, to a lesser degree, *M. alba* are grown for silkworm culture. The plants, propagated by cuttings, are close-planted to form a high-density orchard and are maintained as shrubs. Trees are severely pruned to force numerous shoots from which leaves are stripped to feed silkworms. The silk is spun by the larvae for preparing their cocoons. The cocoons are boiled, and the fine silk thread is then unwound and made into fabric.

The ornamental weeping mulberry, grafted about 2 meters high on an erect stem, assumes the shape of an umbrella. A fruitless mulberry clone is planted as a shade tree because it grows rapidly and does not shed fruits that stain walkways and cars.

Feijoa, *Feijoa sellowiana*

The feijoa or pineapple guava is one of two South American evergreen species that have edible fruits and belong to the myrtle family, Myrtaceae; the other is *Psidium cattleianum* or strawberry guava (Bailey, 1930). Although the feijoa is subtropical in origin, ranging from northern Argentina to southern Brazil, it requires more than 50 hours of winter chilling to bear well. It is rather cold hardy, tolerating winter temperatures as low as $-9°$ C ($-15°$ F).

The plants bear attractive flowers with an inferior ovary. On the outside the thick petals are whitish and tomentose, covered with densely matted hairs. On the inner side they are purplish, with extensively exserted dark red stamens. The sweet petals are added to fruit salads. Most cultivars are self-fruitful, but they are more productive with cross-pollination. The flowers, borne on short shoots, develop into oblong green fruits about 7 centimeters long and 4 centimeters in diameter. The fruits that mature in early winter develop a flavor reminiscent of pineapple.

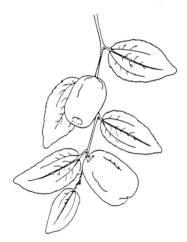

Figure 15.6
Fruiting shoot of a jujube.

Plants are propagated by seeds, layerage, cuttings, or by whip-grafting scions on seedling rootstocks. They may be propagated by suckers from own-rooted cultivars. The bushy plants require little or no care because they are relatively free of diseases and insects.

Zizyphus, *Zizyphus jujuba*

There are some 40 species in the genus *Zizyphus*, which has 24 chromosomes in the diploid somatic cells and belongs to the family Rhamnaceae. Of the numerous species the Chinese jujube or date is the most widely cultivated. Linnaeus named it *Rhamnus zizyphus*, but Lamarck gave the species its current name. The name of the genus is also spelled *Ziziphus*, contributing to taxonomic confusion. The species grows in the wilderness from China to southeast Europe; it thrives under semiarid conditions along the Great Wall of China. Several cultivars are grown for their fruits in China and Japan, but in the United States and elsewhere, the deciduous, thorny, willowy jujube tree is grown as an ornamental. The plant is propagated by grafting scions from desirable cultivars onto seedling rootstocks.

The axillary flowers, borne in clusters, develop into small, elongated or roundish, brownish-green drupes (Fig. 15.6), rich in ascorbic acid (vitamin C). The fresh fruit at maturity is about the size of a small egg; it has a crisp, somewhat dry texture. The dried fruit resembles that of the palm date, from which its name, Chinese date, is derived. The Chinese make a glazed candy with the fruit.

References Cited

BAILEY, L. H. 1930. *Hortus, a Concise Dictionary of Gardening and General Horticulture.* The Macmillan Co. New York.

CHANDLER, W. H. 1957. *Deciduous Orchards*, 3rd Ed. Lea and Febiger. Philadelphia.

CONDIT, I. J. 1947. *The Fig.* Chronica Botanica Co. Waltham, Mass.

TUFTS, W. P. 1946. The rich pattern of California crops, pp. 113–238. In *California Agriculture*. Ed. C. B. Hutchison. University of California Press. Berkeley and Los Angeles.

APPENDIXES

APPENDIX A

Natural Hormones and Synthetic Growth Regulators

Natural Hormones

Various naturally occurring hormones are mentioned in this textbook. These substances cannot appropriately be discussed under any one chapter heading because they participate in nearly all physiological processes. For those who wish to delve into the more basic aspect of plant hormones, the textbook *Biochemistry and Physiology of Plant Hormones* by Thomas C. Moore (1979) and the review article by Yang and Hoffman (1984) are recommended.

By definition, plant hormones are compounds that are synthesized in one part of a plant and translocated to another part where they educe certain growth responses. There are currently five classes of natural plant hormones: auxins, gibberellins, cytokinins, abscisic acid, and ethylene. Except for ethylene, which is evolved as a gas, the others exist in "free" or "bound" forms. The bound ones occur as peptides and glucosyl esters of a hormone. As analytical techniques are improved, additional classes of hormones and different kinds within a class will undoubtedly be discovered.

AUXINS

Indoleacetic acid (IAA) was the first hormone to be isolated, purified, and described by plant scientists. The hormone is found in such tissues as the shoot apex and endosperm, where cell division is occurring rapidly. It is transported basipetally or polarly in stem tissue. Some of the roles attributed to this hormone are expanding the cell walls, nutation, geotropisms and phototropisms, tactile responses, inducing the evolution of ethylene, governing apical dominance, promoting flower and fruit set, and initiating roots.

Charles Darwin, the author of *Origin of Species*, also wrote *The Power of Movement in Plants*, published in 1880. In this book, he described the bending of etiolated grass tips toward the light source. Frits C. Went (1928) pursued this phototropic behavior for his graduate thesis and contributed a considerable amount of knowledge about auxin physiology.

The biosynthetic pathway of IAA begins with the amino acid tryptophan (Fig. A.1), although other pathways may exist. After extraction from plant tissues and separation by paper and thin-layer chromatography, IAA and its conjugated forms are ascertained to be biologically active by using the oat mesocotyl and Avena coleoptile curvature or straight-growth tests. Detection and quantification are also possible by high-pressure liquid chromatography and fluorospectrophotometry. Biochemists are currently attempting to synthesize a monoclonal antibody with a marker to identify and quantify auxins in tissues as well as in extracts.

Ascertaining levels of IAA in plant tissues is tedious and difficult because IAA, which occurs in very minute concentrations, is unstable, especially in the presence of light and air. Even after isolation, it breaks down nonenzymatically to scatole. It is readily oxidized by IAA oxidase, an isozyme of peroxidases, to methyleneoxinodole and other intermediate by-products. IAA exists as a free acid, but when applied to tissue slices, it is readily conjugated with amino acids or glucose. These conjugations are considered to be (1) an auxin-sparing reaction, (2) a detoxifying mechanism, or (3) a means of temporary storage.

Two other auxins, 4-chloroindoleacetic acid and phenylacetic acid, have been isolated and identified from legumes and other plant species. Their presence has not been reported in extracts of tissues and organs of fruit tree species.

GIBBERELLINS

The first gibberellin was isolated from rice seedlings infected with a fungus. Japanese rice

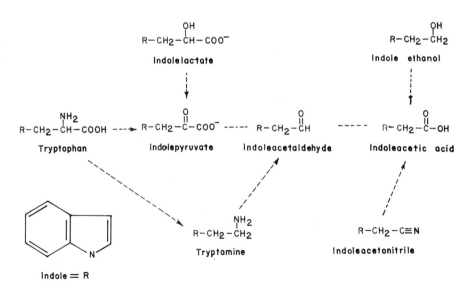

Figure A.1
Biosynthetic pathways of indoleacetic acid, an auxin.

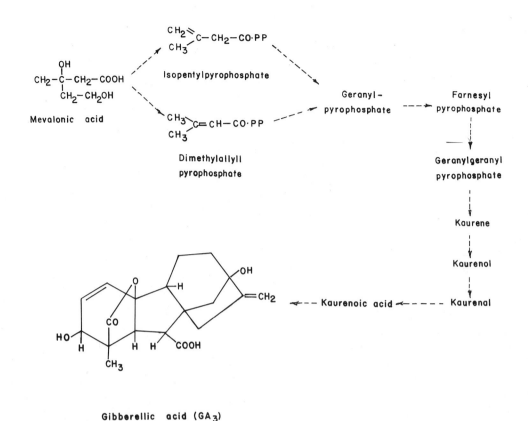

Figure A.2
Biosynthetic pathway of gibberellic acid from mevalonic acid in wild cucumber nucellus–endosperm. (Adapted from Graebe et al., 1965)

growers called the diseased seedling *bakanae* or crazy plant, because infected stem cells enlarged, resulting in erratic growth. The taxon of the fungus was keyed as *Fusarium moniliforme* by European pathologists, but the same pathogen was classified as *Gibberella fujikuroi* in Japan. Professors Yabuta and Sumiki of Tokyo University crystallized the growth-promoting ingredient from the fungi in 1938 and named it gibberellin. Since then, more than 50 substances with ent-gibberellane structure, known as gibberellic acid (GA), have been isolated from diverse representatives of the plant kingdom. Some GAs are biologically active and others are not.

C. A. West and his group at the University of California at Los Angeles discovered that the developing endosperm of the wild cucumber was a rich source of gibberellin. The tissue was used to establish the biosynthetic pathway of GA (Fig. A.2), which starts with mevalonic acid (MVA) (Graebe et al., 1965). The decarboxylation of MVA yields an isoprenoid, a building block which serves as a substrate for the synthesis of abscisic acid, steroids, carotenes, phytol, and even for a part of the cytokinin molecule. Thus, the interruption of gibberellin biosynthesis by the use of growth retardants promotes flowering, enhances or hinders production of chlorophyll (depending on the chemical) and prolongs the rest period of seeds.

The hormone is found in the free and bound

forms in many tissues and organs but principally in those that are rapidly undergoing cell division and enlargement. One bound form is hydrolyzed by glucosidase to yield free GA.

The functions of GA are related to cell elongation and *de novo* induction of enzymes such as alpha-amylase in the barley aleurone layer. These two effects have been quantified so that the elongation of the lettuce hypocotyl, growth of dwarf corn and rice, and hydrolysis of starch in embryoless barley seeds are used to estimate GA levels in tissue extracts.

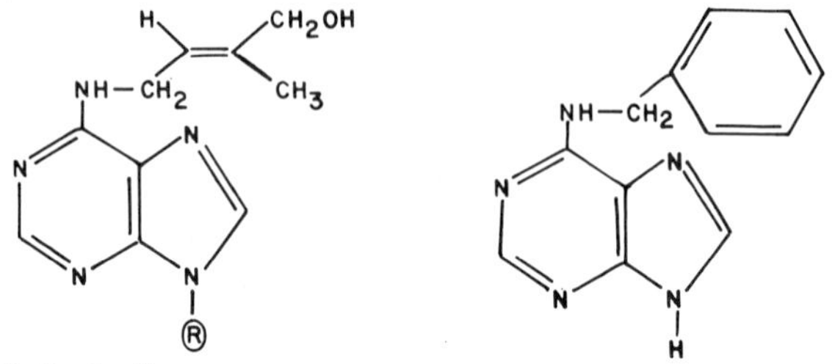

Figure A.3
Structural formula of zeatin and its riboside (left) and benzyladenine (right).

CYTOKININS

Pioneers in tissue culture research observed that coconut milk (liquid endosperm) and yeast hydrolyzates contained constituents that favor cell division and differentiation. Folke Skoog and Carlos O. Miller (1957) had found that tobacco callus enlarges in the presence of these components in tissue culture media. D. S. Letham (1963), working at the Fruit Research Station, DSIR, in Aukland, New Zealand, isolated and identified the first naturally occurring cytokinin. He named the crystalline compound extracted from young corn kernels "zeatin." C. O. Miller (1961), working at the University of Indiana, had also isolated a kinetinlike substance from immature corn kernels but had not characterized it. Since these discoveries were made, the riboside and nucleotides of zeatin have been isolated from many plant tissues.

Kinetin was discovered before zeatin, but it was found to be an artifact because the compound was detected only in autoclaved samples of DNA and not in fresh material. Cytokinins have the isopentyladenosine structure (Fig. A.3) and promote growth. Tissue culturists use zeatin or one of its less expensive synthetic substitutes, benzylaminopurine (BAP) or benzyladenine (BA, Fig. A.3), to induce shoot morphogenesis from callus. Cytokinins do not induce root initials.

The exact synthetic pathway for cytokinins has not been established. Zeatin is presumably formed from adenine, which has a slight growth-promoting potential, ribose, and isopentyl pyrophosphate. The presence and concentration of cytokinins in extracts are assessed and estimated by growing callus derived from tobacco pith and soybean hypocotyl segments. Other assays utilize retention in the dark of chlorophyll in leaf strips or disks or synthesis in the light of chlorophyll by cucumber and cocklebur cotyledons.

ABSCISIC ACID

The history of abscisic acid, a naturally occurring growth retardant and abscission-inducing agent, goes back to Liu and Carns (1961) who isolated from cotton bolls a compound that accelerated defoliation. The material was tentatively named "abscisin" but was later changed to "abscisin I." After the first discovery, Ohkuma, Lyon, Addicott, and Smith (1965) at the University of California at Davis isolated a crystalline substance from young cotton fruits and named it "abscisin II." In the meantime, P. F. Wareing at the University College of Wales at Aberystwyth isolated a substance from birch leaves that kept buds dormant. This material was named "dormin." Ohkuma and Wareing with the help of chemists Cornforth, Milborrow, and Ryback (1965) at the Shell Research Laboratory in England, found that the molecular structures of abscisin and dormin were identical (Fig. A.4). To settle the confusion of names, these workers agreed to call the compound abscisic acid (ABA).

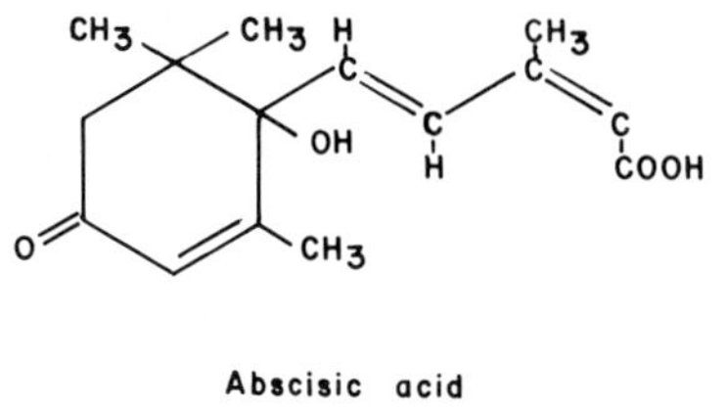

Figure A.4
Chemical structure of *cis*-abscisic acid.

Robinson and Ryback (1969) demonstrated that ABA is a 15-carbon sesquiterpenoid, derived from mevalonic acid. Abscisic acid also occurs as a glucose ester which is thought to be a storage form of this hormone. Another pathway of ABA synthesis may be from cleavage of a carotenoid, violaxanthin (Taylor and Smith, 1967).

Abscisic acid inhibits GA-induced *de novo* synthesis of alpha-amylase by the barley endosperm. Since ABA stems from mevalonic acid metabolism, it is not surprising that the application of growth retardants to inhibit gibberellin biosynthesis would promote ABA synthesis and alter the synthetic pathway of cytokinins.

ETHYLENE

The gas ethylene is known to cause epinasty and chlorosis of leaves, and subsequently their abscission, to induce aerial roots on stems of tomato, and to hasten the ripening of fruits. Membrane permeability is influenced by this gas, so that some substrates and enzymes are allowed to react, whereas normally they would be compartmentalized.

The biosynthesis and mode of action of this hormone could not be accurately assessed until gas chromatographic techniques were refined to a point that allowed ethylene to be measured in parts per billion (ppb). Yang and Hoffman (1984), working in the Mann Laboratory at the University of California at Davis, elucidated the biosynthetic pathway (Fig. A.5).

Ethylene is evolved during almost any stressful event, such as wounding of plant tissues, or bending of branches by winds, or the application of any mechanical force causing tissue distortion. Treatment of plants with synthetic auxins, such as 2.4-dichlorophenoxyacetic acid (2,4-D) or its analogue, 2,4,5-trichlorophenoxypropionic acid (2,4,5-TP), causes plant tissues to generate the gas. Bases of cuttings treated with indolebutyric acid evolve ethylene proportional to the auxin concentration. According to Yang, auxin stimulates the production of S-adenosylmethionine (SAM), which is metabolized to aminocyclopropanecarboxylic acid (ACC), the immediate precursor of ethylene. Yang and his coworkers demonstrated that when tissues are treated simultaneously with aminoethoxyvinylglycine (AVG), a synthetic inhibitor of ethylene generation, and ACC, ethylene evolution is not interrupted. This indicates that AVG inhibits the conversion of SAM to ACC but not the breakdown of ACC to ethylene.

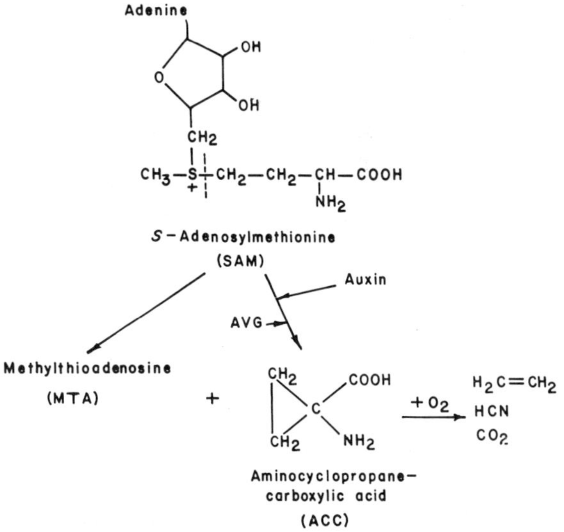

Figure A.5
Biosynthetic pathway of ethylene from S-adenosylmethionine. The dotted line indicates the point of cleavage; the arrows indicate where auxin and aminoethoxyvinylglycine (AVG) induce and inhibit, respectively, the formation of ACC. Hydrocyanic acid (HCN) is presumably metabolized into a amino acid immediately. (Adapted from Yang and Hoffman, 1984)

The precursor ACC was originally discovered in apple cider, but the biological significance of its presence was unknown until the reactions described were elaborated. This compound is found in plants growing in water-saturated soils, which cause an anaerobic condition in the root zone. Its translocation to the aerial portion of the plant where it is converted to ethylene may explain why such plants become chlorotic.

The complexity of the interaction among the different hormones was demonstrated by Lau and Yang (1973) with hypocotyl segments of mung bean. They found that IAA applied exogenously to these pieces was rapidly conjugated with aspartic acid while it induced ethylene evolution. When these hypocotyl segments were co-treated with IAA and kinetin, the auxin-sparing reaction was suppressed and ethylene evolution was enhanced. They concluded that the level of free IAA is regulated by cytokinin, which in turn moderates the amount of ethylene in tissues.

Synthetic Growth Regulators

Structural formulas of widely used growth-retarding compounds whose usages are described in the text are shown in Fig. A.6.

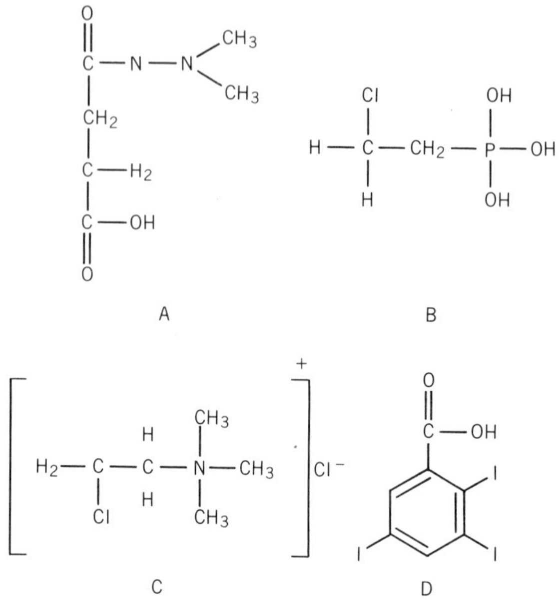

Figure A.6

Synthetic growth regulators: (A) Daminozide, butanedioic acid mono(2,2-dimethylhydrazide); also known as succinic acid, 2,2-dimethylhydrazide (SADH); Alar and B-9 are trade names. (B) Ethephon, (2-chloroethyl)phosphonic acid (CEPA); ethrel is a trade name. (C) Chlormequat, (2-chloroethyl)trimethylammonium chloride (CCC); cycocel is a trade name. (D) 2,3,5-triiodobenzoic acid (TIBA); Regim-8 is a trade name.

References Cited

CORNFORTH, J. W., B. V. MILBORROW, G. RYBACK, AND P. F. WAREING. 1965. Chemistry and physiology of "dormins" in sycamore. Identity of sycamore "dormin" and abscisin II. *Nature* 205:1269–1270.

DARWIN, CHARLES. 1880. *The Power of Movement in Plants.* D. Appleton and Co. New York, N.Y.

GRAEBE, J. E., D. T. DENNIS, C. D. UPPER, AND C. A. WEST. 1965. Biosynthesis of gibberellins. I. The biosynthesis of (−)-kaurene, (−)-kaurene-19-ol, and *trans*-geranylgeraniol in endosperm nucellus of *Echinocystis macrocarpa* Greene. *J. Biol. Chem.* 240:1848–1854.

LAU, O-L., AND S. F. YANG. 1973. Mechanism of a synergistic effect of kinetin on auxin-induced ethylene production. *Plant Physiol.* 51:1011–1014.

LETHAM, D. S. 1963. Zeatin, a factor inducing cell division, isolated from *Zea mays. Life Sci.* 2:569–573.

LIU, W. C., AND H. R. CARNS. 1961. Isolation of abscisin, an abscission accelerating substance. *Science* 143:384–385.

MILLER, CARLOS O. 1961. A kinetin-like compound in maize. *Proc. Natl. Acad. Sci. U.S.A.*

MOORE, THOMAS C. 1979. *Biochemistry and Physiology of Plant Hormones.* Springer-Verlag. New York, Heidelberg, Berlin.

OHKUMA, K., J. L. LYON, F. T. ADDICOTT, AND

O. E. SMITH. 1965. Abscisin II, an abscission-accelerating substance from young cotton fruit. *Science*. 142:1592–1593.

ROBINSON, D. R., AND G. RYBACK. 1969. Incorporation of tritium from ((4R)-4) tritiated mevalonate into abscisic acid. *Biochem J*. 113:895–897.

SKOOG, F., AND C. O. MILLER. 1957. Chemical regulation of growth and organ formation in plant tissues cultured *in vitro*. *Symp. Soc. Exp. Biol.* 11:118–131.

TAYLOR, H. F., AND T. A. SMITH. 1967. Production of plant growth inhibitors from xanthophylls; a possible source of dormin. *Nature*. 215:1513–1514.

WAREING, P. F., C. F. EAGLES, AND P. M. ROBINSON. 1964. Natural inhibitors as dormancy agents. In *Regulateurs Naturels de la Croissance Vegetale*. Ed. J. P. Nitsch.

WENT, FRITS C. 1928. Wuchstoff and Wachstum. *Rec. Trav. Bot. Need.* 25:1–116.

YANG, S. F., AND N. E. HOFFMAN. 1984. Ethylene biosynthesis and its regulation in higher plants. *Ann. Rev. Plant Physiol.* 35:155–189.

APPENDIX B

Table of Equivalent Weights and Measures

Metric	Equivalents	U.S. Equivalents
Lengths		
1609.3 meters	5280 feet	1 mile
1 kilometer	1000 meters	3281 ft
1 meter	100 cm	39.37 inches
0.9144 meter	3 ft	1 yard
30.48 centimeters	12 in	1 foot
2.54 cm	25.4 millimeters	1 inch
1 micron	0.001 mm	1/25,400 in
Area		
4050 square meters	43,560 sq ft	1 acre
1 hectare	10,000 sq meters	2.47 acres
1 sq meter	100 sq decimeters	10.76 sq ft
Weight		
28.375 grams		1 ounce (oz)
454 grams	16 oz	1 pound (lb)
907 kg	2000 lb	1 ton
1 kilogram (kg)		2.2046 lb
1 metric ton (MT)	1000 kg	1.102 ton
1 MT/ha	1784 lb/acre	0.892 ton/acre
Volume		
3.785 liters	4 quarts	1 gallon
3.785 kg water	8.337 lb water	1 gallon
1 cubic meter	1,000,000 cu cm	35.3 cu ft
28.32 kg water	7.481 gallons	1 cu ft
1 liter	1000 ml	1.057 qt
0.946 liter	946 ml	1 quart (qt)

APPENDIX C

Some Common Units of Expression

Nutrient concentration
: millimole of element per kilogram dry matter
percentage on dry weight basis
milligrams per unit leaf area or leaf nitrogen
parts per million (ppm)
  1 ppm = 1 mg per liter (W/V)
       1 µl per liter (V/V)

Tissue and organ composition
: grams per kilogram dry matter
percentage on dry weight basis
parts per million
grams per unit organ, e.g., fruit, seed

Moisture content
: grams per kilogram fresh or dry matter
grams per organ
percent on dry weight basis

Crop yield	metric tons per hectare
	grams per unit plant
	grams per unit leaf area
	fruits per linear length of shoot
	fruits per cross-sectional area of branch or trunk
Fruit set	percent of flowers at anthesis
	fruits per unit length of shoot
	fruits per 100 clusters or spurs
Growth rate, linear	millimeters per unit time, e.g., day
volume	cubic centimeters per organ per unit time
weight	grams per organ per unit time
Relative growth rate	increment of growth per initial size per unit time, as a percentage
Photosynthetic rate	micromole of carbon dioxide per square meter of leaf area per second
Transpiration rate	micromole of water per square meter of leaf area per second
Respiration rate	milliliters of carbon dioxide per kilogram fruit per hour
Ethylene evolution	microliters of ethylene per kilogram fruit per hour
Water tension	megapascal (MPa)
	(equivalent to 0.101 bar or atmosphere)

APPENDIX D

Laboratory Exercises and Objectives

A recent survey of horticultural departments in the United States revealed that students are exposed to numerous horticultural principles and practices. The following laboratory exercises and objectives dealing with fruit trees and vines are designed to familiarize students with the practical arts of pomology. Most are accompanied with questions; their answers should reinforce the objectives of the exercises.

Fall Term

OUTDOOR PRACTICES

1. Harvesting techniques
    A. Objectives
        1. By applying various maturity indices, determine when to begin har-

vesting such crops as apples, grapes, pears, and kiwifruits.

2. Learn how to use the following instruments in the orchard to estimate harvest date.

    a. Hand refractometers for determining soluble solids and percent Brix.

    b. Magness–Taylor pressure tester to measure flesh firmness.

    c. Pull gauge for measuring fruit removal force.

    d. Color charts for assessing ground and surface colors.

    e. Size gauge or minimum-diameter rings.

3. Observe and participate in hand and mechanical harvesting operations.

B. Questions

1. What are the advantages and disadvantages of hand and mechanical harvesting techniques?

2. How might cultural practices and equipment be changed to reduce bruising, improve quality, and increase yield?

C. Materials: ladders, picking buckets, bags, and boxes.

2. Propagation

A. Objectives

1. Become familiar with sexual propagation by collecting and stratifying seeds.

2. Learn different methods of asexual propagation.

    a. Graft different species using various methods.

    b. Prepare layerages of blackberry and grape canes.

    c. Collect branches and canes for making hardwood cuttings.

B. Questions

1. Why must seeds be soaked for two to three days before stratification?

2. The basal part of a walnut shoot is preferred as a scionwood source by grafters, but the same part on sweet cherry and plum is discarded. Why?

3. Why would a grower want own-rooted trees? What are some disadvantages of planting own-rooted trees?

4. What are the advantages of clonally propagating rootstocks?

C. Materials and tools: peat moss and plastic bags for storing seeds and scions, grafting tools, grafting wax, adhesive and/or plastic tape, wide rubber bands.

3. Pruning and training of dormant trees and vines

A. Objectives

1. Distinguish the different approaches to pruning and training by species and cultivars and by age.

    a. Spur bearers versus shoot-bearing species.

    b. Hand versus mechanically harvested crops.

    c. Based on training systems, e.g., vase, palmette, spindle bush.

2. Relate canopy configurations to leaf area indexes.

B. Questions

1. Why does heading back of apple shoot

terminals promote lateral spur formation?

2. Why does pruning when the xylem is exuding sap limit the growth of grape and kiwifruit vines?

3. What is the basis for the horticultural axiom, "Pruning is a dwarfing as well as a fruit-thinning process"?

4. What are the advantages and disadvantages of a vertical fruiting wall such as a palmette? Of a flat, horizontal bearing surface, e.g., a pergola?

C. Tools: hand pruning shears and long handle loppers.

4. Lay out an orchard and a vineyard; plant trees and vines

A. Objectives

1. Become aware of the topography and cultural practices when planning and laying out an orchard.

a. Restriction on tree row direction by irrigation and cultivation practices.

b. Placement of cross-pollinizers of crops that are wind-pollinated and/or mechanically harvested.

2. Relate planting distances to scion–stock combinations, bearing habits, and training systems.

3. Become aware of plant and soil conditions.

a. Observe conditions of graft union, health and relative sizes of scion top and roots, and presence of suckers.

b. Note soil conditions and preplanting preparations, e.g., soil texture and moisture content, chiseling, back-hoeing.

B. Questions

1. Why do growers plant trees and vines on berms? What are some disadvantages of this practice?

2. What are some dangers of planting when the soil is too wet? For example, the auger glazes the wall of the hole.

3. What is the ratio of the pollinizer to the main cultivar? What are some considerations when selecting a pollinizer?

C. Equipment and materials: surveyor's transit, planting wire, stakes, tractor with an auger, shovels, water for settling soil around roots after planting.

5. Field trips: visit orchards, packing sheds, and cold-storage facilities

A. Objectives

1. Compare the cultural practices of orchards and their bases, e.g., flood versus sprinkler irrigation, sod versus clean cultivation.

2. Assess the efficiency of packing sheds and cold-storage plants, e.g., the rate at which the fruits are handled and cooled, precautionary measures taken to minimize bruising.

B. Questions

1. What are the advantages and disadvantages of a sod culture?

2. What purposes do the berms serve?

3. What is field heat and how is it reduced?

4. What are the hazards of storing different commodities in the same cold-storage rooms?

5. What system is used to "scrub" the atmosphere of the cold-storage rooms to remove undesirable volatile gases?

INDOOR PRACTICES AND DEMONSTRATIONS

6. Harvest or maturity indices and their usages

   A. Objectives

   1. Relate harvest indices with stages of fruit maturity.

   2. Familiarize students with various pieces of laboratory and field equipment for measuring maturity standards.

      a. Table model refractometer for measuring sugar concentration.

      b. pH meter for determining titratable acidity.

      c. Shear press for determining flesh firmness.

      d. Field equipment, see 1A.

7. Propagation

   A. Objectives

   1. Learn various grafting techniques for different species and combinations.

   2. Collect and prepare hardwood cuttings.

   B. Questions

   1. Why are grape cuttings allowed to form callus tissue before they are planted in the nursery?

   2. What are some reasons for graft incompatibilities?

   C. Tools and materials: same as 2C.

8. Flower and fruit morphology

   A. Objective

   1. Learn morphological characteristics as a means of classifying and identifying flowers and fruits.

   B. Questions

   1. Why is the point of insertion of floral parts a key character in describing flowers and fruits?

   2. What are the distinguishing characteristics of a berry, drupe, pome, and nut?

   3. What are some of the favorable and undesirable economic characteristics of the cultivars being studied?

   C. Equipment and tools: dissecting microscope and knives.

9. Vegetative characteristics for species identification

   A. Objective

   1. Learn to identify species by stem and foliar characteristics.

   B. Questions

   1. Why is it important to distinguish between juvenile and mature characteristics?

   2. How do the growth habits of terminal bearers differ from those of lateral bearers?

   C. Equipment: same as 8C.

10. Determine the end of rest by inducing budbreak of shoots and seeds

    A. Objective

    1. Decide when chilled plants and stratified seeds are no longer in the resting stage.

    B. Questions

    1. Why should bases of stems brought in from the orchard be recut under water?

    2. Why is hypochlorite or silver thiosul-

fate added to the water in a container holding dormant shoots?

C. Equipment: pruning shears, jars, and antiseptic solutions.

Spring Term

OUTDOOR PRACTICES

1. Pollen compatibility and breeding techniques

    A. Objectives

    1. Determine self- and cross-compatibilities.

        a. Emasculate and bag flowers.

        b. Collect pollen and self- and cross-pollinate known compatible and incompatible partners.

    2. Ascertain the time when embryos of early-ripening cherry and peach cultivars abort.

    B. Questions

    1. Why should reciprocal pollinations be made between two parents? What would a breeder learn from these progenies?

    2. Why is there no need to cover emasculated flowers?

    3. How does a breeder obtain early-ripening cultivars if their seeds abort?

    C. Equipment and materials: paper bags, twine, vials, glass rods for applying pollen to stigma.

2. Treatments for enhancing fruit size at harvest

    A. Objectives

    1. Learn how some cultural practices influence harvest size.

        a. Thin flowers and fruits, girdle branches to enhance fruit size; combine with Exercise 5 whenever appropriate. Follow fruit development by measuring tagged fruits.

        b. Ascertain the leaf:fruit ratios of individual limbs on compact, spur-type, and normal, standard apple trees.

        c. Compare shoot and leaf sizes on thinned and unthinned branches.

    2. Establish reference dates for estimating harvest sizes and dates of peaches and prunes.

    B. Questions

    1. Why is blossom thinning more effective than fruit thinning in obtaining better fruit sizes at harvest?

    2. The Compact Red Delicious/M9 combination partitions more assimilates for fruit production than the standard Red Delicious/seedling apple rootstocks. Why?

    3. Why does the reference date vary from one season to the next?

    C. Equipment: chemical blossom thinners, hand sprayers, calipers, tags, knives.

3. Propagation

    A. Objectives

    1. Learn the methods of asexually and sexually propagating plants that mature their crops in the spring.

        a. Collect bud sticks for spring and June budding.

        b. Graft with stored dormant scionwood by using different methods.

        c. Examine ovules of early- and late-maturing cultivars.

d. Layer suitable species by using various techniques.

   e. Collect and stratify cherry and apricot seeds.

   f. Mist-propagate leafy cuttings.

   B. Questions

   1. Why is the basal portion of a dormant sweet cherry shoot discarded as a source of scion wood?

   2. When does an embryo abort in an early-ripening peach?

   3. Why do seeds germinate faster when the seed coats are removed? What role does the endocarp play during germination?

   C. Equipment and tools: Same as Fall Term, 1C.

4. Pruning and training: summer pruning

   A. Objectives

   1. Learn the fundamental bases for summer pruning of young nonbearing and older bearing trees.

   2. Achieve early bearing through coordination of dormant and summer pruning.

   B. Questions

   1. What is meant by "delayed heading"?

   2. Why is summer pruning more debilitating than dormant pruning?

   3. What alternative practices can be used to obtain better fruit color other than summer pruning?

   C. Equipment: hand shears.

5. Agricultural chemicals

   A. Objectives

   1. Familiarize students with the usage of agricultural chemicals, their advantages and disadvantages.

      a. Dissolve and apply growth regulators, blossom and fruit thinners, and antitranspirant and antisprouting agents.

   2. Assess some side effects, such as the uptake and translocation of herbicides and antisprouting agents.

   B. Questions

   1. How does a wetting agent improve penetration of a chemical through the cuticle?

   2. What are some precautionary measures to observe when applying agricultural chemicals?

   3. What is meant by a systemic pesticide?

   C. Equipment: hand sprayers, DANGER DO NOT TOUCH labels, rubber gloves.

6. Frost protection devices

   A. Objectives

   1. Observe and/or demonstrate numerous means of preventing frost damage.

      a. Wind machines.

      b. Orchard heaters.

   2. Become familiar with various thermographs and thermometers.

      a. Insert temperature probes in various parts of trees and in the soil at different depths and record temperatures with a thermograph.

      b. Observe how the high–low thermometer functions.

   B. Questions

   1. What are some limitations of orchard heaters? Of sprinkler systems?

2. What is meant by a strong inversion?

C. Equipment: temperature probes, thermometers, and thermographs.

7. Root distribution

    A. Objectives

    1. Observe how rootstock species, cultivation, and mode of rootstock propagation influence rooting depth and distribution.

    2. Evaluate the influence of soil texture on external root morphology, i.e., the lack of fine roots in sandy soils.

    B. Questions

    1. Why are roots on cuttings shallower than those on seedlings?

    2. Sod culture promotes shallow rooting. Why?

    3. Shallow-rooted species growing under sod are less likely to show symptoms of nutrient deficiency, provided nitrogen supply is adequate. Explain.

    C. Equipment: tractor with trenching tool.

8. Soil–plant–water relationships

    A. Objective

    1. Expose students to various apparatus and tools for measuring different environmental and plant parameters.

        a. Tensiometer and neutron probe for determining amount of soil moisture.

        b. Schollander bomb for estimating water tension of plants.

        c. Autoporometer for measuring relative humidity, transpiration, and photosynthesis rates.

        d. Sling psychrometer for determining relative humidity.

    B. Questions

    1. Stomatal closure results from high temperature, wind, and lack of soil moisture. Explain.

    2. Define the terms *saturation* and *compensation points* with respect to photosynthesis.

    C. Equipment: tensiometers, neutron probe, Schollander bomb, autoporometer, and sling psychrometer.

9. Harvesting techniques and maturity indices for spring crops

    A. Objectives

    1. Familiarize students with early season crops and the need for special handling.

        a. Use field and laboratory equipment listed in Fall Term 1A for determining degree of ripeness of sweet cherry, apricot, plum, and other fruits.

        b. Group fruits of different species by ground color and apply maturity criteria in the laboratory.

INDOOR EXERCISES

10. Bud, flower, fruit, and seed morphology

    A. Objectives

    1. Learn the classifications of buds, flowers, and fruits by their morphological characteristics.

        a. Dissect organs under a microscope and notice the characteristics used by taxonomists to classify plants and fruits.

    2. Trace the seasonal development of the ovule and its parts, especially the embryo.

B. Questions

1. Why are floral characteristics rather than fruit traits better suited for taxonomic purposes?

2. When does cytokinesis occur in the endosperm?

C. Equipment: dissecting microscope, knives, tweezers.

11. Pollen germination

    A. Objectives

    1. Ascertain pollen viability.

    2. Observe the effects of temperature, varying sugar contents, and boron and calcium salts on germination and growth of pollen grains.

    B. Question

    1. Why does a pollen grow better *in vivo* than *in vitro*?

    C. Equipment: Petri dishes with different agar and sugar concentrations, controlled-temperature growth chamber.

12. Species and cultivar identification

    A. Objective

    1. Learn to identify species and cultivars by their vegetative characteristics.

    B. Questions

    1. What are several external and internal stem characteristics used for identifying species?

    2. What are some leaf characteristics used by taxonomists for classifying species?

    C. Equipment: dissecting microscope and knives.

13. Harvest indices

    A. Objectives: same as Fall Term, 6A

    B. Questions

    1. Fruits intensely red with anthocyanin pigments tend to pose more postharvest problems than do those with small amounts of red pigments. Why?

    2. Why do flesh firmness and pigmentation of apples harvested from tall trees tend to vary more than do those of apples harvested from short trees?

    3. Why do early-ripening cultivars generally contain smaller amounts of soluble solids than late-ripening ones?

    C. Equipment: same as Fall Term, 1A.1 and 6A.2

APPENDIX E

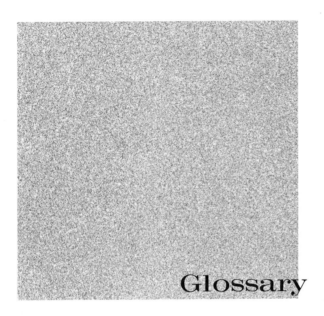

Glossary

This glossary defines botanical and horticultural terms used in the text. Dictionaries and other textbooks may define the same words somewhat differently.

Abaxial. Side of a lateral organ situated away from the axis.

Abscise. To separate (n. abscission).

Abscisic acid. A plant hormone inhibiting growth, promoting abscission, and extending dormancy.

Abscission layer or zone. Specialized layer of cells at which leaves and fruits separate from stems and pedicel and/or peduncle, respectively.

Accessory tissue. An edible tissue associated with but not a part of the pericarp, e.g., the sepal of the mulberry fruit, the receptacle of the strawberry, or the appendicular tissue (an enlarged, fleshy floral tube) of an apple.

Acclimation. The natural process of becoming inured to climate.

Achene. A small, hard, dry, indehiscent fruit with a single seed.

Adjuvant. A chemical added to a solution to enhance the action of the main ingredient, e.g., a wetting agent to lower surface tension of a spray mixture so that the droplets spread, thus improving penetration.

Adnate. United with or adhering to (n. adnation).

Adventitious. Developing in an unusual or irregular position, e.g., buds arising at places other than the node, and roots arising from stems and leaves.

After-ripening. A dormant period of seeds before germinating.

Aggregate fruit. A fruit consisting of two or more individual pistils borne on a single flower, e.g., blackberry, strawberry.

Air layerage. A method of stimulating root formation on a stem in place by wounding and then keeping the zone moist.

Aleurone layer. The outermost layer of the endosperm of grass seed. Source of alpha-amylase which hydrolyzes starch; the layer which develops color in "Indian" corn. Used for gibberellin bioassay.

Androecium. The stamens collectively (pl. androecia).

Anemophilous. Having wind-disseminated pollen, e.g., kiwifruit, walnut.

Anion. Negatively charged ion, e.g., sulfate and nitrate ions.

Annular. Ring-shaped, e.g., spring and summer wood of the xylem.

Anther. Pollen-bearing part of the stamen.

Anthesis. The pollen-shedding period; blooming period.

Anthocyanin. Class of water-soluble flavonoid pigments which range in hue from red, blue, to violet; synthesized by fruits, stems, and leaves.

Apetalous. Lacking petals (n. apetaly).

Apical dominance. Inhibition of lateral bud development by the apical bud on the shoot terminal.

Apical meristem. Actively dividing mass of cells at the tip of a shoot or root.

Apomict. An individual plant which arises from the ovary without fertilization of the egg, e.g., a nucellar embryo.

Apomixis. Seed production in the ovary without fertilization of the egg; i.e., asexual embryogenesis.

Appendage. A secondary part attached to a main structure (adj. appendicular).

Ascorbic acid. Vitamin C, an essential vitamin for human beings.

Asexual reproduction. Propagation by asexual or vegetative means, such as cuttings, budding, grafting, cloning.

Assimilation. Transformation of organic and inorganic materials into cellular components, e.g., protoplasm and cell wall.

Auxin. A class of natural hormones such as 3-indoleacetic acid, 3-chloroindoleacetic acid, and phenylacetic acid, which promote cell division and participate in root initiation, fruit set, nutation, tactile response, and phototropisms and geotropisms.

Axil. The angle created by an organ and its axis, e.g., the upper angle between a leaf and the shoot to which it is attached.

Axillary bud. A bud formed in the axil of a leaf.

Bark. The corky tissue of woody stem, external to the lateral cambium.

Berry. A simple fruit in which the pericarp is succulent.

Biennial bearing. (syn. alternate bearing) Cropping every other year.

Bitter pit. A physiological disorder of apple and pear caused by a lack of calcium.

Black end of pear. A physiological disorder of European pears (*Pyrus communis*) when grafted onto selected Asian pear rootstock (*P. serotina*). The epidermis and flesh in the region of the calyx of the fruit is blackened.

Blight. A disease or injury to plants which causes withering, cessation of growth, and death of parts without rotting.

Bloom. A flower; also a whitish, glaucous, waxy covering of fruits, especially noticeable in plums and grapes.

Bract. A modified leaf; e.g., sepallike structure of strawberry and covering of inflorescence.

Breba. The first or spring crop of edible figs.

Bud. The rudimentary unexpanded shoot or flower. Meristematic tissue protected by scales, having the potential to develop into a flower, an inflorescence, or a reproductive or vegetative shoot.

Budbreak. Emergence of a bud.

Bud scale. A modified protective leaf covering a bud.

Bud sport. A mutant arising from a bud.

Bud stick. A short branch used as a source of buds for budding.

Bur, burr. A spiny covering of a fruit, e.g., the chestnut involucre.

Burrknot. A concentration of preformed root initials on a stem.

Callose. A complex carbohydrate which forms over sieve plates of phloem during winter; also forms in pollen tubes and styles of gametophytic and sporophytic incompatible combinations.

Callus. A wound-healing meristematic tissue; mass of undifferentiated parenchyma cells.

Calyptra. The cone-shaped corolla of grapes which dehisces from the receptacle at anthesis.

Calyx. The outermost whorl of floral parts, the sepals collectively.

Calyx tube. Lower fused portion of calyx, corolla, and stamens; also termed the torus or hypanthium.

Cambium. A thin layer of meristematic cells between the xylem and phloem. Cork cambium is the phellogen found in the outer region of the bark.

Cane. Stem of vines and brambles in the dormant stage.

Capillary movement. Movement of a liquid through fine pores owing to cohesion or adhesion, e.g., water between clay particles.

Caprification. Pollination of the female fig flower with pollen from a caprifig transported by the wasp *Blastophaga psenes*.

Caprifig. A monoecious fig tree or its fruit.

Carbohydrate. A class of compounds composed of carbon, hydrogen, and oxygen.

Carotene. A class of fat-soluble, long-chained C-40 compounds which range in color from red to yellow; synthesized in chromoplastids.

Carpel. A modified leaf enclosing a seed, a megasporophyll. Fusion of carpels forms a compound pistil.

Catkin. A spike or spikelike inflorescence bearing staminate or pistillate flowers. The terminal catkins of chestnuts are bisexual.

Cellulose. A beta-1–4-linked polymer of glucose; a principal constituent of the cell wall.

Chalaza. The base of the ovule which is attached to the funiculus; the point from which integuments arise.

Chemotaxis. Movement of plants or their parts in response to some chemical stimulus.

Chemotropism. A bending movement of plant parts in response to some chemical stimulus.

Chimera. A tissue consisting of two or more genetically different layers of cells.

Chlorophyll. The green pigment associated with photosynthesis.

Chloroplast. A plastid containing chlorophyll.

Chlorosis. Lack of chlorophyll, thus yellowing of leaves; a diseased condition (adj. chlorotic).

Chromosome. One of the chromatin-containing structures of the cell nucleus which transmits hereditary characteristics through gene action.

Clone. A group of plants vegetatively propagated from a single mother plant or a mutant.

Columella. The central white tissue of kiwifruit to which the funiculi are attached.

Companion cell. One of the nucleated cells adjacent to sieve tubes in the phloem that participate in bilateral transport of food.

Compatible. Capable of coexisting in harmony, e.g., two graft partners capable of uniting or the stigma and style of an ovary allowing a pollen tube to grow.

Complete flower. A flower which possesses all four whorls: calyx, corolla, stamens, and pistil(s).

Compound leaf. A leaf consisting of two or more separate leaflets.

Compound ovary. An ovary in which bases of two or more carpels are fused.

Cork. An external, secondary tissue, impermeable to water and gases, derived from the phellogen, the cork cambium.

Cork cambium. The meristomatic cells in the outer bark (phellogen) which give rise to the cork tissue.

Corolla. The collective name for petals.

Cortex. Primary tissue of a stem or root bounded externally by the epidermis and internally by the phloem (adj. cortical).

Corymb. A cymose cluster of flowers in which the outer flowers have longer pedicels than the inner ones, resulting in a flat-topped or convex inflorescence (adj. corymbose).

Cotyledon. The embryonic leaf of a seed (adj. cotyledonary).

Cross-pollination. Transfer of a pollen from an anther of one clone to the stigma of a different clone.

Crown gall. A tumorous enlargement caused by the bacteria *Agrobacterium tumefasciens*.

Cultivar. In fruit culture, a group of similar plants under cultivation that are propagated asexually. A term shortened from the words cultivated variety and abbreviated *cv*.

Cuticle. A waxy layer on the epidermis of higher plants, impermeable to water.

Cutin. A water-insoluble mixture of waxes, fatty acids, soaps, and resinous compounds in the cuticle.

Cyme. An inflorescence in which the center or terminal flower opens first (adj. cymose).

Cytokinin. A class of natural hormones which stimulate cell division and delay senescence; also a mobilizing factor.

Deciduous. Referring to plants which drop their leaves in the fall.

Deep rest. State of a tree or seed, usually in midwinter, when the organism is least metabolically active and minimally susceptible to winter injury.

Dehisce. To split along a natural line to discharge contents within (adj. dehiscent, n. dehiscence).

Dichogamy. Condition in which the stigma is not receptive when pollen is being shed; ensures cross-pollination (adj. dichogamous).

Differentiation, floral. The appearance and growth of individual flower parts at the apex, which now becomes the receptacle.

Dimorphic. Having two forms, as juvenile and adult leaves (n. dimorphism).

Dioecious. (*di* = two; and *oecium* = home, Gk.) Condition in which pistillate and staminate flowers are borne on different plants.

Diploid. A plant of the sporophytic generation having double the number of chromosomes of the haploid generation.

Distal. The end opposite the point of attachment.

Dormant. A bud or seed remaining in the quiescent state because of unfavorable growing conditions (n. dormancy).

Dorsal. The back or outer surface of an organ.

Double fertilization. Simultaneous fertilization of the egg by the sperm nucleus to form the zygote and of the two polar nuclei by the second male nucleus to form the $3n$ endosperm.

Drupe. A fleshy, indehiscent fruit with a stony endocarp, a fleshy mesocarp, and a thin exocarp.

Drupelet. A diminutive drupe, e.g., that of a raspberry fruit.

Emasculate. To remove stamens before anthesis (n. emasculation).

Embryo. The rudimentary plant, consisting of a radicle, cotyledons, and plumule, within a seed.

Embryo sac. A mature female gametophyte, typically with eight nuclei.

Endocarp. The innermost tissue of the pericarp, the pit or stone.

Endogenous. Originating within.

Endosperm. The triploid tissue within the embryo sac that surrounds the embryo in the seed which originates from fusion of the two polar nuclei with one of the male gametes.

Epidermis. The outer layer of cells, the skin.

Epigynous. A flower type in which the gynoecium is below the floral parts, hence, a flower with an inferior ovary (n. epigyny).

Epinasty. An abnormal outward bending or growth of leaves or stems, often caused by exposure to herbicides or growth regulators.

Espalier. A French word for the system of training fruit trees flat against a wall or trellis.

Ethylene. A gaseous plant hormone which regulates senescence, fruit ripening, chlorophyll degradation, etc.

Etiolation. Spindly achlorophyllous growth caused by lack of light.

Evapotranspiration. Vaporization of water through evaporation and transpiration, e.g., loss of soil moisture through evaporation at the surface and transpiration by plants.

Exine. The outermost coat of the pollen grain.

Exocarp. (syn. epicarp) The epidermal layer of the pericarp.

Exogenous. Originating from without rather than from within.

Fertile. Capable of reproduction, in contrast to sterile. Said of soils rich in essential elements (n. fertility).

Fertilization. Union of the male gamete with the female gamete to form a zygote. Act of adding nutrients to soil.

Field capacity. Moisture content of a soil held against the force of gravity after a rain or irrigation, expressed as a percentage on dry weight basis.

Filament. The threadlike structure to which the anther is attached.

Fire blight. A bacterial disease caused by *Erwinia amylovora* that attacks mainly pomaceous species, blackening leaves, forming cankers on the trunk, and discoloring flowers and fruits.

Floret. One of the small flowers in a dense inflorescence.

Fruit. A mature or ripe ovary and other structures associated with it.

Fruit set. The retention and development of the ovary following anthesis.

Gall. A large swelling on plant tissues caused by bacteria, fungi, nematodes or insect parasites.

Gametophyte. The haploid individual or plant generation which bears gametes, in contrast to the sporophyte generation which bears spores. The male gametophyte is the germinating pollen grain; the female gametophyte is the embryo sac.

Gene. Portion of the chromosome that carries hereditary traits.

Geotropism. Growth response induced by gravity.

Gibberellins. A class of diterpenoid hormones which evoke various responses, especially cell elongation.

Girdle. A physical restriction of a stem, usually the removal of a ring of bark to interrupt movement of food in the phloem.

Glabrous. Lacking hairs or trichomes; smooth and shiny.

Gland. A cell or group of cells which secrete substances to the surface; e.g., nectaries.

Glaucous. Covered with a whitish waxy substance, the bloom.

Globose. Like a globe; bulbous.

Grafting. Act of uniting a piece of scion with the rootstock.

Growth regulator. A natural or synthetic compound which regulates plant growth.

Growth retardant. A chemical which retards growth.

Gynoecium. (*gynaeka* = woman; *oecium* = home, Gk.) The pistil, simple or compound (pl. gynoecia).

Haploid. Having one set of chromosomes as in the gametophytic generation or half the number of chromosomes of somatic cells.

Hard end disorder. A physiological disorder of some European pear cultivars when grafted on seedlings of selected *Pyrus serotina*, an Asian pear species. The calyx end of the fruit is hard and shiny.

Hardpan. A compacted or cemented layer of soil impervious to water and impenetrable by roots.

Hardy (plants). Characteristic of plants which are adapted to cold or other adverse climatic condition.

Heartwood. The dense, inner nonliving xylem, usually rich in tannins which give it the brown color, in contrast to sapwood.

Hermaphroditic. Said of flowers which possess stamens and pistals, a perfect flower.

Hormone. A naturally occurring substance which is produced in one part of a plant and is translocated to another part where it induces a growth response.

Hybrid. An offspring from two genetically different plants.

Hypanthium. Fused basal portion of sepals and petals; the floral tube or torus. Enlarged torus often adnate to the fruit.

Hypogynous. Condition in which calyx, corolla, and stamens arise from the receptacle below the pistil (n. hypogyny).

Imperfect flower. Flower which lacks either stamens or pistils.

Incompatibility. Condition in which two organs fail to unite; e.g., failure of sperm nucleus to fertilize the egg after pollination, incapacity of a stock and scion to unite after graftage.

Indehiscent. Remaining closed, not opening by a definite line or pores.

Indigenous. Native to the area.

Induction, floral. Physiological condition required for floral differentiation. The stimulation of a bud to flower at the biochemical level; no morphological evidence of floral parts.

Initiation, floral. Same as floral induction.

Inferior ovary. Ovary which is adnate to the hypanthium or calyx tube; the floral parts arise above the ovary.

Inflorescence. The disposition of flowers on the floral axis, incorrectly applied to flower clusters.

Integument. Two- to three-layered envelope of an ovule; seed coat.

Internode. The portion on a stem between two nodes.

Intersterile. A condition in which two or more cultivars cannot cross-pollinate and produce seeds.

Intine. The inner coat of a pollen grain.

Involucre. A whorl of distinct or fused leaves or bracts subtending an inflorescence, a flower, or a fruit.

Juvenile. A period during which a seedling plant cannot be induced to flower.

Lanceolate. Referring to a lance-shaped leaf; several times longer than wide, broadest toward the base, and tapering to an apex.

Lateral. Borne on the sides of a structure or organ.

Latex. The milky juice of some plants, e.g., fig.

Layering. (syn. layerage) Practice of inducing roots by wrapping or burying shoots or shoot tips still attached to the mother plant.

Leaflet. A subdivision of a compound leaf.

Lenticel. A structure of loose corky cells on young bark that permits gaseous exchange.

Lignification. Deposition of lignin in the secondary cell wall.

Lobe. Any segment of an organ, usually a rounded projection.

Locule. The cavity of an ovary or anther.

Megaspore. The spore that gives rise to the embryo sac, usually larger than the microspore.

Meiosis. Reduction division of chromosomes in the ovule and anther (adj. meiotic).

Meristem. Undifferentiated tissues, the cells of which are capable of dividing indefinitely, and the products that differentiate into definitive tissues.

Mesocarp. The middle layer between the exocarp and endocarp of a pericarp.

Mesophyll. Interior parenchyma cells of a leaf.

Metabolism. The sum of processes of breakdown of complex compounds in living cells (catabolism), by which energy is provided for vital activities, and of synthesis and assimilation of new substances (anabolism).

Metaxenia. An effect of pollen on maternal fruit tissue.

Microspore. The spore that gives rise to pollen grains, male gametophytes.

Middle lamella. The layer, rich in pectin, between primary cell walls.

Midrib. The main or central vein of a leaf.

Mitosis. Nuclear division in which duplicated chromosomes are divided between two daughter cells.

Mixed bud. Bud containing floral and vegetative structures.

Monoecious. (*mono* = one; *oecium* = home, Gk.) A plant bearing both staminate and pistillate flowers.

Multiple fruit. Fruit formed from several flowers, e.g., mulberry, fig, and pineapple.

Mutation. A heritable change in an individual, e.g., bud sport.

Mycorrhiza (pl. mycorrhizae). A symbiotic association between the mycelium—a mass of filaments—of a fungus and the roots of higher plants. The mycelium may be within (endo-) or on the surface (ecto-) of roots; the filaments may extend the activity of the roots.

Nectary. A gland or tissue for secreting nectar.

Node. A place on a stem where leaves and buds originate.

Nucellus. The nutrient tissue in which the embryo sac develops.

Nut. A hard, indehiscent, and woody fruit, usually one-seeded and associated with an involucre.

Nutation. A spontaneous, spiraling movement of a growing plant part, usually of the elongating plumule of a germinating seed; caused by variation in the growth rates on the different sides.

Oak root fungus. A disease of fruit trees caused by the fungus *Armillaria mellea*. The fungus forms brown stringy rhizomorphs, aggregations of densely packed threads intertwined like a rope, which cause destructive rot of the roots of the tree.

Osmosis. Diffusion of fluids through a semipermeable membrane.

Osmoticum. The dissolved substance within a semipermeable envelope that make up the osmotic concentration.

Ovule. The structure which develops into a seed.

Palmette system. A synonym for espalier, a method of training fruit trees to a vertical flat shape on a three- or four-wire horizontal trellis.

Panicle. Compound racemose inflorescence with the younger flowers at the apex or center.

Parenchyma cell. A thin-walled cell, capable of division; often unspecialized but may store food, e.g., ray parenchyma cell.

Parthenocarpy. Fruit formation without fertilization and seed development.

Parthenogenesis. Development of an individual without fertilization of the egg, e.g., nucellar embryo.

Pedicel. The stalk of a single flower of an inflorescence.

Peduncle. The stalk of a solitary flower or of an inflorescence.

Perfect flower. A flower with functional stamens and pistils; hermaphroditic flower.

Perianth. Collectively, the calyx and corolla.

Pericarp. The fruit wall, developed from the ovary wall.

Pericycle. Layer of primary cells immediately within the endodermis and surrounding the stele of most vascular plants.

Perigyny. A flower with a superior ovary in which calyx, stamens, and corolla arise from the floral tube or torus.

Permanent wilting percentage. The range of soil moisture at which plants remain wilted, even when the atmosphere is saturated with water vapor.

Petal. One of the modified leafy parts of the corolla.

Petiole. The stalk to the leaf blade or compound leaf.

Photoperiodism. Response of a plant to day and night lengths, in relation to flowering.

Photosynthesis. The process by which plants convert carbon dioxide and water into sugar and oxygen in the presence of chlorophyll and light.

Phototropism. Directional growth of a plant in response to light.

Phyllotaxy. The pattern of leaf arrangement on a stem.

Phytochrome. A pigment responsive to

red and far-red light in the cytoplasm of green plants.

Pistil. The seed-producing organ, consisting of the stigma, style, and ovary.

Pistillate. Provided with pistils, used to describe flowers when stamens are vestigial or lacking.

Pit-burning. A disorder in which the mesocarp tissue surrounding the pit or endocarp becomes brown and/or gelatinous; usually caused by high temperatures.

Pith. The central, parenchymatous tissue of a stem.

Plumule. The stem- and leaf-producing structure in the embryo of a seed.

Polar nucleus. One of two nuclei, centrally located in the embryo sac. The polar nuclei unite with a sperm nucleus during double fertilization to form the triploid endosperm.

Pollen. The male spores produced by the anther.

Pollination. The transfer of pollen to a stigma.

Pollinator. The pollen-transferring vector or agent.

Pollinizer or pollenizer. The plant which supplies pollen.

Polyembryony. A condition in which two or more embryos are formed in a single ovule.

Polyploid. Having more than twice the haploid number of chromosomes per nucleus.

Pome. A fleshy indehiscent fruit with an inferior ovary having a lignified endocarp.

Primordium. The earliest stage of organ development (p. primordia).

Protandry. An asynchronous condition of a flower in which the anther sheds pollen before the stigma is receptive.

Protogyny. An asynchronous condition of a flower in which the stigma becomes receptive before the anther sheds pollen.

Pubescent. Covered with short, soft hairs.

Pyriform. Pear-shaped.

Raceme. An inflorescence with progressively younger flowers borne near the apex.

Rachis. The central elongated axis of an inflorescence or compound leaf.

Receptacle. The expanded portion of a flower stalk that bears the organs of the flower.

Reference date. An arbitrary date, ten days after the tip of the peach endocarp lignifies, that is used as a physiological marker to estimate fruit size at harvest.

Respiration. The oxidation of carbohydrates during which oxygen is utilized and carbon dioxide and water are evolved.

Rest. Internal conditions that prevent the growth of seeds and buds even though environmental conditions are favorable.

Ringing. Removal of a ring of bark; girdling.

Ripening. Chemical and physical changes by which a mature fruit becomes palatable.

Rootstock. Root material upon which scions or buds are grafted.

Rosette. A dense cluster of leaves arranged in a circular fashion about a stem.

Sapwood. The young, living, and functional xylem.

Scarification. Scratching of seed coverings to allow gas exchange and water uptake to enhance germination.

Scion. The stem section used in grafting that becomes the aerial part of a tree.

Seed. The fertilized mature ovule, consisting of an embryo, seed coats, and food supply.

Self-fertile. Capacity of a plant to produce seeds by self-pollination.

Self-fruitful. Capacity of a plant to produce fruits parthenocarpically or by self-pollination.

Self-incompatibility. Inability of a plant to produce viable seed following self-—pollination.

Self-pollination. Pollen transfer from stamens to the stigma of the same flower, other flowers on the same plant, or flowers on other plants of the same clone.

Self-sterility. Failure to produce viable seed after self-pollination.

Self-unfruitful. Inability of a plant to produce fruits without cross-pollination.

Senescence. Physiological aging in which an organism deteriorates.

Sepals. The outermost whorl of a flower, usually green and leaflike. Sepals collectively form the calyx.

Septum. Any kind of partition, e.g., that of the silique and walnut fruit.

Sigmoid. S-shaped form.

Simple fruit. A fruit derived from a single pistil.

Species. A group of similarly classified organisms potentially capable of interbreeding and different from other groups within the same genus.

Split pit. Cracking of the endocarp during its lignification process caused by an excessive rate of fruit growth.

Sporophyte. The individual or generation which bears asexual spores in a plant that alternates generations, in contrast to the gametophyte of the gamete-bearing generation.

Stamen. The pollen-bearing structure of a flower, consisting of a filament and anther (adj. staminate).

Stem. Main vegetative axis of a plant that produces nodes and buds.

Sterile. Inability to produce viable seed.

Stigma. The enlarged and/or feathery distal part of the pistil that receives the pollen (adj. stigmatic).

Stipule. An appendage at the base of a petiole.

Stoma or stomate. A small aperture operated by guard cells in the epidermis of a leaf, stem, or fruit through which gaseous exchange occurs (pl. stomas, stomata, or stomates).

Stratification. Placing seeds between layers of moist material, usually sand and peat mixture, and exposing them to cold temperature to overcome the rest period so that seeds will germinate.

Style. The stalklike part of the pistil separating the stigma from the enlarged ovary.

Sucker. A shoot arising from an adventitious bud on a root.

Superior ovary. An ovary with the perianth inserted below it.

Suture. A junction or seam of a union where dehiscence may occur.

Syconium. The fig fruit derived from many flowers lining the interior of a hollow fleshy peduncle.

Tetraploid. An individual with four sets of chromosomes.

Tissue. A group of cells of similar morphology that performs a particular function.

Topworking. Changing the cultivar of a tree by grafting over the trunk or large scaffold branches.

Translocation. Movement of water and dissolved substances through the vascular system of plants.

Transpiration. Loss of water vapor through the stomata, cuticle, or lenticel.

Triploid. An individual with three sets of chromosomes.

Tropism. Movement of plant parts in response to an external stimulus.

Turgor. Degree of turgidity resulting from absorption or loss of water from cells.

Variety. A group of similar plants ranking below a species; synonymous with clone in fruit culture.

Vascular. Pertaining to elements and tissues for conducting food or water.

Venation. Arrangement of veins in a leaf.

Vernalization. Exposure of plants or seeds to cold temperatures to induce flowering.

Water sprout. A vigorous shoot arising primarily from latent buds on the trunk or from older scaffold branches, in contrast to suckers which arise from roots and trunk below the ground surface.

Xenia. The effects of pollen on the embryo and endosperm.

Xerophyte. A plant adapted to arid habitats.

Zygote. The diploid cell formed by fertilization of the egg by the sperm nucleus.

APPENDIX F

Color Plates
Showing
Chimeras
and Nutritional
Deficiency
Symptoms

PLATE 1

A Bartlett pear tree with shoots having variegated and albino leaves on the right scaffold branch, indicating a sectorial or mericlinal chimera.

PLATE 2

An albino peach shoot tip arising from a branch bearing both green and variegated leaves.

PLATE 3

A variegated Bartlett pear, a result of mutation.

PLATE 4

A Red Delicious apple that partially reverted back to the Delicious cultivar from which it was derived.

PLATE 5

Pale nitrogen-deficient prune leaves (top row) and normal leaves (bottom row).

PLATE 6

A prune tree severely deficient in potassium in midsummer, showing scorched leaves and dieback of shoot tips.

PLATE 7

French prune leaves ranging from an extreme potassium deficiency to normal (left leaf on upper row). Affected leaves are smaller and chlorotic and develop necrotic areas.

PLATE 8

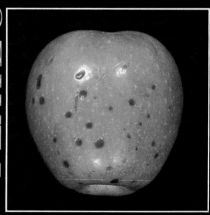

A Granny Smith apple showing symptoms of bitter pit, a disorder caused by calcium deficiency. (Courtesy of W. C. Micke, Department of Pomology, University of California)

PLATE 9

Iron deficiency in prune, which produces pale, yellow leaves with a characteristic fine network of green veins.

PLATE 10

Typical "little-leaf" symptoms of zinc deficiency in prune (left) and a normal shoot (right). Affected shoots have small, pale leaves and poorly developed buds.

PLATE 11

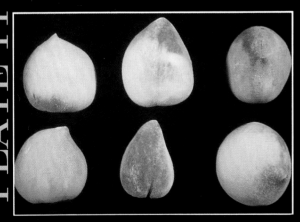

Zinc-deficient peaches with a typical flat, pointed appearance. (Courtesy of O. Lilleland)

PLATE 12

Magnesium deficiency in prune, which causes chlorosis along the terminal and lateral margins of leaves, and an inverted V-shaped green area at their bases.

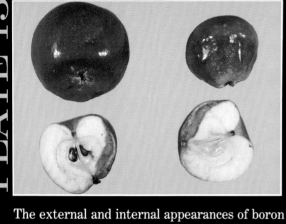

The external and internal appearances of boron deficiency in apples.

Copper deficiency in prune, which causes chlorotic and misshapen leaves and dieback of shoot tips.

Manganese-deficient prune leaves revealing interveinal chlorosis extending from the midrib to the margin.

INDEX

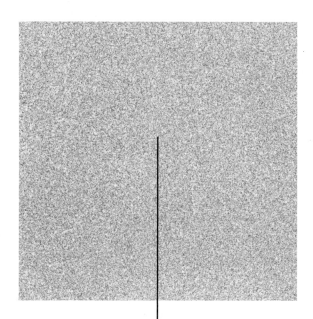

Abscisic acid, *see* Growth Regulators
Abscission:
   of fruits prematurely, 153
     prevention with synthetic chemicals, 154
     promotion with ethylene-generating compounds, 154
   of inflorescence buds of pistachio, 127
   of organs and causal factors, 87
Achene, 20
Acid:
   indoleacetic, *see* Auxin
   indolebutyric acid, 238
   gibberellic, *see* Gibberellic acid
   organic:
     in developing fruits, 135
     metabolism of citric, malic, and tartaric, 139
     seasonal fluctuations, 139
Adult phase, *see* Juvenility, characteristics and heritability
Aerial or air layerage, 240
Aggregate fruit, 20
Air drainage, role in reducing frost damage, 186

Alcohol, ethyl or ethanol:
  secretion by persimmon seeds, 97
Allelopathy, 103
Almond (*Prunus dulcis*) 259–261
  cross-pollinizers, need for, 260
  rootstocks, 260
Alternate (biennial) bearing, 72
  early fruit thinning, as influenced by, 75
  leaves and seeds, roles of, 78
  of pecan, 126
  of pistachio, 127
Alternate system of tree arrangement in orchard, 180
Androecium, (stamen), 10, 11–14
Anthesis, 89
Anthocyanin:
  biosynthesis, 140
  factors that favor formation, 141–142
Apical dominance (correlative inhibition):
  auxin, role of, 31
  cytokinins, overriding effect of, 31
  terminal bud, role of, 31
Apomixis (parthenogenesis), apogamy, 97
Appendicular *vs.* receptacular theory, origin of the inferior ovary, 15–16
Apple (*Malus domestica*), 247–252
  accumulation of mineral elements, 133
  fruit growth curves, 110–111
  rootstocks, 250–251
Apricot (*Prunus armeniaca*), 263–264
  rootstocks, 263

Aromatic compounds, *see* Flavor, and aromatic compounds
Arsenic, effect of excess on trees, 200
Asexual propagation, 224–225
Astringency or puckeriness:
  means of removal from persimmon fruit, 293
  and tannins, 143
Auxin (indoleacetic acid) *see also* Growth Regulators
  apical dominance, role in, 31
  biosynthesis in seeds, 76
  flower bud, role in formation of, 77
  fruit growth, role in, 117
  rooting hormone, 238
  synthetic:
    as abscission inhibitors, 154
    as chemical thinners, 122
    as promoter of flowering, 77
    as promoter of rooting, 238
  tension wood, role in formation of, 34
Available water, 174

Backflow phenomenon in senescing leaves, 58
Bandera (marchand), 215
Bark grafting, 230–231
Basin irrigation system, 176
Bees, honeybees:
  factors governing their activity, 94
  sideworkers *vs.* topworkers, 94
Benzyladenine, a synthetic kinetin, *see* Growth Regulators
Berry, 22

Biennial (alternate) bearing, 72
Bilateral cordon, 216
Bitter pit or stippen (German), 161
Blackberry (*Rubus* spp.), 288
  propagation, 289
  pruning and training system, 220
Black end and hard end of European pear, 162–164
Black line of walnut, 236
Bloom or cuticular wax, 19
Blossom-end rot of peach, 160–161
Blueberry (*Vaccinium* spp.), 286–287
Boron, role in nutrition, 196, 199–200
Branch orientation, effect on flower bud formation, 84–85
Bridge grafting, 234
Buckskin of sweet cherry, 237
Bud:
  abscission, bases for, 87–89
  adventitious, 4, 5
  classifications, 4
  dormancy and rest, 27–28
  latent, 4, 5
Budsport, *see* Mutation, bud sport
Bur (burr) of chestnut, 21
Bur (burr) knots of preformed root initials, 9

Calcium:
  and bitter pit of apple, 161
  and cork spot of pear, 165
  an essential element, 195–196
  symptoms of deficiency, 198
Callose, deposition in the style and pollen tube, 91

Callus, 225
Calyx:
  floral whorl, 14
  role in persimmon fruit growth, 134
Cambium:
  cork or phellogen, 33
  lateral or vascular, 4
  stem girth, 33
Cane pruning, 217–218
Canopy Area Index, (CAI), 54–56
Canopy configuration, 54
Caprification and caprifig, see Fig
Carbohydrates:
  accumulation by developing fruit, 125–127, 137
  metabolism, 138–139
  seasonal fluctuation in stem, 36
  storage:
    in fruits, 35–37
    in leaves, 57–59
    by roots, 35–36, 63–64
    in seed coats, 100–101
Carotene, carotenoid, and xanthophylls, 142–143
  role in photosynthesis, 42
  accumulation in maturing fruit, 142
Carpel, 11, 19
Cation exchange capacity (CEC) of soils, 192
Catkin:
  an inflorescence, 12, 18
  carbohydrate and mineral composition, 273
  competitor or sink for food, 126–127
Cell division and differentiation, region of and duration:
  in buds, 4
  in fruits, 108
  in leaves, 5
  in roots, 8–9

Central leader tree, 211–213
Chemical looseners as harvesting aids, 154
Chemical thinners, 121–123
Cherry, sour (*Prunus cerasus*), 270
Cherry, sweet (*Prunus avium*), 267–270
  gametophytic incompatibility, sterility factor, 90–91
  rootstocks, 66, 268
Chestnuts (*Castanea* spp.), 276–277
  blight (*Endothia parasitica*), 277
  chemical composition of embryo, 100
Chimera, 37
  bud sport, 37
  by graftage (graft chimera), 38
  mericlinal, periclinal, and sectorial, 37
  variegation, fruit and leaf, 38
Chip budding, 229
Chlorophyll:
  in fruits, 140
  role in photosynthesis, 42, 49–50
Chlorosis:
  lime-induced, 199
  nutrient deficiency, symptom of, 196
  prevention, means of, 53
Citric acid, see Acid
Cleft grafting, 230
Climacteric rise, 143–144
Cobnut, see Filbert
Cold hardiness (cold hardy), 184–185
Compaction, soil, effect on root growth, 68
Compatibility, pollen see Gametophytic incompatability; Sporophytic

incompatibilities in pollination
Compensation point, 44
Contour planting, 181–182
Controlled atmosphere storage, 144
Copper, 196, 200
Correlative inhibition, see Apical dominance
Cover crop, see Sod culture
Cracking of fruit skin, 157–158, 162
Cranberry (*Vaccinium corymbosum*), 286–287
  fruit classification, 23
Crop coefficient, 178
Cross-pollination, 89
Crown division, 241
Crown gall bacteria (*Agrobacterium tumefasciens*), 268
Currants (*Ribes sativum; R. rubrum*), 287
  fruit classification, 22
Cuttings (cuttage), 237–238
Cytokinesis and fruit thinning, 123
Cytokinins, 304
  effect on budbreak or dormancy/rest, 31
  promotive effect on flowering, 77, 81
  roots as sites of synthesis, 64

Day length, see Flower bud formation, in strawberries
  vs. temperature and fruit shape, 118
Defoliation and flower bud development, 75, 81
Delayed foliation, 32
Delayed heading, 207–208
Dewberry (*Rubus* spp.), 288–289
  pruning of, 220

Dichogamy, 95
Dioecious plants, (dioecism), 18
Disorders: nutritional, symptoms and treatments, 196–200
  physiological:
    growth-related, 158–161
    maturity-related, 161–162
    rootstock-related, 162–165
    temperature-related, 155–157
    water-related, 157–158
Dormancy (quiescence), 27
  auxins and cytokinins, influence on, 31
  bud scales, effect on, 31
  and chilling, 29
  rain, effect on, 29
  vs. rest, definitions, 28
  rootstock, effect on scion, 29
Dormant budding, 229
Dormant pruning, 207
Double fertilization, 92
Drip (trickle) irrigation system, 177–178
Drupe, 22–24
  growth curve of, 109
  morphological, physical, and chemical changes, 128–131
Drupelets, 25
  development and maturation, 25

East Malling (EM) apple rootstocks, history and types, 250–251
East Malling–Long Ashton (EMLA) apple rootstocks, 251
Eco-dormancy, ecto-/para-dormancy, and endo-dormancy, *see* Dormancy

Effective pollination period, 92
Embryo:
  abnormal development as a cause for poor set, 94
  abortion, 98
  ontogeny, 99
  rescue, a means of culturing immature embryos, 101
  sac, the female gametophyte, 91
Endocarp or pit:
  changes, chemical and physical, 128–130
  lignin deposition in, 128
  morphology, 19, 22
  splitting in stone fruits, 158–160
Endodermis, 9
  and Casparian strip, 9
  as primary tissue, 9
  nutrient uptake, role in, 63
Endosperm:
  as source of hormones, 115–116
  as storage organ in persimmon seed, 99
  triploid tissue, 92
Epicarp (exocarp), 19, 130
Epidermis, orgin, 4
Espalier, methods of tree training, origin, 213–214
Ethylene, 305
  autocatalytic production, 144
  evolution by auxin, 306
  evolution induced by indolebutyric acid, 239
  removal from storage atmosphere by scrubbers, 145
  ripening or senescence hormone, 144
Evapotranspiration rate, 177–178
Exotherm, 185

Fats, *see* Lipid
Fertilization:
  of the egg, 92
  influence of endogenous and environmental factors, 94–95
Field capacity, 174
Field heat, 148–149
Fig (*Ficus carica*), 293–296
  caprification, caprifig, 293
  flower morphology, 20
  fruit morphology, 25
Filbert (*Corylus avellana*), 21, 277–278
Fire blight bacteria (*Erwinia amylovora*), 227
Flavor, and aromatic compounds, 129, 146
Florigen:
  formation in leaves, 74
  mobility in the phloem, 74
Flower bud formation:
  cultural practices which influence, 82–85
  factors which govern, 82
  in grapevines, 81
  hormonal concept, 72–75
  influence of fruit, 75
  influence of seeds, 78
  in kiwifruit, and reversibility, 82
  leaves as source of stimulus for, 75
  nutritional concept, 73–74
  ontogenic stages, 71
  in strawberry, 80–81
  vernalization, 79–80
Flower morphology, 10–14
  and honey bee visitation, 94
  and flower classification, 14–19
Fox (eastern) grape (*Vitis labrusca*), 283
Foxy, flavor, *see* Fox (eastern) grape

Frost:
- factors that favor, 185–186
- means of avoiding and/or reducing damage, 186–189

Fruit:
- cumulative growth and growth rates, 110–113
- factors that affect, 113–118
- for size and harvest predictions, 112
- morphology, 19–20
- physiological disorders, 154–165
- ripening and maturation processes, 137–140
- set, 107–108

Fruit classification, 20–26
Fruit removal force, 152
Furrow irrigation system, 176
- and frost prevention, 186–189

Gametophyte:
- female (mature embryo sac), 92
- male (germinating pollen grain), 13

Gametophytic incompatibility, 90–91
Geneva double curtain, 216–217
Gibberellic acid (gibberellin), 302–304
- and elongation of filament of stamen, role in, 11
- and flower bud formation, 70, 77
- fruit growth, effect on, 114
- and heat stress in grape berry, 117
- seeds, as a source of, 75–76
- as thinner for grape berry, 124

Gooseberry (*Ribes hirtellum*), 22, 287–288
Graft incompatibility, 234–237
Grafting, 230–234
Grand period of growth, 32
Grape (*Vitis* spp.), 281–283
- flowering in, 81–82
- pruning systems, spur *vs.* cane, 216–218
- thinning, cluster and berry, 124–125

Growth regulators, 301–306
Growth retardants, seed germination, influence on, 102
Ground meristem, 4
Gynoecium, 10

Hand pollination, 96–97
Hardiness, *see* Cold hardiness
Hardpan:
- origin, 68
- perched water table, 174
- root growth, effect on, 68

Hardwood cutting, 238
Harvest:
- technology, 146–153
- criteria, 149–153

Hazelnut, *see* Filbert
Head pruning, 216
Heater, orchard, 187–188
Hedgerow tree planting system, 181
Hermaphrodite, hermaphroditic or perfect flower, 17
Hexagonal system of tree arrangement in orchard, 179–180
Honeybees, *see* Bees
Hormones, phytohormones, *see* Growth Regulators
Hydrocooling, as a means of removing field heat, 149

Hypanthium, floral tube (torus), 14–15
Hypogyny, hypogynous flower, 14

Ice nucleation bacteria and basis for frost damage, 188–189
Imperfect flower, 17
Inarching (approach grafting), 234
Indoleacetic acid, *see* Auxin; Growth regulators
Indolebutyric acid, 238
Inflorescence, classification, 18
Insect visitation:
- environmental factors that influence, 95
- influence of floral morphology, 94

Integument, seed coat, aril (testa), 99, 101
- allelopathy, role in, 103
- dormancy, role in, 101
- as protective biochemical barrier, 101
- as temporary storage organ, 100

Inversion, and inversion layer, an atmospheric condition, 188
Iron, an essential element, 196, 198–199

Jujube (*Zizyphus jujuba*), 297
June budding, bases for, 228–229
"June" drop (abscission of immature fruit), 87, 107
Juvenility, characteristics and heritability, 69–70

Kaki, *see* Persimmon
Kiwifruit (*Actinidia* spp.), 283–286

Kiwifruit (*cont.*)
  pruning, 218, 284
  training, 284

Layerage, methods of asexual propagation, 239–241
Leaching:
  of excess fertilizers and salt, 179
  of germination inhibitors in integuments and pits, 101
Leaf:
  morphology and ontogeny, 5–8
  functions:
    biosynthesis of:
      amino acids, 56
      florigen, 75
      phenolics and tannins, 56
      terpenoids and hormones, 56
      carbohydrate and mineral storage, 57
    nitrate reductase, 56
    as photosynthetic apparatus, 41
    transpiration, 46–49
Leaf Area Index, (LAI), 54–56
Leaf variegation, a chimera, 38
Light:
  fruit size and shape, effect on, 118
  quality on photosynthesis, 51
  intensity at saturation and compensation points, 51
  measurements, 44–46
  chilling of buds, negative effect on, 29
Lime (calcium carbonate), a soil amendment, 202

Lipid (oil), as reserve food of embryo, 136
Loquat (*Eriobotrya japonica*), 257

Magnesium:
  roles in plant nutrition, 196
  symptoms of deficiency, 199
Mahaleb (*Prunus mahaleb*):
  cherry rootstock, 66, 267
  root distribution, 66
Malling–Merton apple rootstocks, 250–251
Manganese:
  roles in nutrition, 196
  symptom of deficiency, 200
Marchand (Bandera), 215
Maturation and fruit harvesting criteria, 149–153
Meadow orchard culture, 212–213
Mechanical practices:
  harvesting and limitations, 146, 147–148
  pollination, 96
  pruning or hedging, 209
  thinning, 121
Mentor or recognition pollen, 91
Meristem:
  apical, 4
  culture, a means of propagation, 241–242
  lateral or lateral cambium, 4
Mesocarp, 19, 130
  role in splitting of stone fruit endocarp, 158–159
  mineral accumulation by, 130
Metaxenia, 97–98
Micropropagation, principles and methods, 241–242

Middle lamella:
  role in fruit softening, 130
  mealiness, in relation to, 132
Mineral nutrition, 193–202
  history, 193–194
Minerals:
  accumulation of:
    in drupe, 130
    in pome, 133
  analysis, foliar/petiolar, 200–201
  deficiency and excess symptoms, 196–200
  elements and their roles, 194–196
  and mycorrhizae, 198
Modified central leader of tree training, 211
Molybdenum, as essential element, 196
Monoecious plant, 18
Morrill Act of 1862, and land-grant colleges, 2
Mound/stool layerage, 239–240
Mulberry (*Morus* spp.), 17, 296
Multiple fruit, 20
Mutation, bud sport, 37. *See also* Chimera
Mycorrhiza, 198

Naphthaleneacetic acid (NAA) and amide:
  as a blossom thinner, 122
  as a fruit abscission inhibitor, 154
Nematode spp., 261
  means of eradication, 174
  mechanism of resistance, 226
  resistance by rootstocks, 226–227

Nitrate reduction and nitrate reductase activity:
  in leaves, 56
  in roots, 64
Nitrogen, as essential element, 194–195, 196–197
Nonclimacteric fruit, 144
Nucellar embryony, 97
Nucellus:
  origin and development, 99
  as related to endosperm development, 99
  site of somatic embryogenesis, 97
Nurse-root graft, 233
Nutrients, see Minerals
Nutrition, mineral, see Mineral Nutrition
Nuts:
  classification and morphology, 21
  species, 271–280

Oak root fungus (*Armillaria mellea*) resistance to, 227
Oil, see Lipid
Orchard planting schemes and their rationale, 179–182
  sites, factors to consider, 171–174
Ovule (immature seed):
  development, seed formation, 99
  placentation, 26
  site of initiation, 10, 99

Palmette system of training trees, 214–215
Panicle, a type of inflorescence, 19
Parthenocarpy, origin of, 98
Parthenogenesis, 97
Patch and ring budding, 230

Pathogen-induced graft incompatibility, 236–237
Peach (*Prunus persica*), 261–263
  growth pattern, 109–110
  heritable traits, 262
  rootstocks, 261
Pear decline, a mycoplasma-induced graft incompatibility, 254
Pear psylla, as a vector for pear decline disease, 254
Pears, 252–256
  Asian (*Pyrus serotina; P. pyrifolia*), 255–256
    growth pattern, 110
  European or French (*Pyrus communis*), 252–255
    growth pattern, 110
Pecan (*Carya illinoiensis*), 274–275
  alternate bearing, 275
  flower abscission in, 88
Pectins, dissolution during fruit ripening, 130
Pergola or tendone system of training vines, 219
Pericarp:
  ontogeny, 10
  morphology, 19
Pericycle and root formation, 9
Permanent wilting percentage, 174–175
Persimmon (*Diospyros* spp.), 22, 291–293
  classification, 292
  role of calyx in development, 134
Petals:
  as attractants to pollen and nectar seekers, 14
  and floral morphology, 14

Petiole and its role in sun-tracking, 54
Phenolic substances, 56–57
Phloem:
  cell initiation and maturation, 33
  unloading mechanism, 100
Phorphorus:
  as essential element, 195
  symptoms of deficiency, 198
Photosynthesis, 41–46
  compensation and saturation points, 44–46
  daily trends, 47
  factors influencing, 44, 49–56
  role of pigments, 42
Physiological disorders:
  bitter pit of apple, 161
  blackening of peach, 155
  black and hard end of European pear, 162–164
  black kernel of Persian walnut, 156
  blossom-end rot of peach, 160–161
  bud failure of almond, 156–157
  cork spot of Beurre d'Anjou, 165
  cracking of apple skin, 158
  cracking of sweet cherry, 157
  doubling of sweet cherry, 156
  end-cracking of prune, 157
  gumming, 160
  internal browning:
    of apple, 162
    of prune, 155
  pink calyx, 156
  pit-burning of apricot, 155
  premature shriveling of prune, 155

Physiological disorders (*cont.*)
  russeting of apple and pear, 158
  scald, 162
  side-cracking of prune, 157–158
  skin cracking, 162
  splitting of the endocarp or pit, 158–160
  water core of apple and pear, 155–156
  wilting and shriveling, 161–162
  yuzuhada disease of Asian pear, 164–165
Phytophthora, rootstock resistance to, 227
Pigments:
  role in photosynthesis, 42
  in developing fruits, 140–143
Pistachio (*Pistacia* spp.), 278–280
  inflorescence bud abscission and alternate bearing, 88
Pistil, ontogeny and morphology, 10
Planting systems of orchards, 179–182
Plums, 264–267
  European or French (*Prunus domestica*), 264
  Japanese (*Prunus salicina*), 266
Pollen:
  binucleate and trinucleate, 13
  germination, 13
  growth, 91–92
  morphology, 16
  sterility, 94
Pollination, 89–96
  factors which influence, 94–96

Pollination-constant-variant, phenotypes of persimmon, 97–98
Pollinators:
  pollen vectors that improve pollination, 94
  feeding preference, 95
Pollinizer selection, ratio, and placement in orchard, 182–183
Polyembryony, 97
Polyphenolic substances and tannins:
  fluctuations in fruits, 133
  polymerization in persimmon fruits, 97–98, 143
  as possible phytoalexins, 57
  as products of leaf activity, 56–57
  role in browning reaction, 143
Pome fruit, 247–252
  classification, 25
  growth curves, 110–111
  mineral accumulation in apple, 133
  morphological, physical, and chemical changes, 131–133
Pomegranate (*Punica granatum*), 15, 257–258
Potassium:
  as essential element, 195
  symptoms of deficiency, 197
Preharvest drop:
  causal factors, 153–154
  chemical means of prevention, 154
Procambium, a primary tissue and its derivatives, 4
Protoderm, 4
Protopectin, 130

Prune (European plum/*Prunus domestica*), 264–266
Pruning:
  of brambles and bush berries, 220–221
  during formative years, 206–208
  of mature trees, 208–210
  objectives and principles, 204–206

Quinces (*Cydonia oblonga*; *Chaenomeles sinensis*), 256–257
Quincunx, tree planting system, 179

Rancidity, of oils in nut crops, 99
Rasberry (*Rubus* spp.), 288
Receptacle, 10, 11, 15, 22
Receptacular Theory, 16
Replant ("soil sickness") problem, 173
Rest:
  chilling requirement, 28
  chilling units, 29
  duration of, 28
  growth inhibitors, effect of rain on, 29
  maintaining and overcoming, means of, 30
  nontransmissible effect, 30
  of scion, as influenced by chilled rootstock, 29
Rest-breaking agent, 96
Rhizocaline, rooting hormone, 238
Ring budding, 230
Root:
  distribution, and factors that govern, 65–67
  functions:
    absorption of mineral and water, 62–63

anchorage and support, 64–65
carbohydrate storage, 63–64
growth periodicity, 62
origin and morphology, 8–10
pruning to promote flower bud formation, 84
root pressure and xylem exudation, 65
Root-knot nematode:
  mechanism of resistance, 226
  rootstock resistance to, 226
  soil fumigation to control, 174
Rootstock:
  resistant to:
    crown gall bacteria, 227
    Phytophthora, 227
    oak root fungus, 227
    fire blight bacteria, 227
    nematodes, 226
  species characteristics, 65–67
  for tree size control, 225–226
  tolerant of flooding, 227
Rosetting of plumule from lack of chilling, 102

Saturation point, of light in photosynthesis, 44
Saw-kerf graft, 231–232
Seed:
  development and physiology, 99
  fruit size and shape, effect of, 114–115
  germination:
    factors affecting, 101
    inhibitors in seed coat and endocarp, 101
    need for stratification/chilling, 102
    as source of hormones, 75
  flower bud formation, in relation to, 75
  stratification, 224
Self-fruitfulness/self-unfruitfulness (self compatible/self-incompatible), *see* Pollination
Sexual propagation of fruit plants, 224
Shoot or stem:
  determinate and indeterminate growth habits, 32
  growth in girth, 33
  ontogeny, 4
  as a storage organ, 35
  spur formation, 32–33
Site selection for an orchard, 171–173
Skin cracking, *see* Physiological disorders
Sod culture, *see* Vegetation management or weed control
  principles of, 184
  and spring frost, 188
Sodium, 200
Softwood cutting, 237–238
Soil:
  acidity and alkalinity, (pH), 172
  amendments, purposes of, 201–202
  availability of mineral nutrients, 172
  classification, 191–192
  fertility and structure as site limiting factors, 172
Soil–plant–water relationships, 174–184

Source-sink relationships, 125–127
Specific Leaf Weight (SLW), fluctuations in, 58
Sporophyte, 92
Sporophytic incompatibility in pollination, 91
Spring budding, 229
Sprinkler irrigation:
  principles of, 176–177
  as a means for reducing frost damage, 186–187
Square and rectangular tree planting systems, 179
Stamen (androecium), 11–14
Starch:
  accumulation by kiwifruit, 134
  accumulation by pome fruits, 132
  annual fluctuation in stems, 35–37, 127
  storage in roots, 35, 63–64
Sterility factor, S allele, gametophytic incompatibility, role in, 90
Stomatal density, 8
Stool/mound layerage, 239–240
Stratification, *see* Seed
Strawberry, flower initiation, 80–81
Sulfur:
  as essential element, 195
  symptoms of deficiency, 199
Summer pruning:
  to enhance fruit color and spur longevity, 209–210
  formative years of trees, 207

Tannin:
- polymerization in persimmon, 293
- and polyphenols, levels in maturing fruit, 133, 143
- synthesis in leaves, 56–57

Tatura and ypsilon systems of tree and vine training, 215

T-bar system of training kiwifruit vine, 219, 284

Temperature:
- maximum and minimum, as limiting factors for establishing an orchard, 171–172
- photosynthesis, effect on, 44, 53
- respiration, effect on, 149
- transpiration, effect on, 53

Tension wood, 34

Thinning, fruit:
- and crop load, a means of relieving, 119
- methods: 119
  - chemical thinners, 121–124
    - efficacy of, 123–124
  - hand, 120
  - mechanical, 121

Tip layerage, 240

Topography:
- influence on cold air drainage, 173, 186
- and selection of irrigation systems, 175–178

Transition phase of juvenility, 69

Transpiration:
- daily trends, 45
- factors that influence, 49–53
- photosynthesis, relationship between, 47
- in soil-plant-water relationships, 174–175

Tree:
- size controlling rootstocks, 225–226
- training systems, 210–211

Trench layerage, 240

Trichomes (epidermal hairs), 7

Triple fusion, zygote and endosperm formation, 92

Vase-shaped (open-center) tree, 210

Vegetation management or weed control, 178–179

Vitamin C (ascorbic acid), 135

Water core, *see* Physiological disorders

Water-holding capacity of soils, 174–175

Water potential, components of, 46–47

Water table:
- influence on rooting depth, 67–68
- as limiting factor on orchard establishment, 172

Whip grafting, 232–233

Wind machines, for frost protection, 188

Winter injury:
- causal factors, 184–189
- care of injured trees, 189

Wood renewal system of pruning, 209

Xenia, 97–98

Xylem:
- initiation and differentiation of elements, 9, 33
- exudation, 65

Ypsilon and tatura systems of tree and vine training, 215

Zinc:
- as essential element, 196
- symptoms of deficiency, 199

Zygote formation, 92

E D'